Systems Engineering with SDL

Systems Engineering with SDL

Developing Performance-Critical Communication Systems

Andreas Mitschele-Thiel
Lucent Technologies, Bell Labs, Germany

JOHN WILEY & SONS, LTD

Chichester • New York • Weinheim • Brisbane • Singapore • Toronto

Baffins Lane, Chichester,
West Sussex, PO19 1UD, England

National 01243 779777
International (+44) 1243 779777

e-mail (for orders and customer service enquiries): cs-books@wiley.co.uk

Visit our Home Page on http://www.wiley.co.uk or http://www.wiley.com

Other Wiley Editorial Offices

John Wiley & Sons, Inc., 605 Third Avenue,
New York, NY 10158-0012, USA

Weinheim • Brisbane • Singapore • Toronto

Library of Congress Cataloging-in-Publication Data

Mitschele-Thiel, A. (Andreas)
Systems engineering with SDL : developing performance-critical communication systems / A. Mitschele-Thiel.
p. cm.
Includes bibliographical references and index.
ISBN 0-471-49875-0 (alk. paper)
1. Telecommunication—Switching systems—Design and construction—Data processing. 2. SDL (Computer program language) 3. Systems engineering—Data processing. I. Title.

TK5103.8 .M55 2000
621.382´16—dc21

00-043948

British Library Cataloguing in Publication Data

A catalogue record for this book is available from the British Library

ISBN 0-471-49875-0

Produced from PostScript files supplied by the author.
Printed and bound in Great Britain by Antony Rowe, Chippenham, Wiltshire.
This book is printed on acid-free paper responsibly manufactured from sustainable forestry in which at least two trees are planted for each one used for paper production.

Dedicated to Angelika, Simon and Martin

Contents

Foreword

Performance engineering means to describe, to analyze and to optimize the dynamic, time-dependent behavior of systems. However, it is common for a system to be fully designed and functionally tested before an attempt is made to determine its performance characteristics. The redesign of software and hardware is costly and may cause late system delivery. There are many dramatic examples of this from both the communication and computer industries, most recently concerning Internet providers. Responsible managers state that performance is the major cause of project failure. Moreover, there are many situations in which the functional properties of a system are directly influenced by the temporal behavior and vice versa. For example, communication protocols usually need time-out mechanisms; whether they deadlock or not is dependent on the selected timing parameter. A second example: analyzing the performance of multiprocessors and distributed systems, functional dependencies and communication between distributed subtasks have to be carefully considered; program execution, efficient resource utilization and speed up may crucially depend on them. Therefore, it is clear that performance engineering has to be integrated into the design process from the very beginning.

The provision of tool-supported methods that can be integrated into commonly accepted and widely used software engineering techniques has a key position in this domain. In the area of development of telecommunications systems, the specification and description techniques SDL and MSC have played a dominate role for years. Both languages are standardized by the ITU. Despite the availability of tools for the generation of code from SDL specifications which target a variety of platforms, systematic investigations of the optimization of the generated code have not yet been performed and published. The book presents an investigation of the SDL-92/96 version of the SDL language. It addresses an improvement of the run- time properties by taking into account the characteristics of the applications (communication protocols) and different process scheduling and management strategies. The author concentrates on the efficient implementation of behavioural concepts. For the treatment of issues arising from object-oriented concepts the author applies the traditional flattening approach of the language standard. It is expected that a direct solution would result in further improvements for the performance of SDL-based applications.

The special value of this work lies in the systematic presentation of different implementation strategies, their constraints and their impact on performance. From these results we can derive concrete rules for the design and implementation of application- and platform-specific SDL compilers. Most of the results can also be applied to the most recent language version, i.e. SDL-2000.

It is a very demanding task to describe both worlds, that is, formal specification techniques on one side and performance evaluation and optimization on the other, to survey their essential points and to summarize a state-of-the-art integrated design process. However, the author presents for the first time a systematic and didactically convincing treatise, which explores the question of how to embed methods of performance engineering in real software development technologies. We believe that this publication will become a standard reference for performance engineers, promoting critical discussions and further research activities in this area.

Prof. Dr. Joachim Fischer, Humboldt-Universität zu Berlin
Prof. Dr. Ulrich Herzog, Friedrich-Alexander-Universität Erlangen-Nürnberg

Preface

Context of the Work

The work reported in this book is based on research conducted at the Chair for Computer Architecture and Performance Evaluation (IMMD 7) at the University of Erlangen, Germany, especially research conducted within the research group on design methodology and automation.

The work has been presented as a partial fulfillment for the Habilitation Degree at the University of Erlangen.

The research has been conducted in the context of several projects, most notably the DSPL programming environment for parallel systems supported by Siemens and the German Research Foundation (DFG) by grant SFB 182, the DO-IT project for the development of efficient parallel implementations with SDL, and the CORSAIR project for the rapid prototyping of HW/SW systems based on SDL, MSC and VHDL, both also supported by the German Research Foundation (DFG). Input to the work also resulted from experience of the author as a system designer for ISDN switching systems at Alcatel, and from cooperation with Siemens and Lucent.

Part of the material is based on previous publications and lecture notes. Chapter 2 is in part based on a course in computer science taught at the graduate level, namely the design and optimization of parallel systems. Chapter 6 is in part based on work conducted within the ITU study group 10, question 6 (SDL) and on a tutorial developed jointly with Bruno Müller-Clostermann.

Several of the author's own publications have provided input to the book, most notably [MiMü99] and [HeKM97].

Acknowledgements

First of all I would like to thank Prof. Ulrich Herzog for giving me the opportunity to do this work, for his technical and personal support and guidance, and for providing a stimulating environment.

I thank Prof. Joachim Fischer for careful proofreading and taking over the job of the second advisor.

I would like to thank the State of Bavaria for financial support in the form of a Habilitation scholarship that has helped a lot in doing my studies.

I thank my colleagues at the chair for Computer Architecture and Performance Evaluation in Erlangen and the people who have been part of my research group for numerous discussions on the topic. Thanks in this respect go to Matthias Dörfel, Klaudia Dussa-Zieger, Nils Faltin, Ralf Henke, Lennard Kerber, Peter Langendörfer, Frank Lemmen, Ralf Münzenberger and Frank Slomka.

I especially thank Prof. Bruno Müller-Clostermann and his team, especially Jörg Hintelmann and Axel Hirche, for many discussions that have provided valuable input for the book. Thanks go also to the members of the ITU-T study group 10 for numerous discussions and interest in the topic.

Special thanks go to Winfried Dulz and Martin Paterok for many discussions on various technical and personal matters. I would like to thank Klaus Buchenrieder from Infineon for introducing me to the codesign world and Klaus Wirth from Lucent for a very interesting cooperation.

I thank the people from Alcatel for introducing me to the problems involved with the development of large, complex telecommunication systems, especially Bernhard Gamm, Bernd Janosch, Rolf Kunkel, Karl-Heinz Legat, Bernd Stahl and Martin Wörner.

I would like to thank the people who did the hard job of proofreading earlier versions of the book, namely Ulrich Klehmet, Peter Langendörfer, Frank Lemmen and Frank Slomka, and also the anonymous reviewers. In addition, I would like to thank the editorial staff at Wiley for their support in preparing the final version, especially Sally Mortimore, Sarah Corney and Patrick Bonham.

I thank the technical staff of the Chair for Computer Architecture and Performance Evaluation for their support concerning the computing environment, namely Armin Cieslik, Gaby Fleig and Carlo Schramm.

Last but not least, I have to thank my family: my wife Angelika for providing the support to do this work and her encouragement and understanding during a difficult time, my sons Simon and Martin, and my parents and parents-in-law.

Andreas Mitschele-Thiel

Chapter 1

Introduction

1.1 Motivation

In developing complex computing systems, performance issues play a major role in the systems engineering process. In fact, performance-related problems are a major cause of project failure [Hess98]. Most of these problems result from poor design of the software of the system. Nevertheless, the integration of performance aspects in the engineering process has not been studied very well. In particular, software engineering and performance evaluation are being considered as two rather independent areas. Domenico Ferrari once coined the term 'insularity of performance evaluation' [Ferr86]; even the phrase 'esoteric cult' is sometimes used in this context. As a result of this, each of the two worlds has its own specialists and uses its own vocabulary, models, methods and tools. Especially annoying is that different models are used to deal with different aspects of the system under development. This adds extra cost to keep the models consistent. The extra effort necessary to derive and maintain a separate performance model in addition to the functional model is often shunned by putting off performance issues for as long as possible in the engineering process. The delayed consideration of performance aspects is depicted in figure 1.1. It reflects the fact that performance is mostly considered too late in the development process.

1.1.1 Performance as 'Afterthoughts'

Performance problems often result from poor designs rather than from implementational details. Unfortunately, these design errors often remain concealed until system testing. As design decisions are not as easy to modify as implementational decisions, high cost for redesign results. Fixing performance-related problems can be even more expensive than fixing functional errors [ChKi97]. In the worst case, the project is canceled altogether, for example where the needed redesign would cause a delay that results in missing the market window of the product.

Another problem that often conceals the merits of performance engineering activities is the understandable reluctance of companies to reveal information concerning failed projects. As companies live from their successful products, these data are rarely made available.

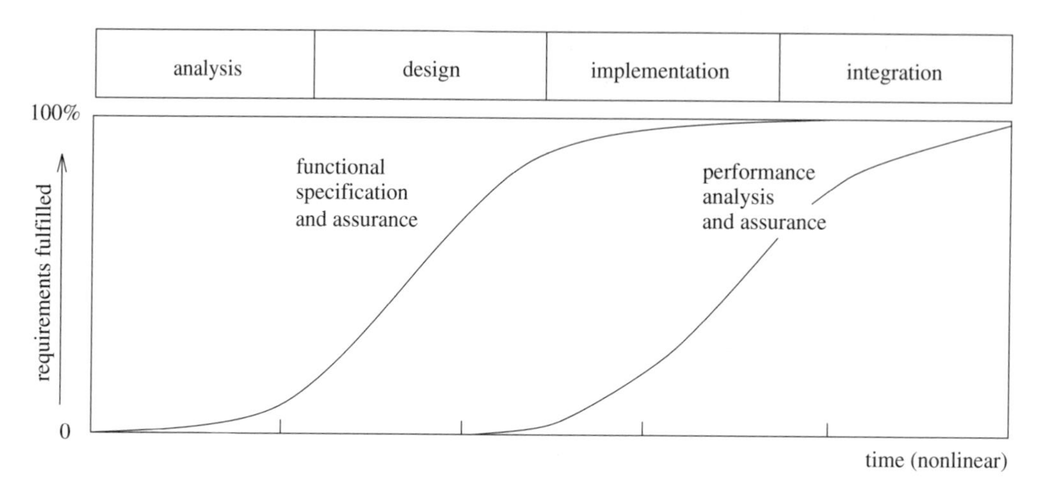

Figure 1.1 Functional and performance assurance in the system development process (adapted from [Herz92])

Considering performance as 'afterthoughts' is especially dangerous with complex systems. For example in the telecommunication sector, product families are offered which evolve over many years and which have to be maintained and updated for a period of 20 years or more. With such systems, the 'afterthought approach to performance' often results in a destruction of the system architecture: in order to quickly meet performance requirements of the system at a stage where the system is already fully implemented and integrated, solutions are selected that are not in accordance with the system architecture (kernel bypasses, etc.). If this approach is taken for several iterations of a product line, complexity is gradually added to the system. This typically results in enormous cost and delays for system integration and testing. In the worst case, the system runs completely out of control due to its enormous complexity.

In order to avoid performance-related pitfalls, the integration of performance-related activities in the early phases of the system development process is of importance. This is the major goal of performance engineering. Early integration allows potential performance problems to be detected where they are introduced rather than when they actually cause a problem.

1.1.2 Engineering Parallel and Distributed Systems

Engineering parallel and distributed applications is inherently more complex than engineering sequential ones. This is mainly due to the asynchronous execution of interrelated activities on different hardware units. Thus, concurrency complicates the system development. In addition, the physical distribution which introduces stochastic problems as erroneous transmission of data adds complexity. The lack of central control makes programming, debugging and testing of such systems extremely cumbersome.

Concerning the performance of the system, the distribution of objects introduces latencies depending on the physical location of the objects. Thus, transparency of location is not achievable for quantitative figures on performance and error rates.

1.1.3 Formal Description Techniques

The motivation for the use of Formal Description Techniques (FDT) is similar to the introduction of performance engineering activities into early development phases. Any redesign causes additional cost. In practice, each phase through which a dormant problem remains undetected adds cost to the product development. Typical figures for software are an increase in cost by a factor of 100 to 1000 for fixing a problem after product delivery, compared to the cost of fixing the same problem during analysis [Boeh81, Pres97]. Even when detecting and fixing a problem during system testing instead, an increase of cost by a factor of 30 to 70 has been reported. Thus, the early detection of problems, functional or non-functional, is crucial for minimizing development cost as well as delay.

Formal description techniques have been devised to support the (early) detection of functional problems and errors. FDTs allow functional assertion to be lifted from the testing phases to a more abstract phase of the engineering process. FDTs allow for taking corrective actions at a higher level of abstraction, i.e. before the actual implementation is derived. Formal description techniques extend system or software engineering concepts by associating a formal definition to the language constructs that allows one to clearly define the semantics of a given system specification or design description. The association of precise semantics allows one to formally reason about the system during early design. Thus, many errors can be detected much earlier in the development process. As argued above, this considerably reduces the cost of corrections.

1.1.4 Specification and Description Language (SDL)

The most widely used FDT is the Specification and Description Language (SDL). In addition, SDL is the main specification and description technique in the telecommunication area [LWFH98]. SDL has been standardized by the International Telecommunication Union (ITU) [ITU93]. In conjunction with tools, SDL is used by the majority of companies in the telecommunication industry, mainly to design communication protocols and distributed applications. In addition, it is employed during the standardization process of new protocol specifications by international standardization organizations such as ITU and ETSI. Besides its use in telecommunication, there is growing interest in using SDL for the design of real-time and safety-critical systems [Gerl97, GoDo95, Morl95].

A major revision of SDL took place in 1992, with the integration of object-oriented concepts into the language. SDL'92 supports the software engineering process from object-oriented specification down to the generation of executable code. Thus, a system specified with SDL may serve as a basis for verification and validation, functional simulation and animation, prototyping, code generation and testing. Typically, SDL is used in conjunction with Message Sequence Charts (MSC) [ITU97]. MSCs especially support the description of execution scenarios and use cases.

Besides a number of proprietary tools and tools from academia, there are two important commercial tools for SDL, TAU [Tele98] and ObjectGEODE [Veri98]. The tools support numerous activities of the development process from formal specification and verification to code generation and testing.

1.2 Goals of the Book

The goal of the book is to provide the knowledge to develop performance-critical parallel and distributed systems, especially communicating systems, with SDL. The book represents the first comprehensive and systematic study of the issues related to performance in this context.

The development of systems that meet the required performance comprises various activities applied during different phases of the systems engineering process. Simply employing a model-based performance evaluation during design to decide the spatial distribution of the entities of the SDL system is not sufficient to arrive at an efficient implementation. Instead, performance has to be kept in mind when making structural as well as behavioral decisions in the SDL description. In addition, code generation and the mapping on the underlying system are important. This is reflected in the book by covering

- a general introduction to performance engineering including performance modeling, model-based performance evaluation and optimization as well as a study of the integration of performance engineering with SDL,
- strategies to describe and structure protocol architectures in such a way that an efficient implementation is possible, and
- strategies to derive efficient implementations from given SDL descriptions.

The book is aimed at engineers developing systems with SDL as well as tool builders.

1.3 Outline of the Book

The book is organized as follows. Chapter 2 provides a general, comprehensive overview on performance engineering. Thus, the chapter provides the basis for the integration of performance engineering activities in the context of SDL. The chapter contains an introduction to systems engineering and performance engineering principles and terminology, a discussion of how performance engineering can be integrated into the system development process, and a discussion of important approaches to performance modeling, performance evaluation and optimization. Unlike most of the books covering this topic, we provide an overview of the whole spectrum of modeling approaches as well as performance evaluation and optimization techniques, from very simple to complex. We will return to the different techniques and show how they are applied in the context of SDL in chapter 6. In addition, the techniques can be applied directly where appropriate.

Chapter 3 is devoted to the design and implementation of communicating systems. Following a general overview of basic mechanisms of communicating systems, the chapter provides a systematic discussion of the important concepts and techniques to efficiently implement communication protocols and protocol architectures. Concerning SDL, knowledge of these concepts and techniques is important for two reasons: first, to understand how to structure SDL descriptions in such a way that an efficient implementation is possible, and second, to build code generators that derive efficient code.

In order for the book to be self-contained, chapter 4 provides an introduction to the formal description techniques SDL and MSC, including methodological issues of system development in the context of SDL.

Chapter 5 discusses the efficient implementation of SDL descriptions. It is the first study that covers the topic in such a comprehensive way and at such a level of detail. The chapter covers the implementation of the extended finite state machines described by SDL and the implementation of the support functions for process management, process communication and buffer management, interfacing with the environment, and timer management. The chapter describes how optimization techniques known from the manual implementation of communicating systems – as described in chapter 3 – can be applied to the automatic code generation from SDL. In addition, measures to deal with non-ideal features and limitations of the underlying system, which are not present in SDL, are discussed.

Chapter 6 discusses the integration of performance engineering in the context of SDL (and MSC). The chapter represents an application of the general concepts and techniques introduced in chapter 2. We provide an in-depth discussion of the extra information to be associated with SDL to support performance engineering, and of the issues concerning the integration with SDL. In addition, we survey important tools that support performance engineering in the context of SDL and MSC.

Chapter 7 concludes the book, providing a short summary and an outlook on further research directions.

Chapter 2

Integration of Performance Aspects in the Development Process

As computing systems become more complex, the development or engineering of these systems has become a major issue. Thus, pure knowledge of the techniques to implement such systems is no longer sufficient. Instead, this knowledge has to be supplemented with knowledge concerning methods and techniques to develop or engineer large and complex systems. The problem was identified during the software crises in the early sixties. Initiated by two NATO conferences on the issue in 1968 and 1969, the term 'software engineering' was coined. Since then, software engineering has received increasing interest in systems engineering. To date, software engineering is often considered the most important topic of systems engineering. This especially holds for systems where the hardware is mainly built from predefined components. This is the case for a large and still increasing majority of computing systems.

As software becomes more complex, performance aspects can no longer be 'tested into the system'. Rather, performance has to be considered from the very beginning and throughout the system development process. Typically, performance-tuning activities started late in the process to rectify performance do not result in a performance achievable with an integrated performance engineering process. In the worst case, it will be impossible to meet the performance goals by tuning, necessitating a complete redesign or even cancellation of the project [WiSm98].

The chapter is organized as follows. In section 2.1, we introduce the basics of systems and software engineering. The basics of performance engineering are introduced in section 2.2. Sections 2.3 to 2.5 discuss major performance engineering activities, namely performance modeling, performance evaluation and performance optimization.

2.1 Basics of Systems Engineering

Performance engineering is a special area of systems engineering. Thus, we start with an introduction to systems engineering of computing systems. In the section, we describe major terms in systems engineering, introduce the phases of the system development process and discuss the major development process models.

2.1.1 Introduction

Before going into the details of systems engineering, we take a closer look at the definition of some major terms used throughout the book. In addition, we describe some basic problems and goals of systems engineering activities.

2.1.1.1 Systems and System Descriptions

System We define the term 'system' as follows (compare also [BrHa93, Thom93]):

> A system is considered as a part of the world that some group of humans considers as a whole. A system consists of a set of interrelated components where each component has some purpose and some properties. A system interacts with its environment, i.e. with a part of the rest of the world and has a defined purpose.

What a system exactly comprises is a matter of definition. Each component of a system may itself be defined as a system. A system is not just an unordered collection of components. Rather, it consists of a structured and interrelated set of components.

An example of a system is a telephone. In this case, the system interface comprises the interface to the service provider, i.e. the subscriber line, or the antenna in the case of a mobile phone, and the user interface comprising push buttons, speaker, microphone, bell and possibly a display.

A more complex example of a system is a telephone switching system. Examples range from small private branch exchanges serving a couple of subscriber lines to large ISDN switching systems, serving several tens of thousands of customers with subscriber lines, and being connected to various other instances of similar or different systems, e.g. a set of other switching systems to support long-distance calls and mobile communication, or to a system that supports remote operation and maintenance.

Another example of a system could also be a part of an ISDN switch, e.g. the call control unit responsible for setting up new connections and handling them appropriately.

The focus of this book is on a specific class of technical systems, namely on *parallel and distributed computing systems* that are 'embedded' in a larger technical context. Thus, the environment of the systems considered here mainly consists of other technical systems rather than of human users.

Parallel and distributed computing system Parallel and distributed systems represent a special class of computing systems (compare [SlKr87]):

> We interpret the term 'parallel and distributed computing systems' as systems that employ multiple resources to process the load imposed on them. With parallel systems, the coupling between the resources is rather tight. Conversely, distributed computing systems consist of a set of autonomous nodes that cooperate to meet some goal. Distributed computing systems may communicate over longer distances and the communication as such has a major influence on the design of the system.

Embedded system The systems we focus on are typically embedded in a larger technical context (compare [Heat97]):

> We interpret the term 'embedded system' in a wide sense, i.e. computing systems that are embedded in a larger technical system and interact with their environment.

We focus on systems where the interactions underlie some time constraints and where performance is a critical issue.

Classic examples of embedded systems are microcontrollers in washing machines and microwave ovens. However, these systems are not the focus of this book. We focus on larger embedded systems with rather stringent time constraints.

Examples of these systems are onboard controllers in aircraft, e.g. to support fly-by-wire, or onboard computer systems in cars that cooperate with many other units, e.g. the engine or the anti-block system. Another example of the kind of embedded systems we have in mind is a telephone switching system as a whole or selected parts thereof, e.g. the components of a call control unit that handle new call requests.

System description There exists a major difference between a system and a system description (compare [BrHa93]):

> While a system is the real thing, a system description describes some aspects of the system. Thus, a system description represents a model of the real system.

Different kinds of system descriptions describe different aspects or views of the system. Examples are structural and functional views. For example, the structural description of the hardware of a telephone specifies the hardware entities as push buttons, microphone, speaker, etc. and the links between them. An example of a functional description is a protocol for call control used by a telephone or a telephone switching system. The functional description of the protocol specifies the functional entities involved in setting up a phone call, maintaining and releasing it.

2.1.1.2 Systems Engineering

Systems engineering deals with the process of developing systems.

Systems engineering Various definitions of the term 'systems engineering' are known from the literature (compare [Thom93]).

> The term 'systems engineering' as used here is defined as the process of building systems such that they meet their intended purpose. We use the term systems engineering to denote the more technical aspects of the system lifecycle. In addition, systems engineering as defined here focuses on the process and the activities involved in deriving the technical product rather than on its use.

However, since the use of the system typically prompts its further development, the difference between use and development is often blurred. The term 'system development' is used in this book as a synonym for systems engineering where the process of developing a system is the main focus.

Methodology Methodologies are concerned with the whole development process of a system. We use the following definition (adapted from [BrHa93]):

> A methodology comprises a set of methods (see below) to support the system engineering process, and procedures and rules for when and how to apply the methods.

Thus, a methodology defines the development lifecycle model employed as well as the specific methods applied in different phases of the development phases.

Method For the different phases of the development process, different methods are typically employed. Our definition of a method is as follows:

> A method comprises techniques and rules to execute a defined activity in the system engineering process, e.g. the requirements analysis or implementation design. Methods define procedures for obtaining the desired results. They are based on *concepts* and on *languages* that support the concept.

In today's practice, most methods are supported by tools to some extent.

Concept Methods are based on concepts and languages.

> A concept is a basic approach to model or describe selected aspects of the system. Each concept represents a specific view of the system, e.g. a functional, behavioral or structural view.

Examples of models supporting specific concepts are data-flow graphs (functional), finite state machines (behavioral) and entity-relationship diagrams (structural). We will elaborate on this further in section 2.1.3.

2.1.1.3 Goals of Systems Engineering

The overall goal of the system development activities is to derive a system that satisfies the needs and expectations of its environment. The environment comprises

- other technical systems interfacing to the system to be developed, and
- humans using, owning or maintaining the system.

Note that the usefulness of a system depends on the environment in which it is used. With respect to the system, humans may take different roles, i.e. use the system, maintain the system or own it. Depending on their role, the humans have divergent interests in the system.

Typical interests of users are ease of use, ease of learning, appropriate functionality, correctness, robustness, and good user documentation. The system administrator is interested in ease of operation and maintainability, including adaptability, extensibility, reliability and low error rate. Also very important for the system administrator is the system documentation. Typical interests of the owner are high productivity, flexibility of the product, the cost of the product and the cost of ownership.

In addition to the environment of the system, the people and company actually developing the system have some interests which are not necessarily in accordance with the interests of the customer. Interests of the company developing the system comprise short-term goals such as making profit as well as long-term goals such as the acquisition of new know-how, entering a new market or enlarging its market share.

2.1.1.4 Problems

The major problems in the system development process are caused by the software parts of the system [Gibb94]. Software is usually blamed for two major problems of the system development process, namely

- the delay in delivery and
- a high number of errors in the delivered product.

Olsen [Olse93] reports that 60% of software projects exceed their schedule by at least 50%. Linger [Ling94] reports 25 to 35 errors per 1000 lines of code before module testing commences. The delivered product typically still contains a number of errors. Typical numbers are up to four errors per 1000 lines of delivered code [Lipo82, Hami86]. Trauboth [Trau93] even reports 8 to 10 errors per 1000 lines after delivery.

Compared to software development, the hardware development process is very well understood. For typical systems, the cost of hardware development is far below the cost of software. In addition, the tendency is to use off-the-shelf hardware components, i.e. processor boards or even complete computer systems. Thus, hardware development is often reduced to hardware configuration and installation.

2.1.1.5 Quality

Satisfying the needs and expectations of the environment of the system has to do with quality. Thus, quality is the central issue in the system development process. Vital to ensuring quality is to understand the needs and expectations of the environment. This is the task of the first phase in the system engineering process, namely the analysis phase. The purpose of the requirements analysis is to study and classify all functional and non-functional requirements of the system to be developed. The results of the requirements analysis are stated in the requirements specification.

Given the requirements specification – assuming its correctness and completeness – the primary goal of the system development process is to derive a system, i.e. mainly a system implementation, which meets the functional and non-functional requirements given in the requirements specification. The requirements include cost and time for the development of the system.

There are two views of quality: the quality of the development process itself and the quality of the result of the process, i.e. the final product.

Product quality The most important quality measures of the product are (see also [Pres97])

- completeness and correctness, i.e. conformance with the functional requirements,
- robustness, i.e. the ability of the system to deal with unforeseen cases,
- reliability, i.e. the frequency and severity of faults,
- user-friendliness, i.e. appropriate user support and guidance,
- performance and efficiency, i.e. appropriate responsiveness and efficient use of system resources,
- maintainability, i.e. easy modification, extension and porting of the system, and
- documentation, i.e. appropriate user and system documents.

Process quality Quality cannot be 'tested into the system implementation' by simply employing appropriate testing mechanisms. Instead, the quality of the system engineering process as such is of vital importance to the quality of the product. Criteria for the quality of the development process are

- method integration, i.e. the ability of the methodology to integrate the used methods,
- practicability of the methods, i.e. the usability of the methods and the development process model in day-to-day practice,
- appropriateness, i.e. how well the process model and the methods are suited for the given application area,

- effort of learning, i.e. how fast the methods and the development process can be learned and adapted by engineers,
- reusability, i.e. the degree of reuse of components in other projects,
- tool support, i.e. the degree of tool support for the process model and the methods, and
- cost effectiveness, i.e. the cost for system development employing the process model and the respective methods.

Quality assurance Due to the quality problems outlined above, quality assurance, i.e. measures that assure the development of products that meet given quality requirements, is of vital importance in the systems engineering process. Quality assurance within the development process is the main focus of the ISO 9000 standards (see [Pres97, OsGl95] for an introduction) and Total Quality Management (TQM) [Arth93, BDPU94].

ISO 9000 is based on the separation of the development activities from quality measurement; these are typically done by different actors. ISO 9000 defines quality mainly as conformance to given measures. Conversely, TQM employs a different approach where quality is not primarily understood as conformance with given standards or measures. Rather, it defines quality as satisfying the expectations of company-internal as well as external customers during the whole development process. TQM is more centered towards the development teams which themselves are responsible for the quality of the product. Unlike ISO 9000, TQM is much more than quality management. It also influences the management philosophy to a large extent.

A detailed discussion on quality assurance can be found in [BaRo87, HaFr89]. Important in assuring quality is also its measurement and appropriate metrics. A comprehensive overview on quality metrics is given in [Fent91]. The evaluation of the quality of the software development process is the major concern of the Capability Maturity Model (CMM) of the Software Engineering Institute [Hump91, PCCW93, Paul95] (see also [Your92] for an overview on the CMM approach and a discussion of issues associated with process assessment).

Verification and validation System verification and validation are two important activities to establish quality assurance. Different definitions of the two terms are used in computer science. In the following, we use the most common definition of the terms in the software engineering area as provided by Boehm [Boeh81].[1] According to Boehm, *validation* is to ensure that the system engineers build the right system, i.e. a system that conforms with the needs of its environment. In contrast, *verification* is to ensure that the system is built right, i.e. to ensure that the outcome of a development phase exactly conforms to the specification provided as input to this phase or an earlier phase. For example, verification is employed to ensure that a design (or an implementation) conforms to the requirements specification. Thus, the scope of validation is product quality while verification ensures process quality.

[1] An alternative definition is to refer to all activities that are related to checking the 'validity' or 'value' of the system as validation [Holz91]. In this context, the term 'verification' is considered as a special kind of validation, i.e. where the validation is based on formal or logic methods.

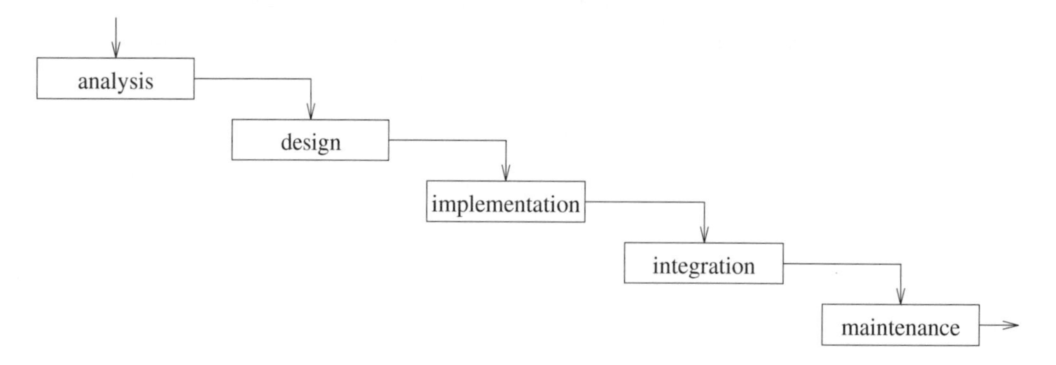

Figure 2.1 The major system development activities (shown in the waterfall model)

The most popular methods of verification and validation are reviews and testing. Alternative approaches are the formal validation and verification of operational specifications, and correctness by construction, i.e. the automatic derivation of the implementation from verified and validated formal specifications.

2.1.1.6 System Development Process

The overall goal of the system engineering process, i.e. to derive an appropriate system that serves its environment, can only be achieved by a coordinated set of measures and activities. Note that only very few systems are developed from scratch. Rather, they represent modifications or extensions of earlier products. Thus, systems engineering is typically a highly incremental process rather than a sequential process as suggested by most textbooks on software and systems engineering.

The selection of the most appropriate development process depends on the kind of systems to be developed. Systems engineering comprises the development of hardware and software. While the development of hardware is well understood, software engineering, especially for large systems, is still a very challenging task. The symptoms that indicate this can be seen in day-to-day practice with complex systems. In those systems, most service interrupts are due to errors in the software rather than hardware problems. Also the effort being put into software development is steadily increasing and in most cases far exceeds the effort for the hardware. In addition, hardware typically has a rather small impact on the performance of the system compared to the impact of software [Hess98, Smit90]. For these reasons, this book is biased towards software aspects.

2.1.2 Phases of the System Engineering Process

The system engineering process comprises five major phases: the analysis, design, implementation, integration and maintenance of the system. This is depicted in figure 2.1 for the most popular development process model, namely the waterfall model [Royc70].

2.1.2.1 Analysis

Purpose The analysis is the first phase of the system engineering cycle. The focus of the analysis is the study of the problem to be solved, i.e. the analysis of the problem domain, before taking some actions. The goals of the analysis phase are

- to identify the purpose and merit of developing the product, and
- to identify the purpose of the product and to understand its exact requirements.

The first question is especially important for the developing company, and is a highly strategic topic. Conversely, the second question is of major interest to the customer and system analysts who have to come up with a specification of the requirements of the system.

The analysis is the hardest task of the system engineering process [Broo87], especially to decide precisely what to build. The analysis is also the most critical part of the system development process since it is the hardest part to rectify later. The errors made in this phase are the most costly to correct later since they influence all other system development activities.

Note that the requirements specification focuses on the needs of the environment rather than on how the needs are provided by the system. Thus, the focus is on the externally visible behavior of the system rather than on its internal structure and implementation. If the requirements specification does not contain internal information on the system, it is referred to as a black box specification. An example of this approach is the NRL software requirements specification method [Heni80]. Others argue that the requirements specification should be potentially executable [Zave84]. In order to support this, the requirements specification has to be operational, i.e. a problem-domain-oriented specification that can be simulated and validated. Thus, internal information is needed in addition to the externally visible information. This approach is supported by formal description techniques such as SDL or Estelle [ISO89].

Subphases The analysis phase is typically divided into three subphases, namely the problem analysis, the feasibility study and the requirements analysis.

The **problem analysis** represents a preliminary study to analyze the important needs of the environment to be supported by the system and discusses principal solution strategies. The problem analysis results in the problem definition. The problem definition is especially important for the management. It describes the project goals, the scope and major directions of the development, and specifies variables and constants of the product to be developed and the resources necessary to conduct the development. The problem analysis provides important input for the development plan and the selection of the appropriate development process model.

The **feasibility study** checks (1) the technical and (2) the economic feasibility of the product development and the product. This comprises the technical feasibility of the product, i.e. the check whether the technological requirements are met by the developing company as well as the question whether appropriate personnel are available at the company or can be acquired. In addition, the economic feasibility of the project has to be checked. Important here is whether the project goals can be met, given the human and technical resources as specified in the problem definition. In addition, a cost/benefit analysis is necessary to decide whether the product and its development pay off for the participating parties, i.e. the customer as well as the developing company.

The output of the feasibility study comprises information on the expected cost and benefit of the project, the resources needed to develop the system, and an evaluation of potential technical alternatives. Thus, the preliminary development plan derived during the problem analysis can be refined with more detailed information. Based on the plan, management decides on further proceedings, depending on whether the development seems feasible and economically beneficial.

The **requirements analysis** is a detailed study of the requirements of the system as seen from its environment. The major tasks of the requirements analysis are to identify, analyze and classify the specific requirements of the product to be developed.

The results of the requirements analysis are documented in the requirements specification. The document forms the legal base for the contract between the customer and the developing company. Unlike the earlier problem definition, the requirements specification should be complete and correct. The requirements specification defines the output of the development process, i.e. the deliverables comprising the product itself as well as product documents.

The requirements specification comprises information on the overall functionality of the product to be developed, the definition of the interfaces to the environment, performance requirements, constraints on the software and hardware, guidelines for the documentation and possibly for the internal structure of the product, and the dates when the deliverables are due. Typically, the requirements are divided into functional and non-functional requirements.

In order to validate the requirements specification, prototypes of parts of the system may be developed. Especially popular are prototypes for the user interface. Depending on the further use of the developed prototypes, there is a distinction between throw-away prototypes and reusable prototypes.

Methods Various methods have been developed that support the analysis. However, as already stated, the boundary between analysis and design is not always very precise. In addition, most of the concepts on which the methods are based are appropriate for the analysis as well as the design. A survey of the basic concepts for analysis and design along with references to important methods is given in section 2.1.3.

2.1.2.2 Design

Purpose The purpose of the system design activity is to decide on the internal structure of the system, comprising its hardware and software components, based on the information given in the requirements specification. Thus, the design deals with the question of how the system meets the needs of the environment as specified in the requirements specification. Entering the design process results in a major change of focus on the system. While the focus of the system analysis is on the system from outside, i.e. focusing on the problem, the focus of the design is on the solution. This change of focus has a major impact on the models employed to specify the system requirements and the design. Often major problems are encountered with the transformation from the analysis to the design model.

The design starts with a static decomposition of the system into a set of components and the definition of their functionality, interfaces and interrelationships.[2] The major purpose of the decomposition is to reduce the complexity of the parts of the system. Thus, a major concern of the decomposition is to maximize coherence within the components and to minimize it between the components. In the following steps, the components are subsequently refined.

The system design has a major impact on the resulting product. It determines to a large extent the quality of the implementation as well as the maintainability of the product. In order to support quality assurance, the verification of the design against the requirements specification is an important issue.

A central issue of the design activities is the selection of the most appropriate alternatives that meet the requirements and minimize development and system cost. A major problem is the complexity of this optimization problem. Typically, the optimum for a subproblem of the design process is not necessarily in accordance with the global goal of the system development activity.

Subphases Three major subphases of the design process can be identified. These are the top-level design (or architectural design), the detailed design and the implementation design.

The **top-level design** comprises the following subtasks:

- the design of the system architecture, i.e. the overall architecture of the system including hardware and software parts,
- the design of the hardware architecture, i.e. selecting the appropriate hardware components and defining their interconnection, and
- the design of the software architecture, i.e. decomposing the software in modules and defining the interactions and interfaces of the modules.

Top-level design documents especially focus on the services provided by the parts of the system (i.e. software modules, hardware components or compound components) and the interfaces between them. As we will discuss in more detail later, partitioning of the system in parts to be implemented in hardware or software can be done at different stages of the development process. Thus, the focus here may or may not be on functional partitioning rather than technological partitioning.

As pointed out already, the software is the most critical part of the design. The goal of the design is to meet the requirements given in the requirements specification, i.e. correctness, robustness, reliability, performance, efficiency, maintainability, etc. The question here is how to translate these rather abstract requirements to guidelines of the design process. The following guidelines are commonly identified:

[2] If the decomposition has been studied or defined in the analysis phase already, this is input to the design phase.

- maximize coherence, i.e. ensure that the module partitioning is made in a way that supports locality,
- minimize interdependence, i.e. minimize the mutual interdependence of the modules and minimize the communication needed between them,
- functional abstraction and reuse, i.e. define modules in such a way that they provide services that can be used by other modules, and
- keep the modules manageable, i.e. keep the modules at a size easily manageable by a software engineer.

Adhering to the principles, top-level design can be viewed as the process of structuring the system in different layers. The lowest layer represents the hardware, the middle layers represent the operating system and the runtime support system, and finally, the application programs reside at the highest levels. In addition to vertical partitioning, horizontal partitioning is applied as well.

Devising a well-structured top-level design is a very challenging task and often considered an art. Thus, there is no universal method or algorithm that allows an appropriate architecture to be derived from the given requirements specification.

The **detailed design** deals with the details within the software modules and hardware components, respectively.

The detailed design of software modules typically comprises the design of the abstract data types and of the algorithms to provide the required functionality. In order to derive a flexible and reusable software design, detailed design should be independent of the target language. Similarly, the detailed design of the hardware components is also technology-independent at this step for the same arguments. The result of the detailed design is a detailed hardware and software description that defines the behavior of the hardware components and software modules as well as their detailed interfaces in a manner independent of the implementation language and the underlying technology.

In the **implementation design** the language-independent software design is refined and transformed to a language-dependent design. Thus, language-dependent decisions are made, as, for example, adaptation to the selected operating system or middleware, and the implementation of the abstract data types. Similarly, the technology-independent hardware design is refined to the technology-dependent design.

Methods Many methods, concepts, languages and tools exist that support system design. At the conceptional level, most design concepts are appropriate for system design, as well as hardware and software design. In addition, since the design is to a large extent an implementation-independent activity, there is a tendency to defer the partitioning between hardware and software to a very late stage in the development process. We will further elaborate on this in section 2.1.4.5. A survey of the concepts for analysis and design is given in section 2.1.3.

2.1.2.3 Implementation

Purpose The purpose of the implementation phase is to derive the software modules and the hardware components from the respective implementation design documents.

Subphases The implementation phase comprises two major activities for the hardware and software parts, respectively. The activities are the actual implementation of the software modules and the hardware components and their test.

During the actual **implementation**, the hardware and software parts are derived. For the software parts, the code is derived from the implementation design employing the selected programming language. Concerning the hardware parts, the actual hardware components are synthesized for the selected technology.

Testing the software modules and the hardware components is an important activity in verifying their correct implementation against the design document and the requirements specification. Both hardware components and software modules are tested in isolation before they are integrated with other parts. Standard works on software testing are Beizer [Beiz90] and Myers [Myer79].

The outputs from the implementation phase are the tested modules and components that constitute the system.

2.1.2.4 Integration

Purpose The last phase before the product is delivered comprises three activities: system integration, integration testing and system testing. Its purpose is to complete the system for delivery and to ensure that the requirements are met.

Subphases System integration and testing are highly interrelated activities.

System integration is the subsequent addition of the hardware components and the software modules to the system until the final system is established which supports the full functionality. System integration deals with connecting different hardware components, integrating different software modules and the integration of software and hardware.

Integration testing is also a stepwise activity employed after new modules or components are integrated into the system.

System testing is employed after all parts that constitute the system have been integrated.

Integration and system testing are vital activities to verify conformance of the system with the requirements specification and to validate that it meets the needs of its environment. Tests can be classified according to their focus:

- the operation test is the most common test and tests the system in normal operation,
- the full-scale test and the stress test run the system at its limits or extreme limits with respect to the system configuration,

- the performance or capacity test measures the performance of the system with different loads,
- the overload test evaluates the behavior of the system under overload conditions,
- the negative test exposes the system to situations for which it is not built, i.e. its incorrect usage,
- the ergonomics test focuses on the man–machine interface,
- the test of the documentation evaluates the quality and usefulness of the user and system documentation, and
- the acceptance test is employed by the customer as a final check.

Testing activities are described in detail in [Beiz90, Myer79].

After completion of the system test, the system goes into regular operation.

2.1.2.5 Maintenance

System maintenance is involved during the timespan starting when the system goes into regular operation until the system is removed from service. Maintenance deals with the adaptation or evolution of the system to ensure continuous service. Maintenance has to deal with changes due, for example, to changing environments, or to changing functional or performance requirements. Maintenance is also needed to remove errors encountered after the system has gone into regular operation. In principle, maintenance revisits all phases of the development process.

Note that the cost of maintenance (cost of ownership) typically surpasses the development cost of the system. For example, a ratio of 2 to 1 for maintenance over development cost for software systems is reported in [LiSw80].

2.1.3 Basic Concepts for System Analysis and Design

Various methods have been developed to support analysis and design. In the following, we concentrate on the basic concepts employed by the different methods, rather than on the methods themselves. Concepts[3] form an essential role in the development process. They provide a means for the abstract description of specific aspects of the system under development. Thus, concepts serve the purpose of

- capturing the engineer's ideas, and
- checking the ideas (verification and validation).

[3]Note that sometimes the term 'representation' is used as a synonym for concept.

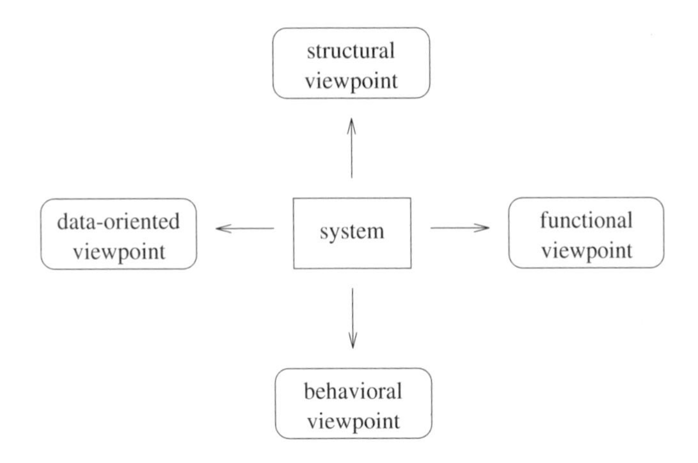

Figure 2.2 Principal viewpoints on the system (adapted from [Budg93][4])

Most methods used in practice are based on one of the basic concepts or on a combination thereof. The basic concepts as described in the following are employed by methods used for analysis as well as design.

Different concepts support the modeling of different aspects (or views) of the system. Thus, a concept can be considered as a means to express a certain viewpoint of the system under development. This is shown graphically in figure 2.2. In the following, we describe the different classes of modeling concepts (see also [Budg93, Flei94, Teic97, ChCo91]).

2.1.3.1 Structural Models

Structural approaches focus on the structure of the system, i.e. its basic building blocks and their interrelation. Thus, the approach focuses on rather static aspects of the system. Typically representative of the structural concept are block diagrams to describe the system structure or the software or hardware structure of a system. Other examples of structural models are Structure Graphs [Buhr84] devised to support the design of Ada systems and to some extent Structure Charts [StMC74]. Both approaches are oriented towards design rather than analysis. Structure Graphs are employed by the HOOD design method (Hierarchical Object-Oriented Design) [Robi92]. Structure Charts are employed by Structured Design methods [Page88, DeMa78].

2.1.3.2 Behavioral Models

Behavioral models focus on the description of how the system interacts to input from the environment. The concept focuses on the causality between events and the responses of the system to

[4] We are indebted to Pearson Education Ltd. for permission to use figure 5.1 from 'Software Design' by David Budgen, published by Addison-Wesley in 1993.

the events. The most important examples of this are state transition diagrams and Petri nets. State transition diagrams are employed by various analysis and design methods, e.g. by extensions of the Structured Analysis and Design methods for real-time systems [WaMe85] and by formal description techniques such as SDL and Estelle. A popular extension of state transition diagrams to support hierarchical modeling is Statecharts [Hare87]. A textual variant of state transition diagrams is state transition matrices. Other examples of the concept are communicating processes and systems [Hoar78, Miln80], known today as process algebras.

2.1.3.3 Functional Models

Functional models describe the system in terms of its tasks. Thus, the concept focuses on the functions provided by the system and possibly on their hierarchical organization. The most popular examples of functional models are data-flow diagrams which describe the flow of data through the system, and pseudo code. With data-flow diagrams, the data flow is modeled as the transformation of input data to output data applying a set of operations. The most important analysis methods based on data-flow diagrams are the Structured Analysis techniques, e.g. SA [DeMa78, GaSa79] and SADT [RoSc77]. Another technique that also supports the specification of the data flow within a system is a Message Sequence Chart (MSC) [ITU97]. Other approaches that support (among others) the functional view are Structure Diagrams, e.g. Jackson Structure Diagrams [Jack75], and Structure Charts [StMC74].

2.1.3.4 Data-Oriented Models

Data-oriented models focus on the data objects maintained by the system including their structure and interrelations. The most popular example of this is Entity-Relationship Diagrams [Chen76]. Another approach that supports (among others) the data-oriented view is Structure Diagrams [Jack75]. Examples of methods based on this paradigm are Jackson Structured Programming [Jack75] and the Warnier/Orr method [Warn74].

2.1.3.5 Object-Oriented Models

Another important class of models is represented by object-oriented approaches. Unlike the basic concepts discussed above, object-oriented approaches borrow several ideas from the above concepts rather than employing a fully disjunct set of new principles. With the object-oriented approach, objects from the real world are mapped on objects in the modeling technique. Object-oriented approaches apply four basic principles, namely abstraction, encapsulation, inheritance and polymorphism. The most important aspect of object-oriented approaches is inheritance which supports reuse at a high level of abstraction. This allows objects sharing some common properties to inherit the properties from a common object class. Popular examples of object-oriented methods have been the Object-Oriented Design method by Booch [Booc91], the Object Modeling Technique (OMT) [RBPE+91], and Object-Oriented Software Engineering (OOSE) [JCJÖ92]. An overview on these and other object-oriented methods can be found in [JCJÖ92].

More recently, the Unified Modeling Language (UML) has been developed and is increasingly adopted by industry. UML [BoJR98, RuJB98] draws from previous experience with object-oriented approaches, i.e. merges OMT, the Booch approach and OOSE. UML is continuously updated and standardized by the OMG [OMG99]. UML comprises a large set of modeling concepts to describe various aspects of the system. The diagram types supported by UML are:

- *object and class diagrams* to model various kinds of objects or classes and their interrelations,
- *use case diagrams* to model the interactions among the system and external objects,
- *sequence diagrams*, i.e. a notation very similar to MSCs (see section 4.2.1), to describe the details of use cases by means of a sequence of messages exchanged between objects,
- *collaboration diagrams* as an alternative form to describe details of use cases with focus on the structure of the collaborating objects rather than the exchanged messages,
- *state diagrams* similar to Statecharts [Hare87] to model behavioral aspects,
- *activity diagrams* as an alternative to state diagrams employing elements known from Petri nets,
- *component diagrams* to describe the structure of software components with their interdependences,
- *deployment diagrams* to model the structure of hardware components with their interfaces, and
- *package diagrams* to describe the structure of models.

The UML standard itself does not cover methodological aspects. In order to provide guidelines on the software development process, the Unified Process [JaBR99] has been proposed.

2.1.3.6 Other Approaches and Further Reading

A different and somehow broader classification is taken by the ODP (Open Distributed Processing) reference model [ITU95]. ODP distinguishes five viewpoints that model different parts of the system under development. The viewpoints are enterprise, information, computation (behavior), engineering (design) and technology. The computation, engineering and technology viewpoints are important for the design. The enterprise view focuses on the purpose of the system. The information view focuses on the information elements handled by the system and their interrelationship, and the computation view models the behavior of the system.

Comprehensive discussions of general software design methods are given in [Budg93, OmLe90]. Design methods suitable for the development of real-time systems are discussed in [Goma93]. The suitability of different software design methods for the development of distributed systems is discussed in [Flei94]. In [Teic97], the focus is on design methods for HW/SW codesign. Object-oriented design methods are surveyed in [JCJÖ92]. A comparison of object-oriented and conventional analysis and design methodologies can be found in [FiKe92] and [Wier98].

2.1.4 Development Process Models

Development process models define the process, i.e. the sequence of steps and activities of the system engineering process. Various models for the system development process have been proposed, and we survey the most important ones.

2.1.4.1 Conventional Life Cycle – Waterfall Model

The waterfall model goes back to the late 1960s. It is based on the development process model employed in traditional engineering fields, e.g. in civil engineering. The lifecycle with the classical waterfall model was depicted in figure 2.1.

The waterfall model describes system development as a sequential process employing the phases of analysis, design, implementation, integration and maintenance (compare [Boeh76]). The waterfall model is the most common lifecycle model used in industry.

Classical waterfall model The classical waterfall model does not expect any feedback from later phases to earlier phases of the development process.

The waterfall lifecycle is very appealing from the management point of view since it divides the development into clearly defined steps. Thus, well-known techniques for planning the phases and controlling the intermediate results can be employed. The partitioning of the development process into a set of well-defined activities is a first step in reducing the complexity of the development process.

However, various drawbacks of the classical waterfall model exist. The model is based on the implicit assumption that each development phase results in a complete and correct document (or product). For software, this assumption is unrealistic. The model ignores the fact that software development is not a straightforward process without any feedback. Each software product contains errors that are recognized in later phases of the development process. Thus, feedback to earlier phases is inevitable. The waterfall model completely ignores the existence of errors in earlier phases of the development process.

The waterfall model assumes customer interaction only during the analysis phase and after the product is complete. Thus, it relies on complete and correct requirements specifications. Errors in the requirements specification typically result in an enormous increase of cost due to their late detection.

With the waterfall model, the derivation of executable code is deferred to a very late time in the development cycle. Thus, many problems and errors are discovered very late in the development process. This again results in high cost for correction.

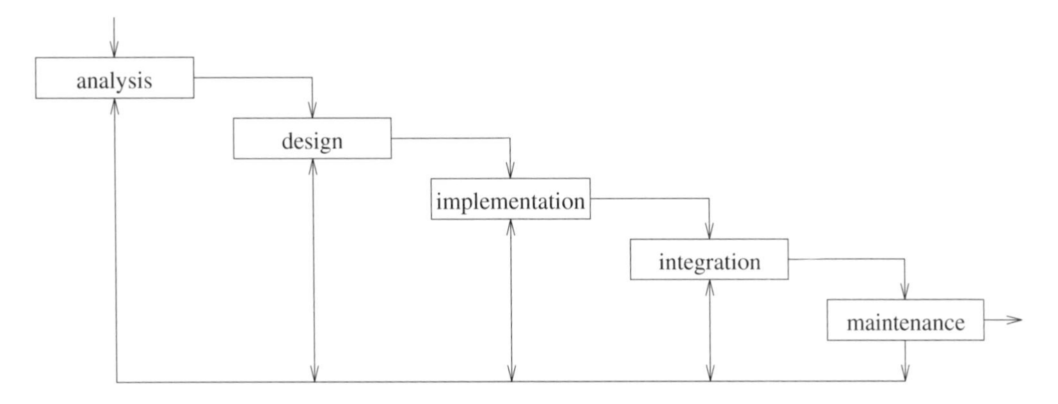

Figure 2.3 The iterative waterfall model

There is also a tendency of engineers to focus on 'paper engineering', i.e. to describe things on paper that could be implemented faster instead.

See also [Your92, Pres97] for a discussion of the problems with the waterfall model.

Iterative waterfall model In order to remedy some of the problems, the iterative waterfall model has been proposed [Royc70]. This is depicted in figure 2.3. It allows the development process to return to an earlier phase in case problems or errors are discovered. Thus, it acknowledges the existence of errors in any phase of the development process and allows one to deal with them right away.

However, the iterative waterfall model is still based on the common assumption of the waterfall model that each development phase is completed before proceeding with the next phase. For example, design is not started before all system requirements are analyzed and documented.

Prototyping Since the correctness of the requirements specifications is so crucial to the waterfall model and major cost are incurred if errors pass undetected, prototyping has been introduced into the analysis phase [Smit90a]. The main purpose of prototyping in the analysis phase is to clarify the needs of the environment and to ensure that the requirements specification is in accordance with the needs of the environment. Two types of prototypes are usually distinguished, throw-away prototypes and reusable prototypes.

2.1.4.2 Evolutionary Model

As discussed above, the waterfall models are only suited for the development of systems for which the requirements are very well understood and well defined. They are not suited for the development of systems with changing requirements or where the requirements are not well defined and understood.

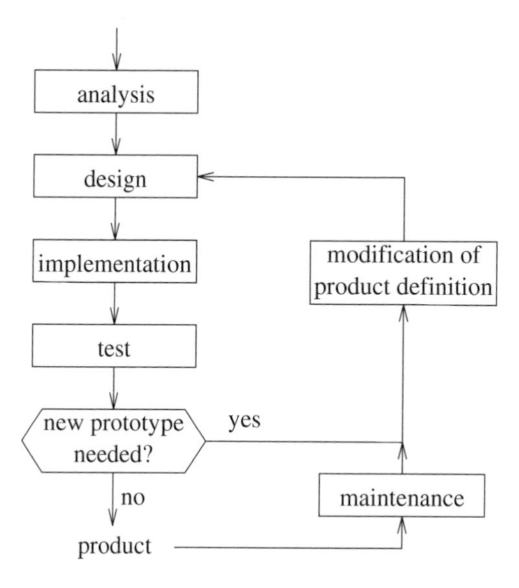

Figure 2.4 The evolutionary model

For these kinds of systems, the evolutionary model has been developed [Gilb88]. The evolutionary model considers system development no longer as a linear process. Rather, system development is considered as an iterative process as depicted in figure 2.4. These ideas have also guided the development of the Unified Process [JaBR99]. The Unified Process provides guidelines on the development of systems using UML.

With the evolutionary model, system development is done in a number of iterations. The result of each iteration is a prototype. With each iteration, the previous prototype is extended. Each iteration comprises an analysis, design, implementation and test phase. The prototype turns into the product when it satisfies the needs of its environment.

With the evolutionary model, maintenance of the system is considered as just an additional cycle in the development process.

A major difference of the evolutionary model from the waterfall model is the frequent interaction with the customer. This is a major advantage, and allows the early detection of problems. In addition, continuous customer involvement improves acceptance of the system by the user. Moreover, the early derivation of prototypes minimizes the risk of encountering unforeseen problems late in the development process and improves the motivation of the engineers.

However, the evolutionary approach also has some serious drawbacks. Evolutionary processes are hard to manage. In particular, resource allocation and planning as well as the evaluation of the progress of the project are very difficult to achieve. In addition, the evolutionary approach relies on easily changeable systems. Thus, the 'quick and dirty' approach often employed for prototyping is not appropriate here. Another problem is documentation. It is not clear how the documentation can keep up with a dynamically evolving system.

2.1.4.3 Transformational Model

The transformational process model can be viewed as an extension of the waterfall model with prototyping. The idea of the transformational approach is to derive a formal and operational requirements specification. Based on the requirements specification, an executable prototype is derived which is verified and validated against the needs of the environment.

The final implementation is derived by subsequent application of transformation rules that preserve the semantics of the system. The transformation rules are intended to transform the requirements specification to an implementation that is also optimized for performance.

The transformational approach has several advantages (see also [McCl89]):

- It supports maintenance at the specification level. Due to the automatic generation of code, changes to the system are made at the abstract specification level only. This radically differs from other approaches, where the documents used in the different development phases as well as the implementation have to be maintained.

- It supports the derivation of well-verified systems. Due to the automatic code generation, inconsistencies between the requirements specification and the design or the implementation are avoided.

- It enables early validation and verification. Due to the formal and operational specification of the requirements specification and, thus, the availability of an operational prototype, early formal validation and verification of the system can be supported.

However, the transformational model is still a major research topic. Many problems have to be solved to support the fully automatic or rule-based derivation of implementations from formal specifications.

2.1.4.4 Spiral Model

The spiral model proposed by Boehm [Boeh86] represents a flexible combination of the above approaches. The development process with the spiral model is depicted in figure 2.5. The spiral model supports a problem-specific combination of various process models under full control of the management. In fact, the waterfall model as well as the evolutionary model can be considered as special instances of the spiral model. The detailed activities employed in each iteration of the spiral are to a large extent a matter of definition.

With the spiral model, system development is an iterative process. Each iteration in the spiral comprises the following kinds of activities:

(1) the definition of the objectives, alternatives and constraints of the iteration,

(2) the evaluation of the alternatives, the identification and resolution of risks,

(3) the development and verification of the (intermediate) result of the iteration, and

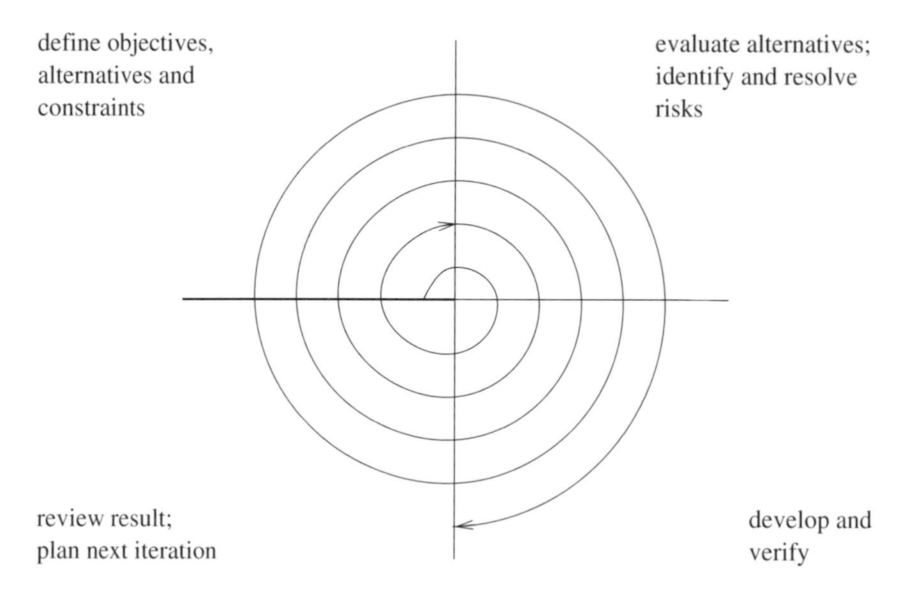

Figure 2.5 The spiral model

(4) the planning of the next iteration.

A review is employed at the end of each iteration. This allows the progress of the project to be evaluated and further proceedings to be determined.

2.1.4.5 A Note on HW/SW Partitioning

So far, the issue of the phase in which the HW/SW partitioning is decided has not been addressed. There are two extremes to this: early and late partitioning.

The traditional system development process employs *early partitioning*, i.e. the partitioning during the top-level design or even the requirements analysis. This is supported by the 'divide and conquer' principle to reduce complexity. Thus, the partitioning is decided before the detailed behavior of the system is defined. Early partitioning allows the most appropriate (technology-dependent) methods and techniques for the development of the respective hardware and software parts to be employed.

The disadvantage of early partitioning is its inflexibility with respect to changing the partitioning later on. Any change in the partitioning typically results in very high cost. Changing the partitioning may be needed if a part of the system implemented in software turns out to be a bottleneck and, thus, should be moved to hardware. Other examples are advances in processor technology that suggest that more functionality should be moved to software for cost effectiveness, or changing requirements.

More recent research proposes instead *late partitioning* in the system development process. The application of this principle is usually referred to as HW/SW codesign [RoBu95], which proposes the design of the system independent of the underlying technology. With this approach, HW/SW partitioning is deferred to the implementation design phase after a complete functional description of the system has been derived. During implementation design, the functionality of the system is mapped on the underlying technology by appropriate tools. Late partitioning is especially popular for smaller embedded systems. The codesign approach shares some common principles with the transformational process model.

Late partitioning more naturally supports the joint validation and verification of functional aspects of the system, independent of the implementation technology used. The major advantage of late partitioning is its flexibility in changing the HW/SW partitioning very late in the development process. Due to the availability of appropriate tools, the cost of changing the partitioning is minimal. Thus, changes due to newly available hardware, changing requirements or other causes are easily made. Altogether, this is a very flexible approach. However, note that so far the codesign approach is applied to rather small systems only.

Methods for the development of systems based on the codesign approach are described in [BeJe95, DöMS00, GaVa95, KaLe93, MiSl97, ThAS93].

2.2 Basics of Performance Engineering

In this section, we motivate the need for performance engineering and introduce basic terminology and activities in the performance engineering world.

2.2.1 Introduction

2.2.1.1 Performance Engineering

Throughout the book, we use the following definition of performance engineering:

> *Performance engineering* is the integration of *performance evaluation* and *performance optimization* activities in the *system engineering process*.
>
> Performance engineering adds new *activities* to the system engineering process. In addition, the results of the performance engineering activities influence the system development process.

Performance engineering comprises a set of interrelated activities. The additional activities are highly associated with the underlying system engineering process. Performance activities may be applied in all phases of the system development process.

2.2.1.2 Performance Evaluation

Performance evaluation denotes the activities needed to evaluate the system under development with respect to performance. Performance evaluation is applied in various phases of the development process and comprises a set of activities. Performance evaluation may be based on models or on measurements, depending on the phases in which the activities are applied.

2.2.1.3 Performance Optimization

Performance optimization denotes activities to derive or improve a design or an implementation with respect to performance. Thus, the major tasks of performance optimization are to identify open parameters of the system design relevant to the system performance and to select values for the parameters such that the required performance figures of the system are met. Since the parameters relevant for performance are numerous and comprise many aspects of the system, performance optimization has to deal with modifying a large set of diverse parameters, e.g. influencing the software and the hardware as well as issues related to the mapping of software on hardware. Performance optimization also has to deal with other constraints of the system under development, e.g. fault tolerance, safety, system cost, etc. Since performance optimization relies on performance figures derived for the system under development, performance evaluation is an integral part of the optimization process.

2.2.2 Myths and Merits of Performance Engineering

2.2.2.1 Myths of Performance Engineering

Before we go into the details of performance engineering, we summarize the major arguments not to integrate performance activities early in the system engineering process (see also [Smit90], section 1.1):

Performance problems are rare In the early days of computers, performance was not an issue with batch systems. However, nowadays computing systems are used online, often in a distributed environment, and have a very demanding functionality. With today's systems, poor performance is a major cause of project failure [Hess98].

Hardware is fast and inexpensive In fact, hardware is often much cheaper than modifying the software for higher efficiency. On the other hand, faster hardware allows only a relatively modest increase in speed compared to the difference of efficiency of different software designs or implementations. The performance of an efficient implementation can easily exceed the performance of a badly designed system by an order of magnitude. Thus, faster hardware may be a remedy only if a modest increase in speed is sufficient.

Building efficient software is very expensive Nowadays there are many methodologies and tools available that support fast evaluation of the performance of the system under development. This adds little overhead to the development of the system. [Hess98] estimates the additional cost as 2-3% of the development cost. On the other hand, performance engineering saves tremendous time and effort compared to the case in which performance problems pass the design phase undiscovered and prompt a redesign of the system later on.

The system can be easily tuned later if there is a need Performance problems – especially with parallel and distributed systems – are typically caused by fundamental problems in the design of the system, and not by an inefficient implementation of given algorithms. Thus, code tuning typically does not remedy the problem. Rather a major revision, i.e. the redesign of major parts of the system, is needed, which is very expensive in most cases.

Efficient software implies higher complexity With the fix-it-later approach, fixing the performance of the implementation often results in an enormous increase in complexity. This is due to the fact that at this late development stage, often system design and structuring principles are violated in order to fix performance problems quickly. Examples of such fix-it-later approaches are the corruption of the layered system architecture by bypassing some of the system layers. While this may work for a number of system releases, there will be a time when the system is no longer manageable due to the subsequently added complexity.

Conversely, with early performance engineering in place, potential performance problems are detected early in the development cycle. Thus, the design can be fixed without adding complexity to the implementation.

2.2.2.2 Benefits of Performance Engineering

The benefits of the early and systematic integration of performance aspects in the system development process are as follows (compare also [Smit90] and [Hess98]):

Increased overall productivity of the development process Performance engineering supports the detection of potential performance problems in very early development phases already. This minimizes cost for redesign and reimplementation of the system due to performance problems.

Faster product development (time-to-market) Performance engineering activities require minimal extra overhead and time in the development phases. On the other hand, the redesign and subsequent reimplementation of a system or parts thereof obviously require much more time.

Reduced risk of uncontrolled cost and delay Early evaluation and detection of potential problems allow control of the system development process to deal with risky parts or aspects of the system first. This also applies to performance problems. It allows the risk of encountering performance problems late in the development cycle to be minimized.

Improved product quality and maintainability Performance engineering starts early in the development process. Thus, the risk of encountering performance problems in the implementation or integration phase is highly reduced. This in turn eliminates the need for 'performance hacking', i.e. ignoring system engineering principles and bypassing parts of the system architecture to improve the performance of the implementation. Obviously, a clean system architecture which follows proper engineering principles supports product quality and maintainability.

2.2.3 Important Terminology

Before discussing performance engineering activities, we define important basic terms used in this context (compare also [Jain91]).

Workload and stimuli The workload describes the load imposed on some entities. A specific workload description may define (1) the load imposed on specific resources of the system, or (2) the load imposed on the system as a whole, i.e. the load imposed at its external interfaces.

Highly related to the workload is the term 'stimuli'. Stimuli specify specific requests for services. Thus, stimuli represent a specific approach to describe the workload.

Use cases A use case denotes a specific usage of the system, i.e. the stimuli provided to the system and the actions performed by the system to provide a specific service. Examples of use cases are the transmission of certain classes of data by a communication system, e.g. a file transfer or some delay-sensitive data for a multimedia application, or operations to update some routing table. Note that different use cases typically impose different workloads on the system.

Available resources Available (limited) resources denote the units which are available for the application, i.e. to handle the load imposed on the system. Important resources are the processors and the communication channels of the system. Further important resources concern the available memory to hold code and data, most notably the size of buffers. Besides the capacity and time characteristics of resources, their service strategies are of importance.

Resource demands Resource demands denote the quantitative requirements resulting from the application or more specifically from the arrival of service requests on some resources of the system. Thus, resource demands specify the cost caused by the implementation (or execution) of the parts of the system on the available resources. Examples of resource demands are the required processing times on processors, or the required memory space. Note that the resource demands depend heavily on the design and implementation of the system, e.g. the actual algorithm employed to solve a problem, or the selected strategy for parallel or distributed processing.

Performance parameters During system design and implementation a large number of decisions have to be made. The potential variables of the system design and implementation that influence the performance of the system are called performance parameters. Examples of performance parameters are the performance of processors or communication links, the spatial distribution of the application on the available processors and the scheduling strategy.

Performance metrics Performance metrics denote the performance figures of the system under development that are subject to the performance optimization either as part of the goal function or as constraints. Important examples of performance metrics are (see [Jain91] for details):

- the response time, i.e. the time interval between a service request issued to some entity of the system and its respective response,
- the throughput, i.e. the rate at which service requests can be served by an entity, and
- the utilization, i.e. the fraction of time a resource is busy.

Performance metrics can be classified as

- application-oriented, and
- resource-oriented.

An example of an application-oriented metric is the response time to process an external service request. Examples of resource-oriented metrics are the throughput or the utilization of some resources of the system.

Performance requirements The performance requirements specify the required values for specific performance metrics, i.e. the values or the range of the values expected from the implemented system. Thus, performance requirements can be considered as constraints of the performance optimization process. For different use cases different performance requirements are typically given. Examples of performance requirements are response time and throughput figures.

2.2.4 Performance Engineering Activities

The section is about the basic activities involved in the performance engineering process. We start with an overview of the basic activities and discuss how the results of performance engineering influence the overall system engineering process. Following this, we outline when and how performance engineering activities are applied. Finally, we discuss the specifics of applying performance engineering in the different phases of system development.

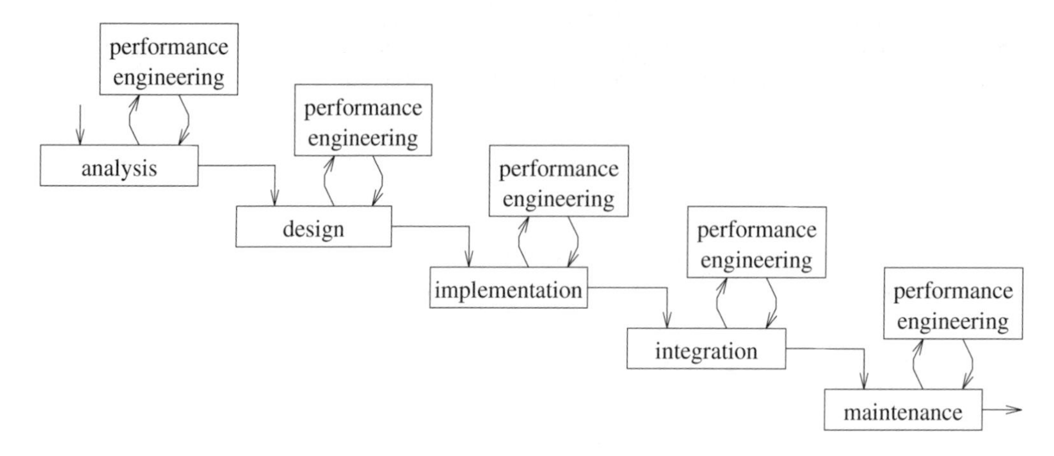

Figure 2.6 Performance engineering subcycles in the system development process (here the waterfall model)

2.2.4.1 Activities of the Performance Engineering Subcycle

As has been pointed out already, performance engineering is applied in various phases of the system development cycle. Performance engineering can be considered as a subcycle in the system development cycle. It adds extra activities and influences the system development process. The influence of the performance engineering activities on the system development process is graphically depicted in figure 2.6. For simplicity of the figure, we assume the waterfall model for the system development process.

In the following, we describe the generic performance engineering activities independent of the specific development phase in which the performance engineering subcycle is applied.

Identify the goals of the performance engineering subcycle Before the performance engineering subcycle is commenced, the goals of the subcycle have to be identified. This involves answering the following questions:

- What is the purpose of the subcycle? The purpose may be
  - to evaluate possible solutions to a decision problem to find the best solution, or
  - to identify the parts of the system critical to performance which have to be studied first.
- What are the performance metrics to be estimated? This depends on the performance requirements imposed on the system under development. For the evaluation, the primary performance requirements as seen by the user should be evaluated (e.g. the response time) as opposed to secondary performance figures such as the load imposed on the resources.

- What is the required accuracy of the evaluation? Depending on the answer, different modeling and evaluation techniques may be appropriate.

- What kind of performance evaluation is performed? Depending on the exact goal of the evaluation, optimistic, pessimistic or average case evaluations (see section 2.2.4.3 for a discussion) may be appropriate.

After the identification of the exact goals of the performance engineering subcycle, the details of the object under investigation have to be studied.

Study the details of the object under investigation Depending on the phase of the development cycle in which the performance engineering subcycle is applied, the object of the investigation may be either an existing system or a system under development (analysis, design or implementation phase).

The activities include

- the identification of the service requests issued to the system (workload), including their frequency, possible outcomes of the service requests and possible performance requirements on the execution of the service requests,

- the analysis of the execution environment (available resources),

- the analysis of the static structure as well as dynamic aspects of the system, and

- the identification of the resources used by specific service requests (mapping).

Decide on the modeling approach After the goals of the performance engineering subcycle have been identified and the object under study is well understood, the modeling approach can be decided on. Different modeling approaches will be described in section 2.3. As we will see in section 2.4, the selected modeling approach highly influences, among others, the kind of performance metrics that can be derived, and the accuracy of the evaluation.

Build the performance model After the modeling approach has beed decided on, the actual performance model for the system or a part of the system can be built. Thus, the 'mind model' of the object under study is mapped into a meaningful or 'equivalent' performance model. This may be a complex task. In particular, the selection of an appropriate level of abstraction is very important here. If the model is too abstract, the accuracy of the evaluation may suffer; if the model is too detailed, the evaluation becomes time and resource consuming.

At this stage, the performance model is a more or less abstract model of the system with a focus on performance aspects. Often performance models represent extensions of one or more of the basic system modeling concepts described in section 2.1.3.

To complete the model and to allow a quantitative analysis of the model, the next step is essential.

Derive quantitative data for the performance model In this step, the quantitative performance data have to be gathered and the performance model has to be attributed with these data. This comprises all factors having a non-negligible influence on the performance of the system. The performance data include the resource demands of the different parts of the application and the performance data of the available resources of the system.

The alternatives to derive these performance data for the system under development will be described in section 2.3.7. If the performance evaluation is applied to the final implementation, measurements can be made instead of a model-based performance evaluation. If parts of the system are implemented, a mixed model- and measurement-based performance evaluation can be employed. Note that this step may be iterated with previous steps to decide on important and unimportant parts to model.

Transform the performance model to an executable or assessable model In this step, the performance model is mapped on a model assessable or executable by performance evaluation techniques. Thus, the limitations and peculiarities of specific performance evaluation techniques are taken into account. Examples of assessable models are task or process graphs and Markov chains. An example of an executable model is simulation models.

Evaluate the performance model In this step, the model is executed or analyzed and the performance figures are derived. The evaluation of performance models is supported by a wide variety of tools. In order to ensure the correctness of the performance evaluation, a validation of the results should be done. This may be based on expert intuition, on a comparison with measurements, or on a comparison with other performance evaluation techniques, e.g. that focus on special cases. The evaluation of performance models is described in more detail in section 2.4.

Verify the performance results against the performance requirements The last step directly related to performance evaluation is the comparison of the performance figures obtained by the performance evaluation with the respective performance requirements. The comparison can either be done during the performance evaluation or afterwards. In case the performance evaluation is embedded in an automatic optimization process, the performance evaluation is done on the fly [Mits94, MiLH96, DöMS00].

A note on performance evaluation methodology and tools The performance evaluation methodology usually presented in textbooks follows a scheme similar to the one embedded in our performance engineering subcycle (e.g. see [Jain91, Kant92, Lave83]).

Performance modelers and tool builders have achieved considerable progress with respect to the later performance evaluation steps. However, they are often heavily challenged when being confronted with the earlier steps. A major problem is the description of the workload. Another problem is to develop a performance model that is consistent with a dynamically evolving design.

2.2.4.2 Using the Results of the Performance Engineering Subcycle

As defined in section 2.2.1.1, performance engineering is highly integrated with systems engineering. Depending on the purpose and the results of the performance evaluation, one of the following actions is performed next:

- System refinement: The parts of the system which have been identified by the analysis as performance-critical or risk-critical are refined first. Thus, the performance engineering subcycle is left and control returns to the main system engineering cycle.
- Model-based optimization: Based on the performance model, some parameters of the system are modified, i.e. different system alternatives are selected, and the performance evaluation is redone with the new parameter values.
- Model refinement: If more accurate or other performance data are needed based on the results of the previous analysis, the performance model is modified or another performance evaluation technique is applied.

System refinement The purpose of this action is to minimize the risk of encountering performance problems late in the development process. Thus, uncertainties have to be resolved and minimized as early as possible in the development process. This can be achieved by selecting the parts for which no accurate performance figures can be derived in the current stage of the system development cycle and refining them first. After the refinement, the selected parts are reevaluated. For the selection for refinement, two criteria are of importance:

- the influence of the part on the overall performance of the system, and
- the degree of uncertainty about the performance of the respective part of the system.

The higher the uncertainty and the larger the influence on the overall performance of the system, the more important is an early refinement.

Note that the discussion is also influenced by the development process model employed.

Model-based optimization If the performance does not meet the expectations, the respective system parameters have to be modified and the performance evaluation process has to be repeated with the modified system. Relevant system parameters that may be modified depend on the development process phase in which the performance evaluation is employed and the severity of the performance problem encountered. For example, if a performance evaluation is done in the implementation phase, small problems can typically be remedied by modifying the detailed design; for more serious performance problems changes to the top-level design or even the system requirements may be necessary.

A prerequisite for the automation of the optimization is that the open system parameters and alternatives as well as the constraints on the optimization have to be identified prior to the optimization. Besides their identification, the open parameters, the constraints and the goal function of the optimization have to be defined formally.

Model refinement In order to minimize the overhead for performance evaluation, we may start the performance evaluation process with a superficial performance evaluation. The results of the first evaluation can then be used to identify performance-critical parts of the system which require a more accurate performance evaluation before further actions in the system development process (i.e. system optimization or refinement) are taken. Thus, we may identify parts of the system which are critical for the overall performance and which should, therefore, be modeled and analyzed more accurately than others. Alternatively, we may identify use cases that are more critical to performance than others and, thus, may need further study.

2.2.4.3 When and How to Employ Performance Engineering

When As discussed in section 1.1, the late identification of problems results in higher development cost. Thus, performance problems should be detected and remedied as early as possible in the development cycle. This includes the analysis phase. Here already potential performance problems can be detected.

The cost of a performance evaluation is negligible (see below) compared to the cost of revisions. This is especially true for decisions taken in early development phases. Thus, a performance evaluation should be done whenever there is a risk of selecting a wrong solution, i.e. a solution that may cause high cost in the case of revision.

Accuracy The accuracy of the performance evaluation depends on the phase in which the analysis is applied, i.e. the detail of the design, and the accuracy of the performance data and models. Thus, the application of the performance evaluation in late development phases definitely allows for more accurate results. Our suggestion is to employ an evaluation for optimistic as well as for pessimistic assumptions on the performance of the system and to start with simple models.

The use of an optimistic and a pessimistic evaluation is essential for early development phases where accurate input to the performance evaluation is hard to derive. If the pessimistic evaluation of the system indicates that the performance requirements are met, no further performance study is needed at this point. If the optimistic evaluation already indicates performance problems, we have to seek alternatives, i.e. some action has to be taken to improve the performance. If the performance of the system is in between, we suggest the identification of the performance-critical parts of the system. The performance-critical parts should be refined first in the development cycle. An alternative, in case a more accurate performance evaluation is possible, is to apply it to the performance-critical parts to get more accurate performance results.

We suggest starting with simple models in the early development phases rather than deriving a very detailed functional model for which no accurate performance data are available at this stage, or which is overkill due to the fact that performance is not a problem for most parts of the system. Based on the results of these simple evaluations, it can be decided what to do next, i.e. to select one of the alternatives described in section 2.2.4.2.

Effort As we will see in section 2.4, many simple evaluation techniques exist ('back-of-the-envelope' type of techniques). This is much cheaper than later revisions, i.e. when considerable effort has been put into the detailed design or even the implementation ([Boeh81, Smit90]). Many simple techniques can be applied by people who are not experts in the field of performance evaluation. In addition, we will show in sections 2.4 and 2.5 how the performance engineering process can be supported by tools and automated to a large extent.

2.2.4.4 Performance Engineering and Development Phases

Even though the basic performance engineering steps identified in section 2.2.4.1 apply to all phases of the system development process, the exact purpose and goals of performance engineering activities vary depending on the phase of the system development cycle in which the activities are applied (compare [Smit90], section 1.3).

Analysis The main purpose of the performance engineering activities in the analysis phase is to

- identify and define the workload imposed on the system along with its performance requirements (requirements analysis), and
- check the principal feasibility of the system (feasibility study) with respect to performance and cost.

In this phase, optimistic and pessimistic estimations are typically derived, e.g. during a feasibility study. Examples of performance evaluations at this stage are the analysis of critical paths or the estimated load per resource (e.g. processor, network, memory, etc.).

In order to estimate the load imposed on the system at this early stage of the development, especially for the feasibility analysis, we recommend the use of the 80/20 rule, i.e. focus on the 20% of the use cases that are likely to use 80% of the resources. However, note that the 80/20 rule may not be appropriate to estimate the response time of the system or for systems with hard real-time constraints.

If there are conflicting requirements or requirements that have a detrimental influence on the performance, the requirements may be renegotiated. While this is obvious for the analysis phase, it may also hold for later phases. In [Smit90] it was pointed out that slight changes to the functional requirements may have a large impact on the resource demands and the performance of the system.

Top-level design Due to the principal differences of top-level design from detailed and implementation design, we deal with them differently.

The main goal of the top-level design is to design the system architecture, i.e. identify the components of the system (especially the services they provide) and the interrelations between the components. The purpose of performance engineering activities in this stage is to support the derivation of a system architecture that meets the performance requirements along with all the other requirements and constraints. The following tasks have to be tackled:

- identify the performance-relevant parameters of the system architecture, i.e. the parameters that are open for a performance optimization,
- evaluate the alternatives for the system architecture, i.e. for the system partitioning, modulization, parallelization, etc., and derive estimates of the performance for the different workload scenarios of the system, and
- identify the parts or system components critical to the performance of the system.

If performance-critical parts have been identified, the same rule as above applies. Thus, the respective components should be detailed first in order to estimate and minimize the risk of potential errors and uncertainties of the performance evaluation.

Detailed design and implementation design As the design proceeds, more design decisions are made which in turn allow a more detailed estimation of the performance of the system. The main tasks of performance engineering activities in these subphases of the design are to

- identify the system parts most critical to the performance of the system (if not done already) and pay special attention to them, i.e. the parts for which performance figures are critical or have a major impact on the overall performance of the system,
- identify the possible design alternatives of the respective subphase,
- decide on the most appropriate design alternatives,
- derive more detailed estimations of the performance of the parts of the system for the different use cases, and
- check whether the global performance requirements are met for all use cases and workloads.

If performance-critical components have been identified, they should be implemented first in order to estimate and minimize the risk of potential errors and uncertainties of the performance evaluation.

Implementation and integration The purpose of the performance engineering activities in these phases is to check whether the predicted performance of the components of the system as well as of the system as a whole are met.

As soon as the components are implemented, performance figures based on measurements can be derived. As long as only parts of the system are implemented, the measured performance figures can be aggregated with the model-based performance figures to estimate the overall performance of the complete system more accurately than is possible in the design phase.

A very important purpose of the performance engineering activities at this point is to maintain a performance data base with detailed performance data on the software and hardware components of the system. This is helpful for various reasons, especially to provide performance figures for later products or later product cycles.

At this stage also, an analysis of the quality of the earlier performance evaluations should be employed in order to measure and improve the quality of the performance engineering process. This allows one to avoid errors and minimize inaccuracies in the performance engineering process.

Maintenance During maintenance, performance evaluation is used to estimate the impact of hardware or software modifications, and to decide on appropriate alternatives. Possible modification may be solely for performance reasons, i.e. to serve more service requests, or to modify or extend the functionality of the system.

2.3 Performance Models

In order to support a model-based performance evaluation or optimization, models are needed that describe the aspects of the system relevant to its performance. From the viewpoint of performance evaluation and optimization, these models of the system can be considered as the input language to describe the system to be evaluated and optimized.

2.3.1 Introduction

2.3.1.1 Purpose of Performance Models

As pointed out already, modeling is an important issue in systems engineering. Models support the understanding of the system. Modeling employs various principles to deal with complexity. As discussed in section 2.1.3, models are used to describe structural, functional, behavioral and other aspects of the system.

The main purposes of performance models are

- to describe, accurately and intuitively, the aspects of the system that influence performance,
- to deal with the complexity involved with evaluating and optimizing the performance of the system, and
- to support the automatic evaluation and possibly the optimization of the system under development.

Note that the actual performance model employed for performance evaluation and optimization may vary, depending on its specific purpose.

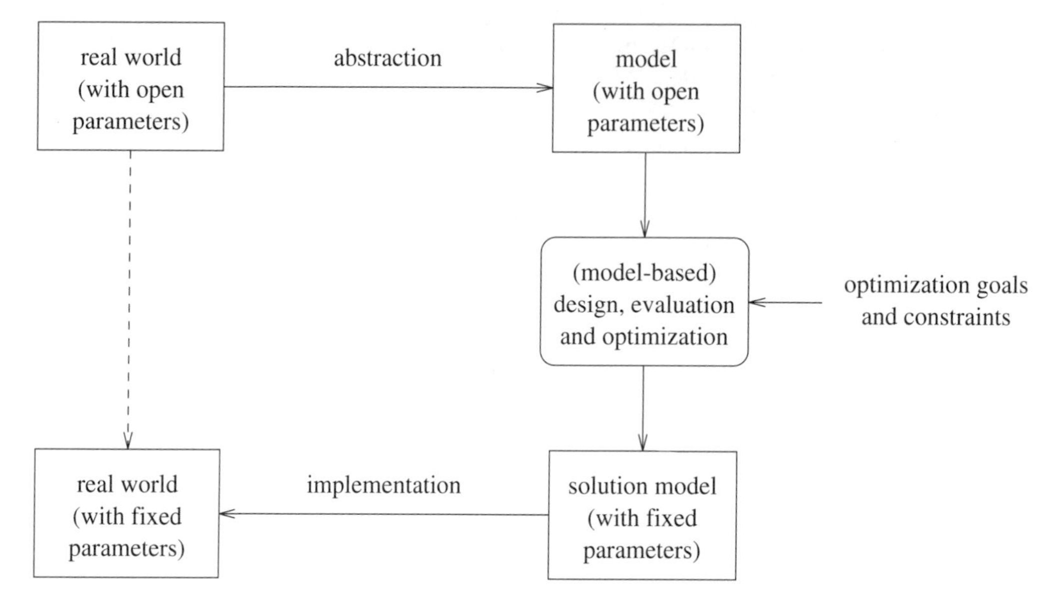

Figure 2.7 Real world and model

Qualitative and quantitative aspects Typical system models employed in the development process focus on qualitative aspects of the system. As described in section 2.1.3, qualitative aspects comprise structural, functional and behavioral aspects, including relations and dependences between the different components or parts of the system, e.g. the communication relationship between parts of the software or precedence relations between the parts during the execution. Unlike typical models for system analysis and design, performance models focus on quantitative rather than qualitative aspects of the system.

Dealing with complexity In order to deal with the complexity when modeling performance-relevant aspects of systems, a set of principles known from systems engineering is applied (see also [BrHa93, Budg93]).

Abstraction is a means of focusing on the aspects important for the specific development task to be solved by using the model. In the case of performance engineering, the models are the basis on which to evaluate the performance of the system and to select the most appropriate design and implementation alternatives to support the required performance. The abstraction can be viewed as a transformation from the real world into the model world. Based on the model world, design decisions are made. The result of the decision-making process, which may also be represented in the model world, is then transformed back to the real world. The back-transformation to the real world is a part of the design, implementation and integration phase of the system development. The relation between the model and real world is depicted in figure 2.7.

Projection allows the system to be viewed from a specific angle, rather than considering the system as a whole as with abstraction. Projection is applied to focus on specific performance phenomena of the system, i.e. to focus on the resource utilization or on the performance figures visible at the external interfaces. Projection is also applied when the system is evaluated for different workload scenarios.

Partitioning and aggregation are especially applied with hierarchical performance models. The idea is that parts or components of the system are separately modeled and evaluated in detail. The respective results are approximated by simpler submodels in the overall model. Thus, the evaluation of the overall model is based on simple models of the parts or components, rather than on the respective detailed submodels.

Generalization and specialization are employed with object-oriented modeling techniques. With these techniques, performance models can be derived by inheritance and specialization of already established classes. This approach is often used to derive simulation models from given class libraries. It saves time in building the model.

Formalization is an essential principle and a prerequisite to support tool-based evaluation and optimization.

Approximation is commonly involved when abstractions from the real system are made. In addition, approximation is typically involved with hierarchization.

2.3.1.2 Issues with Performance Modeling as Part of the System Engineering Process

In order to derive an appropriate and useful performance model, several issues are of importance. They can be classified as issues concerning the modeling approach as such and issues concerning the specific model instance to describe a system.

The most important questions concerning the modeling approach in general are as follows.

Expressive power Expressive power allows for the accurate specification of the system by the selected modeling approach. Thus, expressive power is important for the derivation of performance models that allow for the accurate modeling of important phenomena of the system.

Ease of use Ease of use defines the simplicity of constructing a model, i.e. whether the parts of the system or different aspects can be described in an intuitive manner and independent of each other. Ease of use allows one to break down the problem in independent parts and to express the solution in an additive manner. This supports flexibility which allows for changes to the system to be limited to small parts of the model. This in turn allows easy incorporation of changes to the system into the model.

Flexibility is especially important when the performance model is supposed to serve for the evaluation of the system under various workloads, to study the influence of a variety of modifications of the system, or when the model is the base for an automatic optimization of the system design or configuration. Especially for the latter case, the model has to be flexible enough to be automatically adaptable to different design alternatives. If an automatic optimization is supported, the parameters to be varied as well as the constraints for the selection of certain decision alternatives have to be modeled explicitly. Depending on the kind of optimization being supported, the description of the constraints on the optimization can be a complex matter.

Automatic model derivation The question whether the automatic derivation of the performance model is supported is important to minimize cost of performance engineering activities in the system development cycle. The requirement that the performance model is to be derived automatically from some design model or implementation has an influence on the modeling approach itself. In this case, the mapping of the entities in the design model or the implementation on the performance model has to be formally defined.

Integration with other models For performance engineering, the degree of integration of the performance model with other models used in the systems engineering process is important. An issue related to the automatic derivation of performance models is the degree of integration of the performance model with the other models employed for system analysis, design and implementation. The question is to what extent performance and other aspects of the system are modeled by a single joint model or a core model. Alternatively, does there exist a single comprehensive model from which the special model needed for a specific activity of the development process (e.g. a performance evaluation) can be derived automatically?

Disjoint models allow special focus on the respective aspects and how specific phenomena can be described as clearly as possible. Conversely, joint models inherently support consistency between the models but may get very complex and hard to understand. An important question when different models are used to model the different aspects is how to keep the different models consistent with each other. For example, if changes to the design model are made, the question of how the performance model keeps up with these changes is important. The issues involved with the integration of different models and views are discussed in [Wood95].

Level of detail With the derivation of the specific model instance, many details are involved concerning the specific mapping of the phenomena of the system on constructs in the modeling approach. However, this depends heavily on the specific problem or application area at hand. The most important general question concerning the specific model instance is the level of detail of the derived model.

Detailed models are costly to build and evaluate. In addition, they require detailed performance data which may be hard to derive. Conversely, coarse models may lack the needed accuracy to reason about the performance of the system. Thus, a reasonable compromise has to be found. Note that the appropriate level of detail also depends on the specific purpose of the model.

2.3.1.3 Reference Model for Performance Modeling

Before discussing specific performance models, we define a reference model for performance modeling. The reference model provides a framework to model and understand performance as-

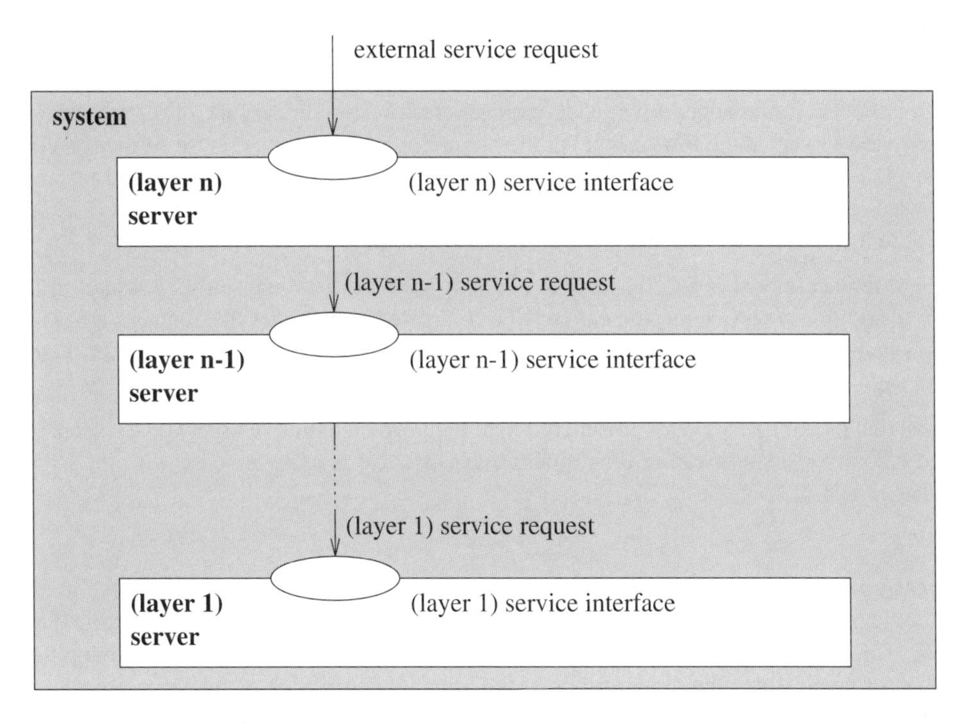

Figure 2.8 Architecture of the reference model for performance modeling

pects of systems. Thus, it should help in understanding the important issues involved with performance modeling.

Architecture of reference model The reference model provides the guidelines to model a system from the performance viewpoint. The architecture of the reference model is outlined in figure 2.8. The model follows a hierarchical structure similar to the ISO reference model on Open Systems Interconnection (OSI) [ISO84].

The reference model is employed to model a **system** from the performance viewpoint. In principle, the system may be a complete computing system or only a part of a computing system, e.g. a computing subsystem, the available resources only, or the computer hardware plus the operating system.

A system in the reference model consists of a hierarchy of **servers**. Each server has a **service interface** where it offers services to upper-layer servers. Services can be requested from a server by issuing **service requests** to the respective server. **External service requests** are the service requests that are issued from the environment to the system.

In order to serve a service request, a server may itself request services from lower-level servers. Thus, services (or service requests) are implemented by a set of service requests issued to lower-layer servers. The manner in which a service request is served by services of lower-level servers is called **implementation**. Thus, the implementation of a server defines how a service request issued to it is mapped on service requests for lower-layer servers. The only exception applies to the servers at the lowest level in the hierarchy, which may not issue further service requests. Note that the mapping may be dynamic. Thus, it may depend on the actual instance of the service request issued to the server, the state of the server, or the results of a service request it issues to lower-level servers on behalf of the service request it is serving.

The reference model is strictly recursive, i.e. service requests may only be issued to lower layers. In addition, a service request can only be completed if all the service requests it has issued to lower layers on behalf of the initial service request have been completed. However, note that this does not necessarily mean that concurrent service requests are served sequentially.

Note that the kind of services considered here focus on quantitative aspects, i.e. system stimuli and resource demands, rather than qualitative aspects, e.g. as is the case with the ISO/OSI reference model.

Application of the reference model In the following, we instantiate the reference model to an example performance modeling approach. The specialized reference model is outlined in figure 2.9. The model comprises three layers (types of servers), the application, the mapping and the resource layer. Thus, the model supports the separate description of the three aspects of the system as proposed in [Herz89].

The architecture of the performance model as shown in the figure is often employed to model parallel and distributed systems. The reference model is especially popular for performance simulation and model-based optimization techniques. As an example of the use of the three-layer model consider a distributed-memory system running a parallel program. In this case, the application layer models the parallel program, the resource layer represents the available processing resources and communication links of the system, and the mapping layer defines the spatial mapping of the application program on these resources.

In the remainder of this section, we focus on the general nature of the three layers. Examples of models for the different layers will be provided in sections 2.3.2 through 2.3.4.

The **application layer** models the application part of the system. Typically, the application layer represents the application software run on the system. The application layer provides services to the environment, i.e. requested by system stimuli, and provides these services based on services provided to itself by the underlying mapping layer. In a sense, the services (or resource demands) requested from the underlying mapping layer are a function of the application layer and the external service requests (system stimuli).

The **mapping layer** associates the services (resource demands) requested by the application layer to the actual resources. Typically, the mapping layer models the distribution of the application software on the available processing and communication resources of the parallel or distributed system. The mapping may be static or dynamic, depending on the specific needs of the application. Note that the mapping layer is a very important part from the optimization standpoint.

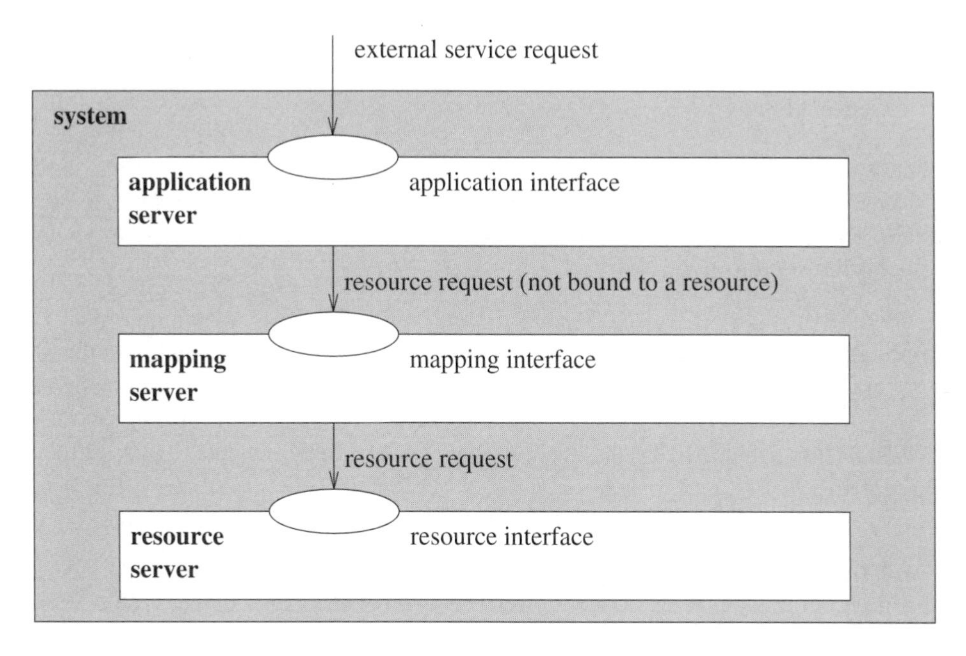

Figure 2.9 Three-layer performance reference model for parallel and distributed systems

Its parameters are important for the optimization of the system performance. Many model-based optimization techniques for parallel and distributed systems focus solely on these parameters.

The **resource layer** models the physical entities of the system that provide the physical resources to serve the requested services. Typical examples of the resources are processing entities as well as communication entities of distributed systems. To what extent system services (as provided by operating systems) are modeled by the resource layer is a matter of definition.

In addition to the three layers that represent the system, a fourth description is needed that models the environment.

The **system stimuli** describe the workload imposed on the system as a whole. The stimuli are external service requests issued to the system.

We highly encourage the use of disjoint models as proposed by the reference model. It allows one to keep the resource demands issued by the application and the available resources as separate as possible. The separation supports flexibility in changing the system and to automatically evaluate different design and implementation alternatives. In addition, the separation supports reuse of the respective descriptions for other product variants or product lines.

In the following, we apply the three-layer reference model to describe approaches to model the different aspects of the system, i.e. the application, the mapping, the resources and the system stimuli.

2.3.2 Application Modeling

2.3.2.1 General Issues

The application model describes the application-specific aspects of the system. Thus, the application model refers to the part of the system that makes the rather generic computing resources accessible to the specific application problem at hand. Typically, the application layer models the application software run on the system.

The application model in the three-layer reference model As pointed out with the three-layer reference model, the application model represents a performance-oriented description of the services the system offers at its external interface. The application model itself uses services provided at the mapping interface, i.e. services provided by the mapping model to provide these externally offered services.

Purpose of the application model The main purpose of the application model is to model quantitative aspects of the application, i.e. aspects relevant to the performance of the system. However, as the performance of systems also depends on qualitative aspects, these also constitute a part of the application model.

- *Quantitative* aspects of the application especially are temporal properties of the application, e.g. resource demands from components of the system (internal services) such as execution cost, communication cost or memory requirements.
- *Qualitative* aspects of the application model specify semantic properties of the application at a more or less abstract level. Qualitative aspects of the application include
  - the structure of the application, i.e. of its implementation,
  - the data flow,
  - the control flow, and
  - the data dependence of the application.

Classification of application models Depending on the data dependences of the application, the execution triggered by the external service requests may vary considerably. In addition, executions of the system may also depend on the actual state of the system at the time of the arrival of the service request. Thus, a use case is a function of (1) the stimulus or stimuli that trigger it, (2) the state of the system on arrival of the stimuli, and (3) possibly other stimuli arriving while the initial stimulus is served.

Applications differ in the possible **variance of the execution**. Thus, application models may be classified according to the variance of the use cases they support.

- *No variance:* The data dependence of the application is negligible from the viewpoint of the performance model. Thus, no data dependence is visible at the abstract level of the application model.
- *Quantitative variance:* The data dependence is influencing the resource demands of the internal services, but has no impact on the kind of internal service requests issued during the execution. Thus, the execution structure of the application is static.
- *Quantitative and qualitative variance:* The data dependence has also an impact on the internal service requests issued during the execution of an external service request. This is the worst case for performance modeling. It is also a very common case.

Note that the variance visible in the model depends on the level of detail in the model. Thus, for a given use case, a detailed model may exhibit quantitative and qualitative variance. Conversely, a coarse-grain model of the same use case may not exhibit qualitative variance or may not even show quantitative variance.

Note that often only the performance of a small set of the possible use cases is of interest. However, this does not necessarily imply that the other use cases are completely negligible. An example is the memory requirement for the program code in an embedded system which is independent of the frequency of the occurrence of the different use cases. Another example relates to hard real-time systems. There, all possible use cases of the system have to be considered that are able to negatively influence the performance of use cases with hard real-time constraints.

A second problem is the occurrence of **mutual interference** of a set of executions or use cases. Due to limitations of the available resources this may be an issue. Mutual interference between use cases may be classified as follows:

- *No interference:* Different executions do not influence each other in any way. For example, no activity of an execution may be waiting for a resource that is held by an activity belonging to another execution. This is guaranteed when the period between two consecutive stimuli arriving at the system is larger than or equal to the execution time caused by a stimulus. While this is a sufficient condition for the property to hold, weaker conditions may be sufficient for specific systems. An example is a strictly pipelined system, where it may be sufficient that the execution time spent on the bottleneck server for a single stimulus is smaller than the period of the arrivals of the stimuli.
- *Mutual interference:* As pointed out above, mutual interference denotes the case where a set of executions is interfering at the available resources. Interference has a negative impact on the performance figures of the use cases. A performance figure especially sensitive to interference is the response time. Conversely, the influence of interference on throughput figures and resource utilizations is usually much smaller.

Note that the issue is highly related to the definition of the workload, i.e. the number and kind of system stimuli issued to the system during some time period. We will deal with these more advanced issues in section 2.3.2.3.

Selection of an appropriate application model Important questions when selecting an application model concern

- which specific aspects of the application are to be modeled and which are to be neglected, and
- the level of detail of the model.

The selection of an appropriate application model depends on many factors, most notably the following:

- the structural and functional properties of the application, e.g. the presence of precedence relations between tasks or the presence of data-dependent behavior,
- the expected accuracy of the model,
- the required quality of the computed solutions of the evaluation or optimization problem, and
- the kind of evaluation or optimization problem to be solved, i.e. the evaluation of the response time for certain use cases or the optimization of the mapping of the processes implementing the application on the processors of a distributed memory system.

2.3.2.2 Basic Application Models

From the literature, a wide variety of approaches for modeling the application of a system is known. We describe the most important ones amenable for a performance evaluation and optimization (see figure 2.10). Surveys of application models can also be found in [NoTh93, Heis94].

Example We use a single example throughout this section to illustrate the different application models. The example is given in figure 2.11. It outlines the design of a simple parallel or distributed program with four processes. Actually, the figure does not only describe the application. In order to support the intuitive understanding of the model, we have attached a time axis to the program. Thus, the figure implicitly models aspects of the resources and the mapping.

In the context of performance modeling, we discriminate between the terms 'task' and 'process' as follows. A *task* denotes some unit of execution that performs a specific task or provides a specific service. Tasks are entities focusing on the functionality of the system. Examples of tasks range from a single processor instruction to a complex sequence of execution, e.g. to partially or completely handle an external service request. A *process* represents a unit that exhibits a certain behavior. Processes may serve several tasks. Examples of processes in our sense are operating system processes or threads provided by thread libraries.

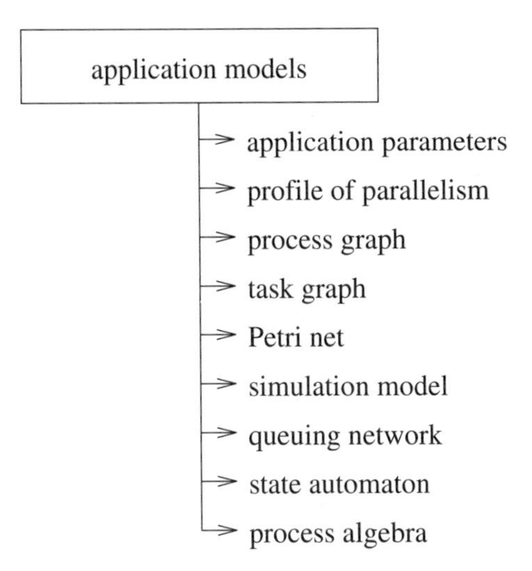

Figure 2.10 Application models

Application parameters A simple approach for modeling the application is to employ a set of parameters. Important parameters are

- the degree of parallelism,
- the processing demands, i.e. the resource demands for tasks or processes of the application,
- the communication demands. i.e. the resource demands for communication between different tasks or processes,
- the length of the longest path in the application,
- the degree of synchronization of the tasks, i.e. the number of predecessors (fan-in) and the number of successors (fan-out), and
- the memory requirements of the processes.

For the different parameters, the average, maximum and minimum values as well as the variance may be specified.

The important parameters of the tasks of the example application are given in table 2.1. The average degree of task parallelism denotes the average number of parallel tasks over the makespan (longest path) of the application (i.e. 45 time units).

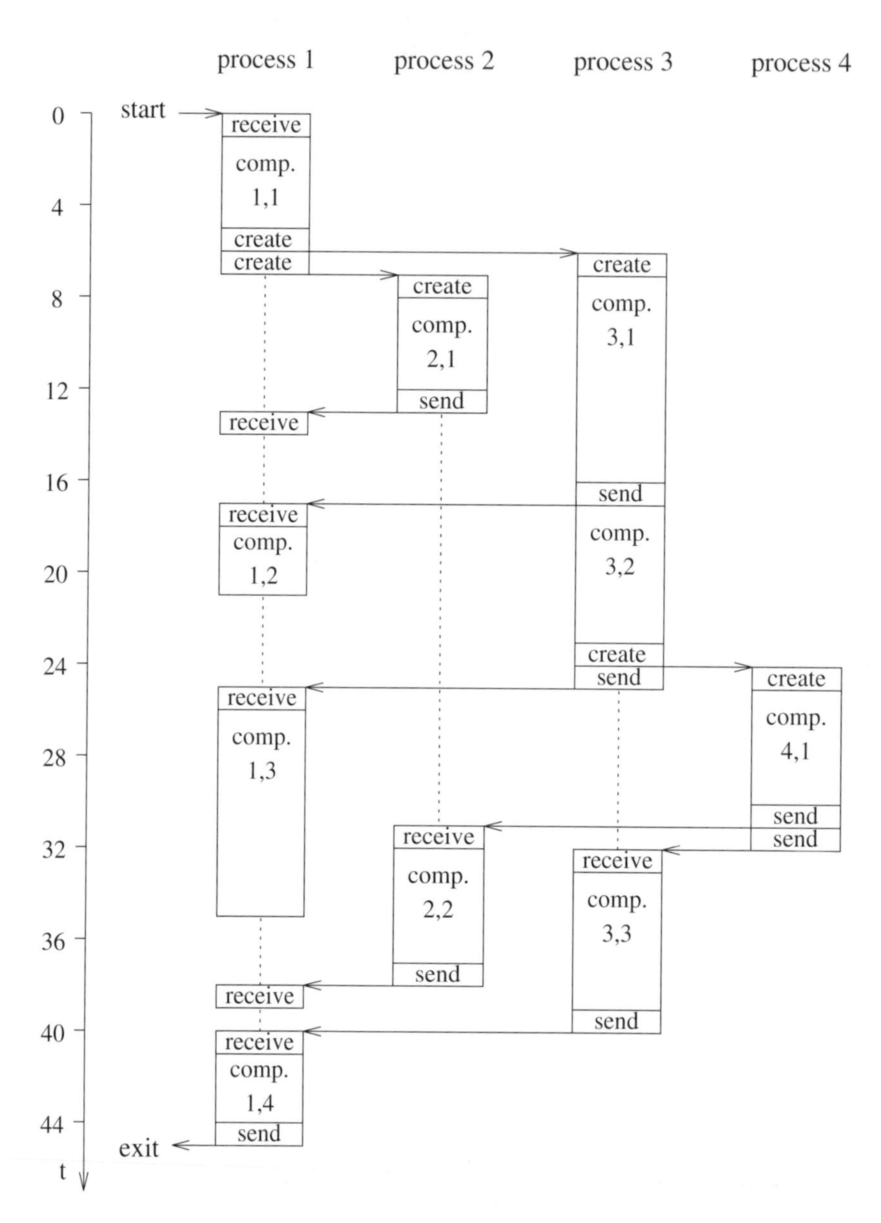

Figure 2.11 Outline of the design of a parallel or distributed program

Profile of parallelism Another approach for characterizing the application is the parallelism profile [Heis94]. In figure 2.12, two profiles of parallelism are given for our example. As we focus on the application, the availability of a sufficient number of processors needed to serve all

Application parameters	Minimum	Maximum	Average
task parallelism	1	3	1.71
processing demand per task	3	9	7.7
processing demand per process	9	28	19.25
communication demand	*not specified*		
longest path	45		
fan-in	1	3	1.4
fan-out	1	2	1.4
memory requirements	*not specified*		

Table 2.1 Parameters of the application

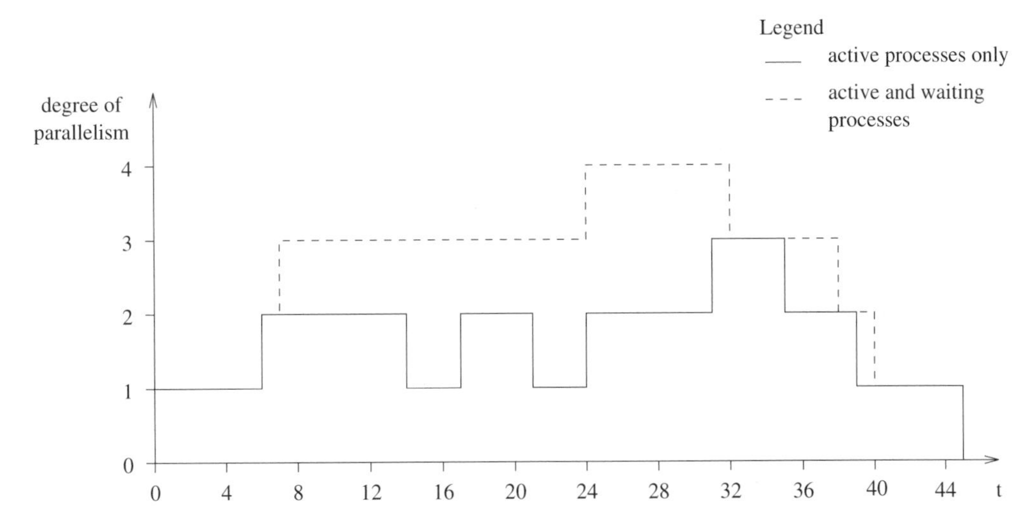

Figure 2.12 Profile of potential parallelism

processes in parallel has been assumed. The solid curve in the figure represents the degree of parallelism with respect to active processes, i.e. processes that perform some computation. Conversely, the dashed curve shows a different interpretation. It describes the degree of parallelism with respect to active and waiting processes. This denotes the fact that any process, whether active or waiting, consumes some resources, e.g. requires memory space or causes overhead for process management.

Process graphs In order to graphically describe communication relations between processes, process graphs (see also [NoTh93]) have been introduced. A process graph consists of processes and communication relations. The process graph for our example is given in figure 2.13. In the

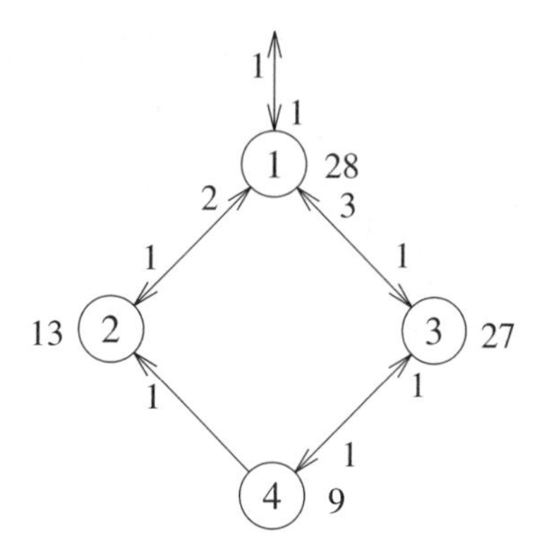

Figure 2.13 Process graph

graph, processes are modeled by nodes, and communication relations are modeled by arcs. Note that the modeled communication relations reflect the spatial distribution of communications between processes rather than any temporal dependences. In fact, there is no information in the process graph that states the time when actions are performed. The graph just states that certain parts of the system represent a certain load for the resources. Especially, the arcs of process graphs do not provide any information on precedence relations between activities.

Typically, the nodes are annotated with information concerning the processing demands of the process. If our three-layer reference model is applied, processing demands are given by the kind and amount of services required from the resources. Under some circumstances, e.g. that the system has a homogeneous set of processors, the processing demands may also be given directly, e.g. the required processing time. For simplicity of the graph, this approach is also taken in figure 2.13. Thus, the figures associated with the processes denote the processing demands of the processes. Note that the given processing demands specify the sum of the demands required by a process to handle some load. In our example, the given cost specifies the processing demands for a single execution of the example program given in figure 2.11.

The arcs of the process graph can be annotated with information concerning communication cost. Examples for the specification of communication cost are the total amount of data transmitted between two processes, the number of communications between them or the absolute time to transmit the data between the processes. Note that the first two approaches are consistent with our three-layer reference model, while the last is not. In our example, we have annotated the arcs with the number of communications between the processes. In general, communication relations may be modeled by directed arcs or by undirected edges. The design choice depends on whether or not the additional information provided by directed arcs is relevant for the performance of the system. This may be the case if links or channels with different characteristics are employed for the different directions.

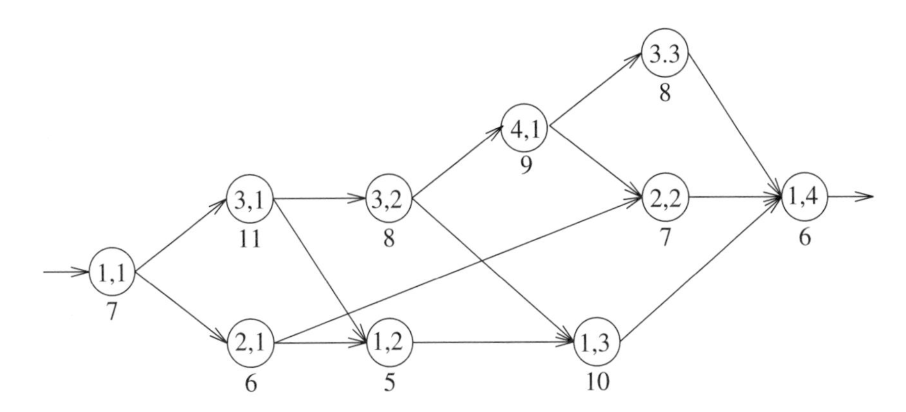

Figure 2.14 Task graph

Task graphs Unlike the more spatial focus of process graphs, task graphs are more oriented towards temporal relations, especially precedence and synchronization constraints. A task graph (see also [NoTh93]) consists of tasks and precedence relations. Because of this, task graphs are also called precedence graphs. A task graph defines a partial order on tasks. The task graph for our example is given in figure 2.14. In the graph, tasks are modeled by nodes, and precedence relations are modeled by arcs. With task graphs, arcs may represent simple precedence constraints that may not require any resource demands, or may model an explicit communication between the respective tasks. In the example task graph, we do not display precedence relations between the tasks of the same process where these are modeled indirectly by other precedence relations, e.g. between nodes 1,1 and 1,2.

The nodes of task graphs are annotated with information concerning the processing demands of the tasks. If our three-layer reference model is applied, processing demands are given by the kind and amount of services required from the resources to process the respective task. Under some circumstances, e.g. that the system has a homogeneous set of processors, the processing demands may also be given directly, e.g. the required processing time. For simplicity, this approach is also taken for figure 2.14. Thus, the figures associated with the tasks denote their respective processing demands.

The arcs of the task graph can be annotated with information concerning communication cost. Examples for the specification of communication cost are the size of the message transmitted between the two tasks, or the absolute time to transmit the message between the tasks. Note that the first approach is consistent with our three-layer reference model. The last approach is not. In our example, we have not annotated the arcs since we have assumed that communication takes zero time. Thus, there is no difference in the arcs which would require an annotation.

Several variants of task graph models exist. Extensions to express disjunctive structures, e.g. to model branches, have been proposed in [Mits99, GiLi90, ChAb82]. Examples of models that integrate aspects of task graphs are execution graphs as defined in [Smit90], MSCs (see section 4.2.1 for details), and dependence graphs [BaGS94] used by compilers to optimize code.

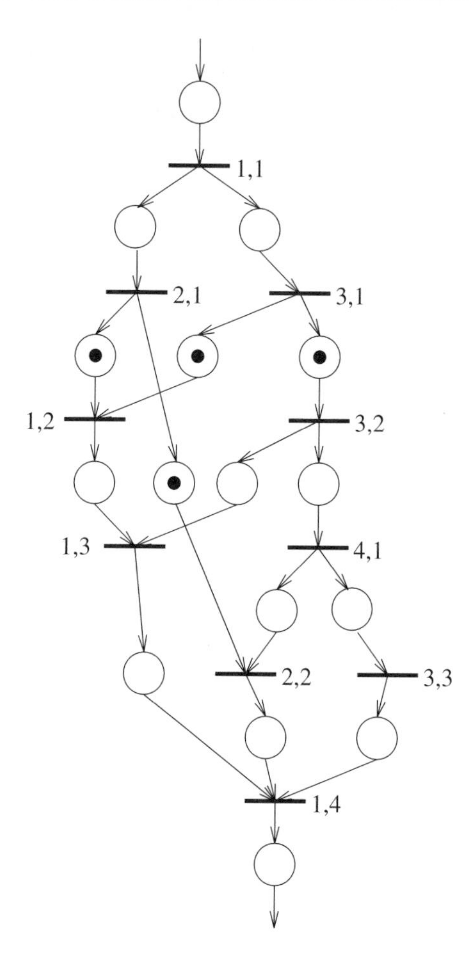

Figure 2.15 Petri net (with tokens marking a specific state of the computation)

Petri nets Petri nets (see also [Pete77]) are strong in modeling synchronization constraints and contention. Petri nets are more expressive than the previous approaches.

Petri nets may be appropriate to model the application. The Petri net model of our application example is depicted in figure 2.15. Similar to the task graph above, we neglect edges between the tasks of the same process where appropriate. Essential parts of Petri nets are places (circles), transitions (bars), tokens (bullets) and arcs. Places store tokens waiting for transitions to fire. A transition may fire if a token is available at each of the places that has an arc pointing to the transition. When the transition is fired, a token is removed from each of the places. In addition, a token is inserted in each of the places to which the transition is pointing.

Performance modeling with Petri nets is typically supported by adding delays to the transitions [Ramc74, AjBC86].

A nice feature of Petri nets for application modeling is that they allow the description of disjunctive as well as conjunctive structures. Thus, conditional branches as well as fork/join structures can be described. In addition, complex synchronization constraints can be modeled with Petri nets.

However, as can be seen in the figure, Petri nets do not necessarily provide a very concise approach to model applications.

Simulation models The basic models introduced so far may lack the expressive power to precisely model some specific phenomena of the system under study, e.g. complex dependences between the modeled entities.

An approach to enhance the expressive power of the performance model is to employ simulation models. Simulation models represent pseudo code with respect to the details of the functionality of the application. The model representing the application is executed on the model of the resources. Thus, some of the constructs in the simulation language consume time on some resources. Simulation models typically support deterministic as well as a variety of stochastic variables to model time. With simulation, details concerning behavioral aspects of the system can be modeled employing data structures and control-flow constructs known from programming languages. For example, this allows one to model dynamic reaction to load figures of the underlying system, a feature needed for overload control in telecommunication systems or to model systems with QoS guarantees.

However, flexibility has its price. Due to the growing expressive power, the modeling approach becomes complex, and the evaluation of the model is typically limited to simulation. Thus, the evaluation of the model may become very time-consuming.

Other models that include application aspects Other models that are not typically used to solely model applications, but rather to model systems as a whole, are queuing networks and timed automata.

Queuing networks [Jain91] are networks of a set of individual queuing stations. Unlike most of the other models, queuing networks typically assume stochastically varying time intervals. Queuing models are typically applied to derive a single monolithic performance model, which describes the complete system including the resources and the mapping on the resources. An exception to this is layered queuing networks [RoSe95] that specify the application as a network of queuing stations (servers) that are mapped on a separate (underlying) queuing network that models the resources.

Performance-relevant aspects of the application may also be modeled by a state automaton. The states of the automaton represent the states of the application. With timed automata [Alur90], time is associated with transitions, or more precisely, clocks and conditions are used to define the time when a transition occurs (see [LaSW97] for an introduction). Timed automata are well suited to modeling temporal requirements.

Process algebras have been extended to model time and performance aspects. With timed process algebras, timed actions are added to model deterministic delays. Thus, timed and timeless

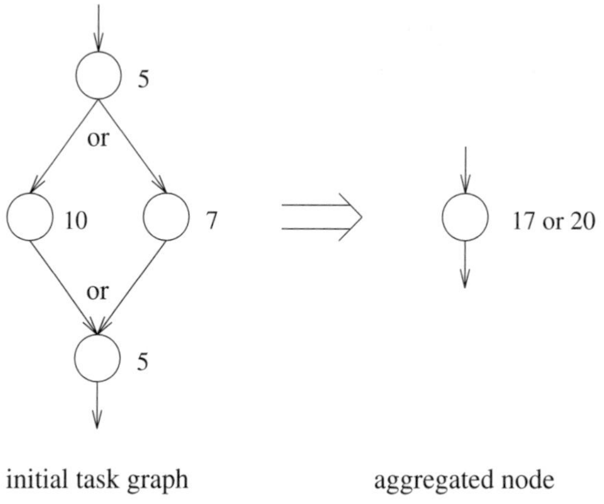

initial task graph aggregated node

Figure 2.16 Application of the aggregation principle to a task graph with disjunctive logic

actions are used to describe the application. An overview of timed process algebras can be found in [NiSi91].

2.3.2.3 Modeling Advanced Aspects

So far we have mostly assumed that the application can be represented by a single deterministic description, e.g. by a single task graph with deterministic processing and communication demands.

In section 2.3.2.1, two issues have been identified, namely (1) the possible variance within a single execution, and (2) the mutual interference between a set of executions or use cases. The first issue, i.e. the problem of dealing with variance in executions, is clearly a modeling issue. The second issue is also related to the performance evaluation.

Several approaches to dealing with variance in the application exist and we will summarize the most important ones.

Aggregation of nodes One approach to dealing with variance is aggregation. Aggregation reduces the level of detail. This may allow one to aggregate a set of data-dependent executions (set of nodes) into a single, more coarse activity (single node). This is graphically displayed in figure 2.16 for a task graph with disjunctive logic. Aggregation works perfectly as long as the influence of the data dependence can be confined to a small set of nodes which can be aggregated to a composite node. If this is not strictly the case, approximations may be applied.

Under the given assumptions, the application of the aggregation principle transforms a model with qualitative and quantitative variance to a model with quantitative variance only. For example, in figure 2.16 the difference in the length of the two paths results in a variance of the execution

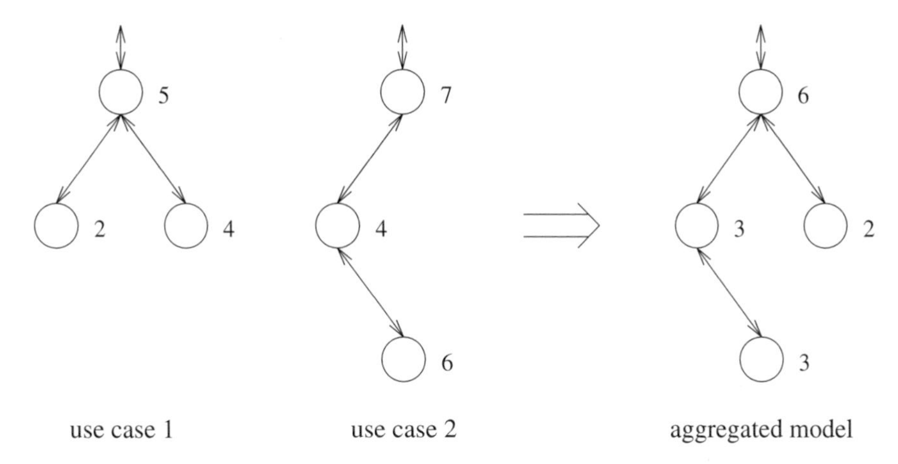

Figure 2.17 Application of the aggregation principle over different use cases

cost of the aggregated node. Depending on the specific case, also quantitative variance may be completely eliminated or minimized to such a degree that it is negligible in the aggregated model, e.g. by concentrating on the worst case.

Aggregation of resource demands An alternative to the aggregation of nodes in the model is to focus on aggregated service or resource demands right away. This approach can be easily applied with process graphs due to the fact that these models abstract from temporal dependences. In this case, each node and edge in the process graph is annotated with the sum of the resource demands that results from the workload rather than the resource demands for a single execution or use case. An example is displayed in figure 2.17. For the aggregation of the resource demands, we have assumed that the two use cases appear with equal frequency. Thus, the resulting resource demands are averaged over the respective resource demands of the use cases. A disadvantage of the approach is that the model may be too abstract for some performance metrics to be evaluated. For example, the neglect of temporal aspects with process graphs does not allow accurate reasoning about the response time of use cases.

Overlaying and enumeration of the evaluation An alternative to dealing with the problem in the model is to pass the problem to the evaluation. In other words, a set of models is used to model different use cases instead of using a single model that represents all use cases. Then the system is evaluated for each of the different use cases independently, or jointly in case the workloads caused by different use cases overlap in time.

As an example of the independent evaluation of different use cases, consider two task graphs that model two use cases that do not interfere in their execution. In other words, the system is never executing both task graphs concurrently. In this case, the response times for the two task graphs can be evaluated independently. The results are combined after the evaluation is complete [Mits99].

Stochastic modeling Instead of using deterministic figures with the models, stochastic values may be employed. For example, probabilities may be associated with arcs to model data-dependent branches. In addition, stochastic figures are used to model various time distributions of resource demands, e.g. processing times or communication cost. The use of stochastic figures is common with queuing models and simulation. Other important examples where this approach is taken are stochastic task graphs [BaJL93, SaTr87], stochastic Petri nets [Moll82, AjBC86, Lind98], stochastic automata [PlAt91], and stochastic process algebras [HeHM98]. However, note that stochastic models are more costly to evaluate than deterministic models.

Worst case assumptions Instead of modeling the application in its different varieties, worst case assumptions may be used. For example, a conditional branch in an application is replaced in the model by the longer path. Thus, data dependence is reduced to its worst case. The problem in this approach is the risk of overengineering the system.

2.3.3 Resource Modeling

2.3.3.1 General Issues

Resources in computing systems are physical resources such as processors, communication devices and storage devices. In addition to physical resources, logical resources may be of importance. Important logical resources to be described in performance models are entities which are not abundantly available. Examples of such logical resources are critical sections which may only be executed by a limited number of processes at a time, or data entities protected by semaphores or similar constructs to maintain consistency.

The resource model in the three-layer reference model As described in the three-layer reference model, the resource model represents a description of the resources offered to the application. The resource layer is the only layer that does not rely on other services.

Purpose of the resource model The purpose of the resource model is to model quantitative as well as qualitative aspects of the available resources.

- *Quantitative* aspects of the resources are especially their number and capacity. Examples include the processing capacity of processors or other devices, their possible degree of parallelism, the bandwidth and latency of communication devices, and the capacity of storage devices.

- *Qualitative* aspects of the resource model comprise their service strategy and information on the configuration and possible combination of the different resources.

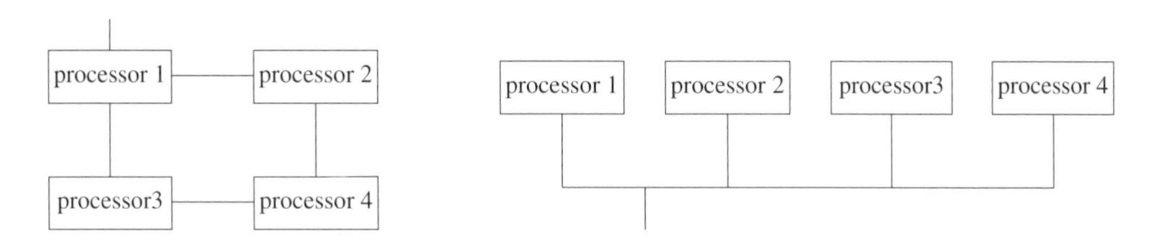

(a) partially connected multi-processor system (b) fully connected (bus-based) multi-processor system

Figure 2.18 Two machine graphs

Selection of an appropriate resource model When selecting the resource model, two aspects are of importance. Besides the level of detail, qualitative aspects of the resource model are important. The extent to which qualitative aspects are considered by the resource model depends on the kind of evaluation or optimization problem to be solved. For example, if the amount of the resource demand of the parts of the application is the only concern, e.g. to evenly distribute the load on different resources, service delays caused by contention can be completely neglected. On the other hand, contention and the strategy to handle it is very important if the system is to be optimized for minimal response times.

2.3.3.2 Basic Models

The graphical description of resource models is only possible to some extent. It is mostly limited to the structure and interconnection of the computing resources. Additional annotations are used to formally specify peculiarities and details of the resources. The most important examples of information typically described separately are the capacity of the resources, the services they offer, and their service strategies.

Machine graphs Graph models represent the most popular class of resource models for the automated model-based optimization. With machine models, two types of graphical symbols exist, namely nodes to model computing resources and edges or arcs to model communication resources. Examples of simple machine graphs are given in figure 2.18. The figure depicts two different configurations of a multiprocessor system with four processors (nodes). The processors are connected to other processors and to the environment by a set of links (edges).

Obviously, additional information is needed to supplement the machine graph. The most important are the capacity of the processors and the communication links, and their strategies to handle contention. Typically, only a single service is provided by each of the resources, namely a computation service for the processor, and a communication service for communication links.

Machine graphs are often chosen as an input representation of optimization algorithms to compute the allocation of the processes or tasks on the processors of the system. Important issues these

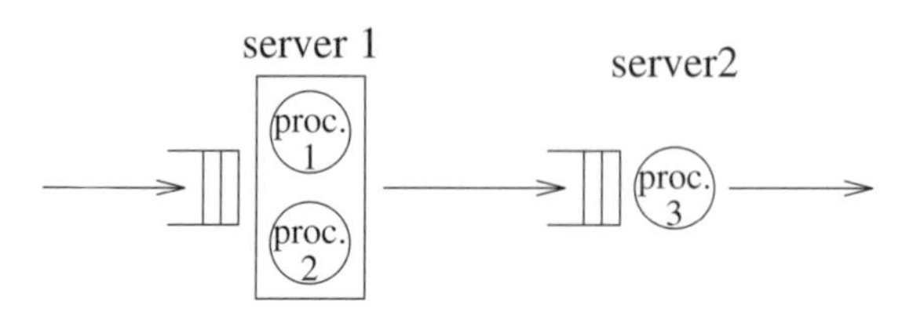

Figure 2.19 A queuing network to model a pipelined multiprocessor system

optimization algorithms deal with are the balancing of the load on the processing resources and the minimization of communication cost.

Queuing networks As noted above, queuing models are typically applied to derive a single monolithic performance model, which describes the complete system including the resources and the application. However, queuing models may also be employed to solely model the resources of the system and their interconnection structure [Mits94]. An example of this is given in figure 2.19. It shows a system with two servers which serves incoming requests in a pipelined manner. The first server is a parallel processing unit with two processors.

If the queuing network focuses on the available resources, the arcs connecting the queuing stations show the principal communication relationships between the servers. Typically, each queuing server only provides a single service. However, extensions to model multiple services exist, e.g. see [PaFi89].

If application-specific information is added, figures describing the probabilities that the different branches of the arcs are taken are typically given.

Petri nets Petri nets are especially suited to model synchronization constraints and contention. Thus, they are often employed to model application aspects rather than the sole modeling of the resources. However, Petri nets are also employed to subproblems related to resource modeling, e.g. to model the access to shared resources as processors or busses. An example of a Petri net model where two processors compete for access to a common bus is shown in figure 2.20.

2.3.3.3 Modeling Advanced Aspects

Service strategies Besides its capacity, the service strategy is probably the most important characteristic of a resource. Service strategies are especially important if the response time of the system is under study.

Examples of service strategies comprise first-come-first-served, last-come-first-served, shortest-job-first, and priority scheduling. In addition to these rather simple (static) strategies, a number of combined strategies as well as dynamic strategies are used in practice.

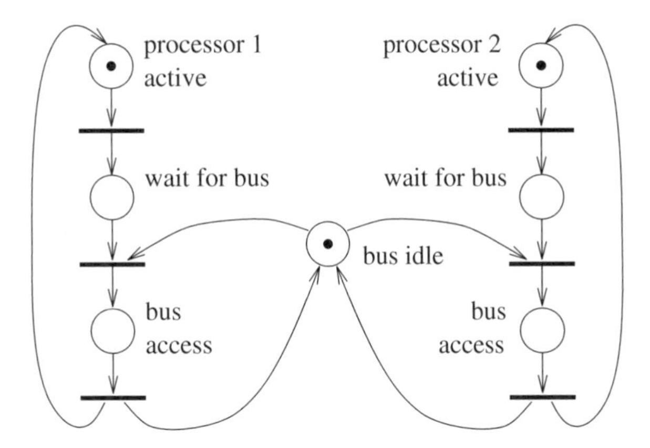

Figure 2.20 A Petri net to model contention caused by limited resources, e.g. a shared bus in a two-processor system [AjBC86][5]

Priority-based service strategies are often classified in preemptive and nonpreemptive strategies. Preemption allows processors or service requests with higher priority to interrupt those with lower priority. With nonpreemptive services or resources, the currently served service request is completed before a second request may gain access to the resource.

The extent to which the scheduling strategy is modeled by the resource model depends on the specific approach. Most flexible are simulation models of the resources.

Caching and swapping So far we have implicitly assumed that the actual resource demand (not the response time) derived from a service request issued to a resource is independent of the state of the resource. With caching and swapping, this is no longer true. Instead, the resource demand is a function of the state of the cache and the main memory.

While swapping is not typically employed in the systems we look at, caching may be a problem. The approach selected to model caching highly depends on the specific application at hand, especially whether the performance requirements imposed are hard or soft. With hard performance requirements, the worst case behavior of the resource has to be estimated. In order to support this, caching strategies should be used that minimize the worst case execution cost rather than the average cost. Conversely, with soft performance requirements. the estimation of the average behavior may be sufficient.

Heterogeneous systems Most resource models for system optimization provide just a single service, e.g. a computation or communication service. The application has to state the amount of the service required from the resource and the mapping model assigns this service request on one of the appropriate resources. As long as the resources are homogeneous, i.e. the actual resource

[5]Reproduced by permission of MIT Press.

demand for a specific resource is independent of the selected resource, this is not a problem. However, the issue becomes more complicated when the resources are heterogeneous. In the easiest of these cases, the resources behave in a linear fashion, meaning that the difference in the resource demands for a specific service request on the different resources is a linear factor. In this case, the respective factors have to be determined for each resource.

More complicated is the case where the actual resource demands on specific resources depend on a larger number of factors. Examples of these factors may be the specific instruction mix that defines the service request. For example, consider a system comprising a processor with a floating point unit and a second processor without a floating point unit. If the second processor has a higher execution rate, the actual execution cost for a service request may be higher on the first processor or vice versa. Thus, the execution cost depends on the specific parameters of the load.

There are two basic approaches to the problem. The first approach measures the resource demands of each service request on all the different processors and uses these values to compute the performance metrics for a specific system design.

The second approach is to identify the important parameters of the load (or service request) that influence the performance. In the above example, we could identify the number of integer operations and the number of floating point operations that constitute a service request. Together with the respective performance figures of the processors for integer and floating point operations, the actual execution cost on the different processors can be derived.

Overhead So far, we have focused on the resource demands directly associated with the application. We have ignored overhead of the underlying system to manage the different service requests issued to it or to provide background services.

There are two principal approaches to modeling cost caused by overhead: (1) to add overhead to each resource request, or (2) to reduce the capacity of the resource to denote the overhead. Depending on the kind of overhead to be modeled with a resource request, the additional cost may be a simple constant factor or a complex function.

2.3.4 Mapping Modeling

2.3.4.1 General Issues

The mapping model defines how the application entities are assigned to the available resources.

The mapping model in the three-layer reference model As already outlined in section 2.3.1.3, the mapping model associates the resource demands requested by the application model with the actual resources.

	Static assignment	Dynamic assignment
Spatial assignment	assignment matrix	balancing parameters and balancing function
Temporal assignment	static schedule	priorities and scheduling function

Table 2.2 Types of mapping models and needed information

Purpose of the mapping model A separate mapping model supports flexibility of performance modeling. It allows one to specify the application and the resources completely independently of each other. The mapping model mainly focuses on qualitative aspects.

The mapping model is especially important from the optimization standpoint. Typically, the parameters of the mapping model are a major subject of the optimization of the system. As pointed out above, many model-based optimization techniques for parallel and distributed systems solely focus on these parameters.

2.3.4.2 Basic Models

Mapping models are typically rather simple. The mapping model may define the spatial as well as the temporal assignment. The specific modeling approach depends also on the time when the assignment is finally decided, i.e. statically or dynamically during runtime.

Static versus dynamic assignment In the static case, the mapping model statically defines how the application components are associated with the available resources. In this case, a simple assignment function is sufficient.

In the dynamic case, the final assignment decision is deferred to runtime and is based on information only available at runtime. Thus, the mapping model itself only provides the general strategy or algorithm that specifies how the final mapping is derived at runtime.

Spatial and temporal assignment The two most important aspects described in the mapping model concern the spatial and the temporal assignment of application components to resources. The spatial assignment defines which application component is assigned to which resource. As described above, this may be done statically or dynamically.

The temporal assignment is concerned with the time instant, or the order, in which the parts of the application are scheduled on the resources.

The information defined by the mapping model for the different types of mappings is given in table 2.2. In the static case, an assignment matrix defines the spatial assignment. The temporal assignment is defined by a static schedule, i.e. the exact temporal order of the execution of the application parts on the specific resources.

In the dynamic case, the spatial assignment is typically called load balancing [HaJo90, KLKZ96]. There are two cases: the application may be distributed once before being executed, or it may be relocated during execution. The information needed to define the mapping in these cases is the balancing parameters for the different application parts and the balancing function. As noted above, the balancing function has as input also some dynamic variables that are not available before the actual execution of the system.

The dynamic temporal assignment is typically called runtime scheduling. If the scheduling is decided dynamically, two things are needed: the parameters of the application components, e.g. their priorities, and a function that defines the scheduling strategy.

2.3.5 Modeling the Runtime System

So far, we have assumed that the application is directly mapped on the underlying resources. In particular, we have not talked about the runtime system, i.e. the operating system and possible runtime support systems as provided by libraries. The important question in this respect is where to model the runtime system.

There are two alternatives. The runtime system may be a part of the mapping model. Above, we have described this approach to modeling the strategies for dynamic scheduling and load balancing. Alternatively, the runtime system may be modeled by the resource model. Thus, the resources as described in the resource model are rather abstract resources providing sophisticated services rather than just physical resources.

Our recommendation is to integrate the parts that are flexible parameters of the design into the mapping model and to keep the parts that are not open design parameters in the underlying resource model or in an additional model, i.e. a separate layer. Thus, the optimization may concentrate on the mapping model for an optimization of design parameters.

2.3.6 System Stimuli

2.3.6.1 General Issues

So far, we have focused on modeling the system with its resources and the application run on it. However, for a performance evaluation, additional information is needed concerning the behavior of the environment of the system. Most essential here is the input to the system, or in our terminology, the external service requests issued to the system.

The important information in a stimuli description is

- the types of the external service requests, and
- their intensity, i.e. the frequency of their occurrence, possibly their exact timing and interdependences with other external service requests.

Given the system specification, an external service request to some extent implicitly defines the execution (or a use case) of the system. However, note that the specific execution also depends on the actual state of the system on the arrival of a service request. The issue has already been discussed in section 2.3.2.1.

Depending on the data dependences of the execution, the actual execution (or use case) triggered by the external service requests may vary considerably. In particular, the execution resulting from a specific external service request may depend on other external service requests issued to the system until the time instant when the specified service request is completed.

Note that the exact variance or dependence of the executions on other external service requests depends on the specific application, especially the data dependences it exhibits.

Various approaches for the specification of system stimuli exist. The selected model to describe the system stimuli depends on the specific information needed to do the performance evaluation. For example, a simulation requires different information on the system stimuli than does a simple critical path analysis.

2.3.6.2 Basic Models

Event traces A simple approach to defining the system stimuli is an event trace. An event trace specifies the exact times when different external service requests (events) are issued along with their type. Thus, an event trace represents a fully deterministic specification of the system stimuli on the time axis.

Tables Another simple alternative that can be found in the literature is the use of simple tables to describe the important characteristics of the arrival process, e.g. the type and period of arriving requests. This approach is especially employed by schedulability analyses which are typically based on very simple models of the system and the system stimuli.

Automata As an alternative to the event-oriented description of the stimuli, the stimuli may be defined by an automaton. The automaton comprises states and transitions. A stimulus is issued upon the execution of a transition. Transitions may be selected based on probabilistic values. The exact time when a transition is fired depends on the transition rate or on time constraints given with the states or transitions of the automaton. Deterministic as well as stochastic arrival processes may be modeled in this way. In the stochastic case, the arrival process is typically a Markov chain.

2.3.7 Model Derivation

So far, we have focused on the different types of models and their appropriateness to describe specific aspects of the system under study. In this section, we describe how to actually derive a specific performance model, i.e. a model that describes the specific system under study. Note that our focus is on deriving models that are input to a performance evaluation of the system. This is different from the direct derivation of the performance figures for the system, e.g. the measurement of response times of the system as seen at its interface.

Since the minimization of the time needed for a performance evaluation is an important concern, the automation of the performance engineering process is important. With respect to the model derivation, we especially focus on the automatic derivation of the application model. The application model is the most costly model to derive since it typically constitutes the most complex part of the performance model.

As described above, there are two aspects a model has to deal with, qualitative and quantitative aspects. Deriving **qualitative information** deals with the issue of deriving the specific instance of the model type for the given system under study, i.e. the question of how the structural, functional or behavioral aspects of the system are described by the model. Deriving **quantitative information** deals with the question of how to derive the quantitative data needed to complete the model, e.g. the actual resource demands of the tasks or processes of the application model.

2.3.7.1 Derivation of Qualitative Information

Before a model can be derived for a given system under development, two prerequisites have to be fulfilled:

- the **model type** used to model the application has to be decided on, and
- the **mapping on the model**, i.e. the exact instructions on how to map each entity of the application under study on an entity of the application model, has to be defined.

If the mapping is intended to be performed automatically, the respective mapping instructions have to be specified formally, i.e. described in a form amenable to tools. Fortunately, this information is to some extent implicitly given by design models as described in section 2.1.3, or by the structure of the code to be modeled.

2.3.7.2 Derivation of Quantitative Information

Several approaches exist for deriving the required quantitative information. The principal approaches are as follows:

- **Measurement** of real software on the real target system (hardware and software) executed with real input data. A prerequisite for this approach is the availability of the implemented system comprising the application, the support system, the hardware, and the input data. Below we describe the application of this important approach to deriving application models.
- **Emulation** (i.e. a detailed simulation) of the execution of the real software on a detailed (complete) model of the target system executed with real input data. A prerequisite for this approach is the availability of exact models of the underlying hardware, the support system and the input data. Emulation is based on the execution of the real software with all its functional and behavioral details.

- **Analysis** of the code instructions or abstract instructions and derivation of the resource demands based on information on the execution times for the different types of instructions. With this approach, the code is not actually executed. Thus, some functional details such as data-dependent decisions have to be decided based on additional information. In addition, some kind of performance data base is needed that provides the quantitative information for each machine instruction or abstract instruction.

- **Empirical estimation** of the quantitative information based on earlier experience. For this approach, neither real software nor hardware, nor the real input data are needed. Instead, it is based on earlier measurements and on experience with the performance of similar systems or system components.

Applicability The different approaches rely on different prerequisites. Measurement and emulation are feasible alternatives only for very late phases in the development cycle, i.e. when the system implementation is close to completion. Analysis is not necessarily based on the code; it may also be based on an abstract description where the execution times for the abstract instructions are available. Thus, analysis is a practical approach for earlier development phases. The same holds for empirical estimations.

Hybrids Hybrids of the given basic approaches are possible. A mixture of different approaches may be employed, especially if the development process model deviates from the classic waterfall model. An example is the use of measurements for the parts already implemented and an analysis or empirical estimation for the other parts. Another example is the measurement of the real software on slower hardware and the derivation of the quantitative information for the real hardware from experience with the performance difference between the two hardware components.

Performance data base Since large systems are seldom built from scratch, quantitative information is typically available or can be derived for an earlier product or earlier product releases. The ideal case is to maintain a performance data base where the relevant data are stored in an organized and structured manner. Based on these data, estimates for the performance of the new product or product release can be made. For this, it is often sufficient to consider the increase in speed achievable with the new hardware technology, or the load added to the system to provide the added functionality.

2.3.7.3 Model Derivation by Measurement

When the prerequisites as defined above are met, a mainly automatic derivation of the application model from an implementation can be achieved.

We focus on the measurement of application- or use-case-oriented aspects of the system only. Other possible aspects or performance metrics are resource-oriented figures of the system. However, these metrics are useful only for the evaluation of a specific system under study and not for possible variants of the system design or implementation.

Derivation of qualitative data The derivation of the model instance of a model type may be done prior to the execution or afterwards. This depends mainly on the kind of model type that is employed and the kind of quantitative information to be derived. For example, with task graph models the actual model instance often depends on the input data for the execution. In this case, the prior derivation of the model instance does not make sense. Conversely, with process graph models this is not a problem at least as long as the process structure of the system is static.

Instrumentation The code of the system is instrumented with additional monitoring instructions that issue an event when executed. The event represents a record of relevant quantitative data of the system execution at this time instant. The event defines the instrumentation point, i.e. the point in the system by which the event has been triggered, and a time stamp. Additional data may be needed to record some context information on the execution.

The instrumentation of application software can be done automatically, if the guidelines on the mapping of the system on the model are given.

Execution and monitoring Next, the system is executed with the respective input data. The specific input data force the system to execute a specific use case for which the performance data are needed. During the execution, the events resulting from the monitoring instructions are recorded by employing a hardware, a software or a hybrid monitoring system.

Trace analysis Typically after the execution of the system is complete, the recorded event traces are analyzed. The major tasks of the task analysis are to identify corresponding events to derive their time duration from the time stamps, and to attribute the performance model with the respective data.

A comprehensive discussion of the measurement and monitoring of systems and the derivation of models from measurements can be found in [KDHM+95]. An overview is also given in [Jain91]. The automated derivation of models from measurements for the case in which the formal description techniques SDL and MSC are used is discussed in [Lemm00].

2.4 Performance Evaluation

While in the previous section we described the approaches to modeling the system with respect to performance, we focus here on the techniques to evaluate performance models, i.e. to derive performance measures for a given performance model. Here we survey approaches to performance evaluation and provide an overview of the respective tools.

As model-based performance evaluation has been a major research area for several decades, many books on the topic exist. Comprehensive information on the performance evaluation of computer systems, and queuing systems in general, can be found in numerous textbooks. An introduction into queuing models and queuing networks is given in [LZGS84]. This book provides a

Figure 2.21 A simple M/M/1 queuing station

comprehensive discussion of performance evaluation techniques for separable queuing networks. Also a very practical book is [Jain91] which covers various aspects of performance evaluation including measurement, simulation and design optimization. A classic in queuing theory is [Klei75] which covers the theory of queuing systems. Focusing more on mathematical aspects of queuing models and networks is [Lave83], which also covers simulation.

2.4.1 Introduction

In order to evaluate the performance of a computing system, information concerning the application, the resources, the mapping of the application on the resources, and the system stimuli must be available. Note that the performance model to be evaluated has to comprise these aspects even though they do not have to be explicit.

2.4.1.1 Performance Evaluation as an Abstract Function

Depending on input parameters describing the available resources, the workload to be served, and the mapping of the application on the resources, numerical values for performance metrics can be determined. In general terms, we have an abstract function f : stimuli $\times$ application $\times$ mapping $\times$ resources $\rightarrow$ performance metrics.

As an example for simple performance functions, consider the M/M/1 model known from basic queuing theory as depicted in figure 2.21, i.e. a single server system where the arrival times as well as the service times are exponentially distributed. Service requests arrive according to a Poisson stream with rate λ (that is, the parameter defining the stimuli), enter the queue and are processed with service rate μ. μ represents the system's processing speed per arrival. Thus, μ jointly models the application and the resource.[6] In this trivial case, the abstract functions f are some simple formulas for the system utilization $U = \lambda/\mu$ and the average response time $R = 1/(\mu - \lambda)$ spent by an external request in the system.

Unfortunately, such simple functions that map system and workload parameters to performance measures are only applicable to a rather limited class of queuing systems. In realistic systems, the models are often large and complex. However, nowadays sophisticated techniques and tools are available that support the derivation of performance functions and their evaluation.

[6] If a network of queuing stations is employed, the application is implicitly modeled by the links between the queuing stations and the probabilities given for branches, i.e. the probability that an output of one station is sent to a certain other station.

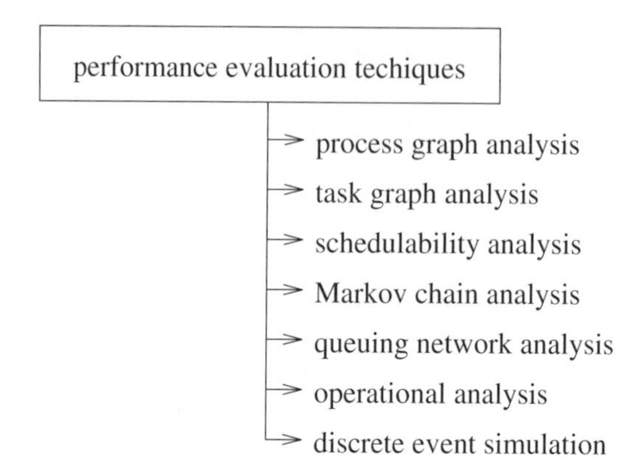

Figure 2.22 Performance evaluation techniques

2.4.1.2 Performance Evaluation in the System Engineering Process

Performance evaluation as part of the system engineering process is especially interested in fast performance evaluations which provide reasonable accuracy. Often the goal of a performance evaluation is to check whether system development is on the right track. Thus, it is often sufficient to roughly estimate whether the performance figures match the performance requirements and to identify cases where special actions with respect to performance are needed. As pointed out in section 2.2.4.2, examples of these actions may be a more detailed performance study or the refinement of some parts of the system design to get more accurate performance estimates.

2.4.1.3 Performance Evaluation versus Verification

An activity highly related to performance evaluation is performance verification. Verification is especially important for systems with hard real-time constraints.

In practice, performance evaluation typically focuses on the derivation of some performance estimates. Conversely, performance verification aims at formal proofs that establish or verify explicit properties of the system under evaluation. Thus, verification allows one to prove that the temporal behavior – as well as the functional behavior – of the system is in accordance with given requirements.

Obviously, a prerequisite for the real system to ensure that the verified requirements are followed is that the assumptions made for the verified model also hold for the real system. Whether a performance evaluation technique is suitable for a formal verification depends on several aspects, including the evaluation technique employed and the kind of assumptions made with the models.

2.4.2 Performance Evaluation Techniques

In this section, we survey the solution techniques shown in figure 2.22.

Traditional model-based performance evaluation is often based on stochastic modeling using queuing networks, Markov chain techniques, stochastic Petri nets and discrete event simulation. More simple techniques which are often sufficient are graph-based analysis techniques, mostly based on deterministic time assumptions. While queuing models offer strong support to describe the available resources, graph models are especially useful to describe the characteristics of the application. Discrete event simulation provides a means to evaluate the system, i.e. the application, the resources and the system stimuli, at an arbitrary level of detail.

After establishing a textual or graphical model of a planned or existing system (as described in section 2.3), it must be transformed to a quantitatively assessable representation that is executable or solvable. In the subsequent step, the assessable performance model is evaluated and the needed performance figures are derived.

2.4.2.1 Process Graph Analysis

The techniques typically employed to evaluate process graphs are very simple. As described in section 2.3.2, process graphs focus on spatial aspects of the application and neglect any temporal relations between the components of the system. Thus, they are only interested in the amount of the load rather than the exact time when the load is present. Graph-based analyses are mainly used by model-based optimization techniques since they provide a fast means of estimating a given system configuration.

Graph analysis techniques typically assume deterministic times (or load). The nodes (and possibly edges) of the graph may be attributed with average, worst- or best-case estimates, depending on the kind of analysis being performed. The analysis techniques are based on a simple summation of the times or load imposed on the different resources of the system.

An example of a simple analysis is given by the following formula computing the load imposed on the resource r:

$$load(r) = \sum_{p \in A_r} load(p, r)$$

where p denotes some process, A_r specifies the set of processes assigned to resource r, and $load(p, r)$ denotes the exact resource demand resulting from the assignment of process p on resource r. As can be easily seen in the formula, the mapping on the resources is defined by the sets A_i. The capacity of the resource r is implicitly modeled by the function $load(p, r)$. Typically, the (number of) stimuli imposed on the system is also reflected by $load(p, r)$, i.e. the load imposed on the resources.

A process graph analysis is limited to the analysis of the load imposed on the resources of the system. The respective resources may be physical (e.g. processors or communication links) or logical (e.g. critical sections).

Process graph analysis typically neglects all influences on the performance caused by temporal constraints. Thus, it is especially useful for systems where throughput is of importance rather than response time, and where the even distribution of the load on a set of resources has to be decided. However, note that it implicitly assumes that contention does not have any negative (dynamic) influence on the load imposed on the resources.

Algorithms analyzing process graphs can be found in virtually any paper devoted to the optimization of the spatial assignment of processes on parallel systems, e.g. see [Bokh81, HwXu90, RaES88, Shen92]. Typically, the cost function for the optimization takes into account processing as well as communication cost and aims at a compromise between the optimal load balance and the minimization of communication cost.

2.4.2.2 Task Graph Analysis

The techniques employed to analyze task graphs typically search for the critical (or longest) path in the graph. The length of the critical path can be used to estimate the response time to compute the application.

Similar to the analysis of process graphs, task graph analyses provide a fast performance estimation. Task graph analyses are typically employed for systems with deterministic resource demands. As a result of this, the analysis techniques are based on simple and efficient sum and maxima formulas to compute the length of the longest path in the graph [Deo74]. The resource demands may represent average, worst- or best-case estimates, depending on the kind of analysis performed.

Some task graph analyses neglect the details of the resources and the mapping. Thus, contention due to limited resources is neglected, which limits the analysis to very optimistic cases. In the most simple case, the analysis focuses on the application only, completely neglecting resource issues. This allows one to find bottlenecks in the application, i.e. in the computation path, independently of the mapping of the application on the resources. However, note that an analysis is meaningless if the capacity of the available resources is heterogeneous or not known at all. In addition, the negligence of the mapping on the resources does not allow one to identify bottlenecks caused by communication.

Resource contention can be modeled by adding additional precedence constraints (arcs) to the task graph that serialize the access of the tasks to resources [Klei82]. This allows one to denote the fact that tasks assigned to a common resource (processor) are processed sequentially rather than in parallel. However, note that this approach employs a single graph model to describe aspects of the application as well as scheduling information.

Extensions of task graph analyses include the mapping on the resources in order to take resource contention into account. Based on information concerning the spatial and temporal assignment of the tasks on the resources, the actual execution order of the tasks (schedule) is constructed. This allows the determination of the response time for the application on the system. The approach is employed by many optimization techniques that optimize the spatial and temporal assignment, e.g. [Duss98, HCAL89, Mits93, PrSa90, WaTs88].

Task graph analysis typically assumes that multiple invocations of a task graph, e.g. as triggered by subsequent system stimuli, do not interfere with each other.

In addition to the analysis based on deterministic times, also techniques for the evaluation of systems with stochastic (general) time distributions exist [BaJL93, Hart93]. If the time distributions are exponential functions, tasks graphs can be mapped on Markov chains and solved by the respective techniques (see below). Fast numerical analysis techniques to derive mean

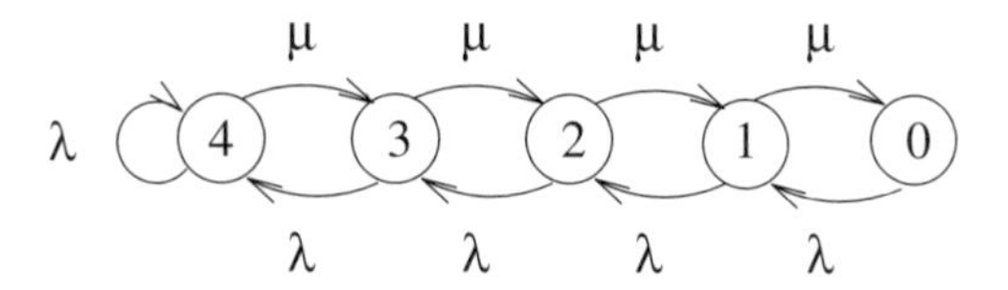

Figure 2.23 Example of a Markov chain to model a queuing station

values for the response times exist for a restricted class of task graphs, namely series-parallel graphs [Hart93]. Approximation techniques and tools exist to estimate the response times of non-series-parallel graphs and graphs with general distributions [Hart93, KDHM+95].

2.4.2.3 Schedulability Analysis

Schedulability analysis is a verification technique that checks whether a system can meet its deadlines. Various variants of schedulability analyses exist. A comprehensive discussion of a number of variants for single processor systems can be found in [BuWe96]. The extension of schedulability to (heterogeneous) parallel systems is described in [Axel96]. However, note that schedulability analysis is typically based on a set of highly simplifying assumptions. The most important assumption seems to be that all tasks of the system are assumed to be independent of each other.

Schedulability analysis typically assumes the deterministic periodic arrival of service requests. Each service request triggers the execution of exactly one task. Typical input to the analysis consists of the period of each task invocation, its computation cost, its deadline and its priority. Schedulability analysis decides whether the deadlines for the different tasks can be met under all circumstances assuming priority-based preemptive scheduling.

Schedulability analysis techniques are based on fixed-point equations for the response time of the different tasks in the system, which are typically solved by a recursive scheme. This subsequently takes into account the delay imposed on lower priority tasks by tasks of higher priority.

Extensions to integrate precedence constraints are described in [SIZL98] and [KaGa93]. However, note that these extensions can result in a considerable increase in the time to verify the system.

2.4.2.4 Markov Chain Analysis

Markov chains represent an explicit (low-level) description of the possible states of the system and the transitions (with rates) between the states. Typically, continuous-time Markov chains are employed for the performance analysis of computer systems. Continuous-time Markov chains are discrete-state continuous-time stochastic processes with exponential time distributions.

An example of a Markov chain to model the queuing station introduced in figure 2.21 is shown in figure 2.23. In the example, the queue is assumed to hold up to three service requests at a time.

The attribute assigned to each state denotes the total number of service requests currently in the station, i.e. the server and the queue.

The theory of Markov chains provides the base for stochastic modeling and evaluation. Markov chains are mostly derived from high-level graphical descriptions, rather than directly developed. Markov chains are often employed as an internal representation for queuing networks, Petri net models, stochastic process algebras and state transition diagrams.

In order to derive performance figures for the steady state, the Markov chain is mapped to a set of linear equations that can be solved by numerical techniques. Here, Gauss elimination or preferably iteration techniques are employed to derive the state probabilities. From the probabilities of the states of the Markov chain, the mean values as well as the distribution functions of various performance metrics can be derived. The theory of Markov chains can be found in [HeSo82, Klei75].

The set of equations can be rather large due to the phenomenon of state-space explosion. With numerical solution techniques, convergence to the steady-state solution may be slow due to ill-conditioned equations. Nowadays models with some million states can be solved, if hierarchical decomposition techniques are used.

In addition to the steady-state analysis, the transient system behavior can be analyzed based on a system of differential equations.

The most important assumption underlying Markov chains is the exponential distribution of the state holding times, or in other words of the arrival and service times of the system. This ensures that the future behavior of the system at any time depends only on its current state. However, also some approximations to describe non-Markov features by Markov chains have been proposed. In spite of their restrictions, Markov chains are a very flexible means of modeling the performance behavior of systems.

2.4.2.5 Queuing Network Analysis

Instead of the explicit description of the possible states of the queuing network by a (low-level) Markov chain and its direct solution, more efficient solution techniques are available for a rather large set of queuing networks. Important is the class of product-form (or separable) queuing networks which can be solved exactly by analytical techniques.

The advantage of product-form queuing networks is that they allow one to independently solve the linear (steady-state) equations for individual stations rather than for the complete network. The state probabilities for the complete network can be derived (by multiplication) from the independent local solutions.

For the exact solution of product-form networks, a set of assumptions has to be satisfied (see [Jain91] for a complete list). Assumptions relate to the service strategy, the service time distribution, the arrival processes and other factors. A variety of rather efficient algorithms has been developed since the late sixties to solve such networks. Examples of exact solution techniques are the mean value analysis for closed queuing networks to directly derive the response time, throughput and number of jobs in the system, and the convolution algorithm. Both, among others, are described for example in [Jain91]. In addition, relatively fast approximative techniques are available for product-form queuing networks as well as for non-product-form models including non-exponential distributions, priorities, and blocking.

2.4.2.6 Operational Analysis

Operational analysis is due to Buzen and Denning [Buze76, DeBu78]. It is based on a small set of simple laws. The advantage of operational analysis is its simplicity which is mainly due to the application of the job flow balance that states that the number of arrivals is equal to the number of completions in a system or subsystem.

The operational analysis does not make any assumptions about the distribution of service and arrival times. It allows one to compute the arrival rates, throughput, utilization, and mean service times of the different stations in the system. Lazowska and colleagues have extended the basic ideas [LZGS84].

2.4.2.7 Discrete Event Simulation

If neither analytical nor numerical techniques can be applied, simulation is often the last resort. The strength of simulation is its flexibility. Although simulation is well established and rather popular in practice, there are many methodological problems. Typical problems are model correctness, inclusion of too many details, vast amounts of statistical data produced by simulation runs, and more. In addition, simulation is an expensive evaluation technique.

There are many textbooks solely devoted to simulation. A comprehensive introduction into the topic can be found in [Jain91]. In addition, numerous simulation tools, both commercial and academic, are available.

2.4.2.8 Hybrid techniques

Hybrid techniques try to combine the benefits of different approaches. Typically, subsystems are analyzed separately under different workloads and the results are aggregated to a substitute representation that has a (nearly) identical performance behavior with respect to throughput and delay. The substitutes are then used as parts of a high-level model which is subsequently evaluated.

2.4.3 Performance Evaluation Tools

In order to relieve the user from the detailed knowledge of performance evaluation techniques, a large number of tools is available to support the evaluation of computer and communication systems. We list some of the typical tool systems.

Since Petri net based tools integrate features to describe functional behavior as well as time aspects, they are very attractive for the formal specification community. Here, we mention DSPNexpress [Lind94], TimeNet [GKZH95], and the QPN tool [BaBK95]. All tools have state-of-the-art graphical interfaces and employ the Markovian solution approach, accompanied by functional evaluation techniques, stationary and transient simulation (TimeNET) or hierarchical techniques (QPN tool).

The tool MACOM (Markovian Analysis of Communication Systems) [KrMS90] provides queuing-station-like building blocks to model telecommunication networks. The models are solved by Markovian numerical techniques.

The tools RESQ2 [Lave83] and QNAP2 [Simu90] are based on the queuing network paradigm,. In particular, model specification is done in terms of queuing networks. Besides analytical algorithms for open, closed and mixed product-form queuing networks, additionally simulation techniques are available to solve extended queuing networks.

A tool especially supporting hierarchical modeling is the HIT tool [BeMW88, BaSc95]. Specification of layered and modular models consisting of components with well-defined interfaces is done with a graphical editor or by a textual description with the HIT language HI-Slang. For model solution, analytical, numerical and simulative techniques are available.

From the commercial field, modeling and planning tools for computer and communications systems such as Best/1, BONeS DESIGNER, COMNET-III, OPNET, and the SES workbench are well established in industry. In addition, we observe that evaluation techniques are increasingly employed in close conjunction with measurement and monitoring. Also frameworks for network management such as HP Openview, Netview and Solstice offer modules for model-based predictions and load forecasting based on statistical evaluation.

Finally, we should refer to classical simulation languages such as SIMULA with its classes SIMULATION and DEMOS, the originally Fortran-based Simscript which is also available with graphics and facilities designed for the development of process- and event-oriented simulation models (SIMSCRIPT II.5), and the transaction- and flow-diagram-based GPSS.

2.5 Performance Optimization

System development is the process of understanding, detailing and deciding. Thus, optimization is a key part of system development. Decisions are made in all phases of the system development process. This also holds for decisions concerning the performance of the system. Thus, performance optimization or optimizations related to performance are employed in various development phases and at various levels of abstraction.

In the section, we describe techniques that support the automation of the decision-making process. Since the performance of systems is typically reflected by the cost function of the optimization and the constraints only, the optimization techniques proposed in the following are not necessarily restricted to performance optimization as such. Rather, the techniques can be employed for other decisions to be made during the system development as well. In order to express this, we focus on general issues. For example, we use the general term ‘quality of the solution’ in the section rather than performance-specific cost functions and constraints.

2.5.1 Introduction

2.5.1.1 Optimization and System Development

The different alternatives for the design decision at hand constitute the search space. We assume that the search space is finite but of considerable size for systems of reasonable size.

Key tasks concerning the performance optimization of parallel and distributed systems are

- the design of the overall system architecture,
- the design of the application-specific parts (software),
- the design of the hardware resources, and
- the mapping of the software on the hardware.

Note that performance is not the only goal. With the development of a system, various other optimization goals and constraints have to be considered, too. As described in section 2.1, typical examples are product cost, development cost, time-to-market, flexibility issues and fault tolerance.

An overview on the design optimization in parallel and distributed systems and the models employed is graphically given in figure 2.24. Models to specify the application, the available resources, the mapping and the system stimuli have been surveyed in section 2.3. The output of the optimization is called 'system model' in the figure. The system model represents a solution of the optimization problem. The system model represents a model of the system at the respective level of abstraction of the input models. In other words, the system model defines the system (design or implementation) at some more or less abstract level. It especially specifies the values for the open parameters which have not been defined in the input models for the optimization, i.e. in the application, resource and mapping models. Thus, the system model defines the software, the hardware, and the mapping of software on hardware.

2.5.1.2 A Definition of Combinatorial Optimization

The decision problems we are dealing with are discrete combinatorial problems. Thus, each solution of the optimization problem can be formulated as a vector $(x_1, ..., x_i, ..., x_n)$, where x_i has a finite range of values. The optimization is typically restricted by a set of constraints, which restrict the possible combinations of the values in the solution vector.

The goal of the optimization is to find the feasible solution with the highest quality, i.e. a solution that conforms to the constraints and results in a maximum (or minimum) value of the given cost function. Examples of possible performance-related cost functions and ways to derive them have been discussed in section 2.4.

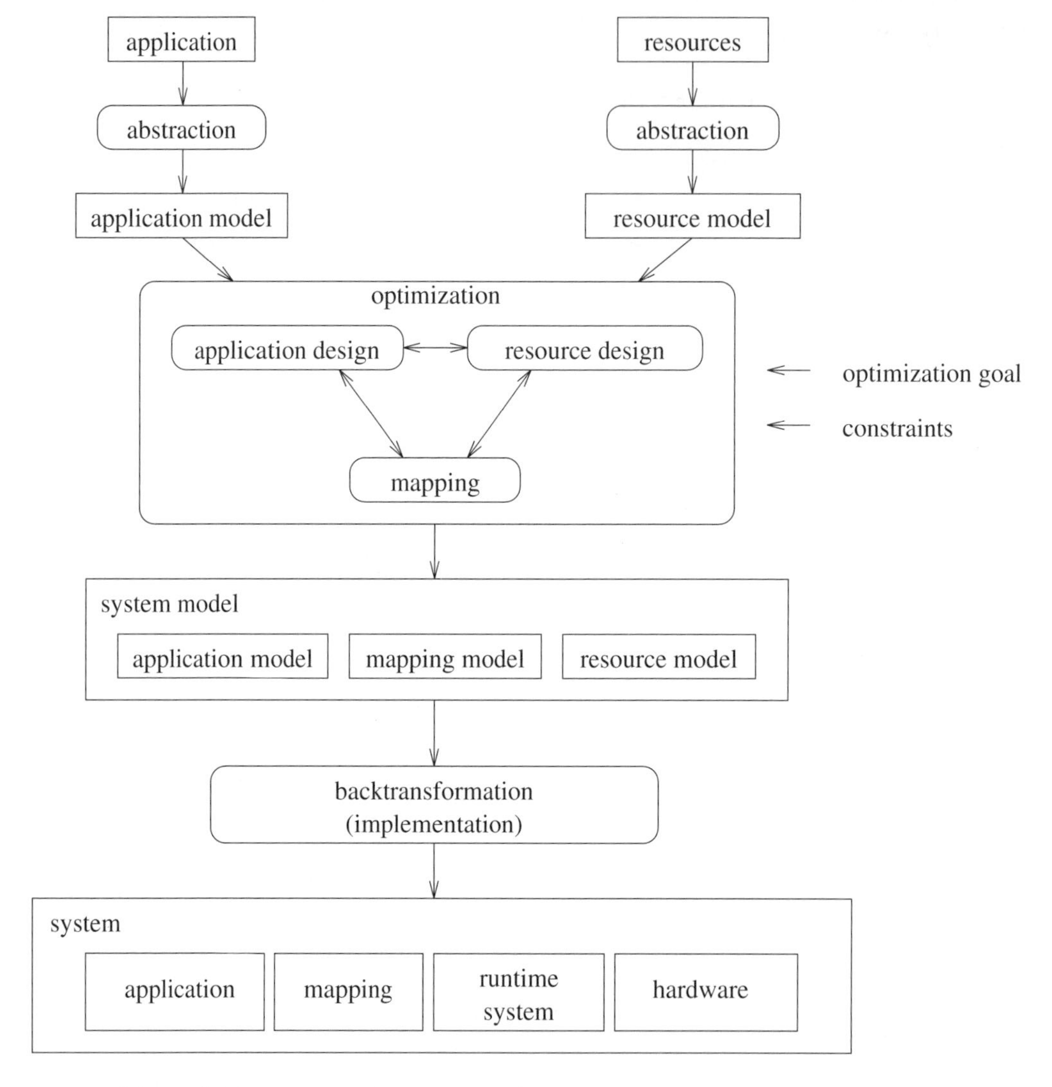

Figure 2.24 Survey of the optimization and the respective models employed

2.5.1.3 Heuristics

In practice, optimization problems can be divided in two classes:

- problems solvable by a deterministic algorithm within polynomial time, i.e. the execution time is a polynomial of the size of the problem, and
- problems not solvable within polynomial time.

Unfortunately, most optimization problems dealing with the optimization of the design and implementation of parallel and distributed systems belong to the second class. Examples are most mapping and scheduling problems. In practice, this means that the optimal solution for the respective problems can only be found for very small problem instances. Or in other words, the computation time to exactly solve a specific problem instance is an exponential function of the size of the problem instance. For a typical multiprocessor scheduling problem, a problem size of about 10 to 20 tasks may already be beyond the capacity of a modern computer system. Thus, heuristic optimization algorithms are often the only choice to solve these problems.

In order to formally reason about the time complexity of problems, the completeness theory has been developed. The class of NP-complete problems contains the related decision problems for which no polynomial-time algorithms are known and which can be solved in polynomial time only if for any of the problems in the class a polynomial-time algorithm can be found. A survey of NP-complete problems can be found in [GaJo79]. A comprehensive introduction into completeness theory is given in [HoSa78]. Optimization problems are highly related to decision problems. Typical decision problems can be solved in polynomial time if the corresponding optimization problem can be solved in polynomial time. If the decision problem corresponding to an optimization problem is NP-complete, the chances of finding a polynomial-time algorithm for the corresponding optimization problem are negligible.[7]

Due to the fact that almost all optimization problems in the area are NP-hard we focus on heuristic optimization in the following.

We use a rather broad definition of a heuristics as proposed in [Reev93a]:

> A heuristics is defined as a technique which seeks good (i.e. near-optimal) solutions at a reasonable computational cost without being able to guarantee either feasibility or optimality, or even in many cases to state how close to optimality a particular feasible solution is.

The definition reflects the typical usage of the term in the recent literature.

2.5.2 Heuristic Search

Most heuristics for NP-hard optimization problems are based on an iterative search. The principle of iterative search is to iteratively

- generate a single (or a set of) solution(s),
- evaluate the quality of the solution(s), and
- search for the next solution(s).

During the search, the best solution is stored, and is output after the search has been terminated.

[7] In fact, the existence of a polynomial-time algorithm for the optimization problem would imply that all NP-complete decision problems can be solved in polynomial time.

2.5.2.1 General Framework

The basic idea of iterative search is to first generate an initial solution, which is then subsequently improved. All optimization algorithms based on the principle of the subsequent improvement of an initial solution exhibit a similar structure. The basic strategy for the search is outlined by the framework given in figure 2.25.

```
(1) solution:= initial_solution
(2) REPEAT
(3)   new_solution:= selection(solution, ...)
(4)   IF acceptance(new_solution, ...)
(5)     THEN solution:= new_solution
(6) UNTIL termination(...)
```

Figure 2.25 Framework for heuristic search

Based on the framework, we provide a general description of the design parameters of this class of optimization algorithms.

2.5.2.2 Basic Elements of Heuristic Search

The different heuristics vary in the details of the search, especially in the specific search strategy, the acceptance criterion and the termination criterion. We first describe the three aspects followed by a classification of the heuristics based on these aspects.

Search strategy The search strategy denotes the approach the optimization algorithm employs to move from one solution to the next. Thus, the search strategy defines where to look for new solutions and how to select them.

The search strategy can be classified as follows:

- The *search area* defines the neighborhood in which the algorithm looks for potential solutions in the next step. The two extremes are
  - *global search*, i.e. the search area consists of the whole search space (all possible solutions), and
  - *local search*, i.e. the search area is restricted to the direct neighbors of the current solution.

  With a local search, subsequent solutions derived by the optimization algorithm only differ by a small degree. Conversely, with a global search, there is no neighborhood relation between two subsequent solutions.

- The *selection strategy* defines the rules employed to select a solution from the identified search area. Alternatives are
  - *deterministic* selection, i.e. the selection of the solution candidate according to some deterministic rules,
  - *random* selection, i.e. the purely random selection of one of the solutions from the search area, and
  - *probabilistic* selection, i.e. the selection of the candidate solution based on a probability function.
- The *history dependence* defines the degree to which the selection of the new candidate solution depends on the history of the search. Alternatives for history dependence are
  - *no dependence*, i.e. the selection of the new solution is independent of the state of the search,[8]
  - *one-step dependence*, i.e. the selection of the new solution depends on the last solution only, and
  - *multi-step dependence*, i.e. the selection of the new solution depends on the results of several previous search iterations.

Most search algorithms are based on one-step dependence, i.e. the case where the search depends on the last previous solution.

Acceptance criterion The acceptance criterion defines under which circumstances a candidate solution is accepted as a new solution.

Alternative acceptance criteria are

- *deterministic* acceptance, i.e. the acceptance of a solution as the new solution depends on some deterministic function, and
- *probabilistic* acceptance, i.e. the acceptance of a new solution is influenced by a random factor.

The acceptance criterion is highly related to the selection strategy. While the selection strategy defines how to select a possible candidate for a new solution, the acceptance criterion decides whether or not to accept this solution or, alternatively, to continue with the previous solution. However, note that continuing with the previous solution only makes sense in the case that the next selection of a candidate solution results in a different solution. This is the case if the state of the search is changed (history) or if a random factor is employed by the selection. Otherwise the two steps are merged to a single step, e.g. with hill-climbing and random search.

[8]Note that this implies that the search area is global.

Termination criterion There are two major alternatives to decide when to terminate the search. The rules for termination may be

- *static*, i.e. independent of the actual solutions visited during the search, or
- *dynamic*, i.e. dependent on the solutions visited during the search.

Examples of the static case are as follows: the search terminates after a fixed number of search iterations has been completed independently of the quality of the solution, or after a certain search time has elapsed. The most important example for the dynamic case is the termination of the search if no improvement of the best solution has been achieved for a given number of search iterations. In this case, the termination condition depends on the goal function, or in other words on the way the quality of the solutions is evolving.

Classification of heuristic search Table 2.3 shows a classification of the most important heuristic search algorithms.

Heuristics	Search strategy								Acceptance criterion		Termination criterion	
	Search area		Selection strategy			History dependence						
	local	glob.	det.	prob.	rand.	none	one-step	multi-step	det.	prob.	stat.	dyn.
hill-climbing	X		X				X		X			X
tabu search	X		X					X	X		X	X
simulated annealing	X				X		X			X	X	
genetic algorithms		X		X			X[a]		X	X	X	X
random search		X			X	X			X		X	

[a]Genetic algorithms consider more than one solution in an iteration. Thus, the history consists of the solutions that constitute the population.

Table 2.3 Classification of heuristic search algorithms

2.5.2.3 Hill-Climbing

Basic idea Hill-climbing algorithms are neighborhood search algorithms that subsequently select the neighbor with the highest quality and continue from there. The search terminates when no neighbor exists that represents an improvement over the current solution. Hill-climbing algorithms belong to the class of greedy algorithms, i.e. the algorithm never goes back to a solution with lower quality. In other words, the climber never goes downhill to finally reach a higher peak.

Classification

Search strategy:	local search; deterministic selection of the neighborhood based on the previous solution
Acceptance criterion:	deterministic acceptance of the neighbor with the highest quality
Termination criterion:	termination when no neighbor exists that improves the current solution

Algorithm The algorithm is outlined in figure 2.26. The hill-climbing algorithm starts with an initial solution which is subsequently improved. The algorithm terminates when no solution of higher quality is within the reachability of the neighborhood.

```
(1) solution:= initial_solution
(2) REPEAT
(3)   new_solution:= select_best_neighbor(solution)
(4)   IF improvement(new_solution, solution)
(5)     THEN solution:= new_solution
(6) UNTIL no_improvement(new_solution, solution)
```

Figure 2.26 Hill-climbing algorithm

With the hill-climbing algorithm, the search can be interpreted as a path consisting of small steps constantly moving to solutions with higher quality.

Discussion The obvious drawback of this simple algorithm is that it is not able to pass a valley to finally reach a higher peak. Thus, it typically settles for a local optimum. This is the reason why pure hill-climbing algorithms are seldom found in the problem domain dealt with in this book. All other algorithms presented in the remainder of the section can be viewed as different approaches to overcome local optima to reach the global optimum. However, greedy ideas are often applied in practice to small parts of the optimization problem, especially where the potential deterrence in the quality of the solution is small in case the algorithm is trapped in a local optimum.

2.5.2.4 Random Search

Basic idea Random search algorithms avoid the potential problem of getting trapped in a local optimum by randomly searching the complete search space. Due to this fully randomized search, these algorithms are often called Monte Carlo algorithms.

Classification

Search strategy: global search of the complete search space; each solution is generated randomly independently of any previous solution

Acceptance criterion: improvement over the best solution found so far

Termination criterion: static bound on the number of iterations

Algorithm The algorithm is outlined in figure 2.27.

```
(1) solution:= initial_solution
(2) counter:= NUMBER_OF_VISITED_SOLUTIONS
(3) REPEAT
(4)   new_solution:= select_random_solution()
(5)   IF improvement(new_solution, solution)
(6)     THEN solution:= new_solution
(7)   counter:= counter - 1
(8) UNTIL counter = 0
```

Figure 2.27 Random search algorithm

Discussion The algorithm is very easy to implement, which makes it an attractive choice for optimization. Easy implementability is mainly due to the fact that no neighborhood relation is needed. Random search can be used as a reference algorithm to estimate the improvements achievable with more intelligent optimization algorithms.

2.5.2.5 Simulated Annealing

Basic idea Simulated annealing algorithms model the annealing process of physical material. The algorithm is based on the phenomenon observed during the cooling process of material. When material is cooled down, it moves from a state of high energy to a low-energy state. In high-energy states, the molecules are highly mobile, thus allowing them to move globally. Conversely, in low-energy states, mobility is highly restricted. With simulated annealing, mobility of molecules can be interpreted in terms of neighborhood of the search. Thus, high energy (or high temperature) indirectly supports a global search by encouraging nonimproving moves. Conversely, low temperature restricts the search by discouraging nonimproving moves. With an appropriate cooling function, material with minimal energy (i.e. high quality) can be obtained. With simulated annealing, high-energy states of the system are interpreted as solutions with low quality. Conversely, low-energy states are interpreted as solutions of high quality.

The application of simulated annealing to optimization is due to Kirkpatrick and colleagues [KiGV83]. The idea is to use the results from physics to optimize the system, i.e. to employ an annealing function that results in a highly optimized solution. A good survey of simulated annealing and a discussion of its application to a large range of problems can be found in [Dows93].

Classification

Search strategy: random local search

Acceptance criterion: unconditional acceptance of the selected solution, if it represents an improvement over the previous solution; probabilistic acceptance of solutions which exhibit a lower quality than the previous solution; in this case acceptance depends on the state of the search (temperature) and the exact quality of the solution

Termination criterion: static bound on the number of iterations (cooling process)

Algorithm The algorithm is outlined in figure 2.28. The algorithm randomly selects a new solution based on the neighborhood of the previous solution. During the search, the temperature is stepwise reduced. The temperature and the quality of the randomly generated solution compared to the quality of the previous solution determine the acceptance of the new solution. In other words, they decide whether or not the new solution replaces the previous solution.

```
 (1) solution:= initial_ solution
 (2) best_solution:= solution
 (3) temperature:= INITIAL_TEMPERATURE
 (4) REPEAT
 (5)   FOR i:=1 TO NUMBER_OF_STEPS_PER_TEMPERATURE DO
 (6)     new_solution:=
           generate_random_solution_in_neighborhood(solution)
 (7)     IF acceptance(new_solution, solution, temperature)
 (8)       THEN solution:= new_solution
 (9)     IF best_solution_so_far(solution, best_solution)
(10)       THEN best_solution:= solution
(11)   ENDFOR
(12)   temperature:= decrement(temperature)
(13) UNTIL temperature = 0
```

Figure 2.28 Simulated annealing

If the new solution has a higher quality than the previous solution, the previous solution is replaced by the new solution. Thus, the new solution builds the new starting point for the next iteration. If the quality of the new solution is below the quality of the previous solution, the selection of the new solution is based on a probability function. The function employed to derive the probability p under which the new solution is accepted is the exponential function

$$e^{-\Delta q/T}$$

where Δq represents the difference of quality between the new and the previous solution, and T denotes the temperature of the cooling process.

It has been shown that if the temperature is decremented logarithmically (i.e. $T := T_0/log(i)$) and the decrement i is approaching zero ($i \rightarrow 0$), then the probability of reaching the optimal solution approaches the value 1 [Leng90]. A survey of theoretical results that may help in the selection of appropriate parameters for the search can be found in [VaAa88, AaKo89].

Unlike pure theory, simulated annealing algorithms often maintain the best solution found so far (lines (9) and (10) in figure 2.28).

Discussion The selection of the parameters of the cooling function, i.e. the decrement function and the starting temperature, is crucial. A small temperature decrement results in large search times. Conversely, a large decrement reduces the quality of the solution. In practice, it has been shown that a linear decrement of the temperature rather than a logarithmic decrement may result in solutions of the same quality within smaller search times [Brau90].

Altogether simulated annealing algorithms are relatively straightforward to implement. This is mainly due to the random search of the neighborhood. A problem may be the selection of appropriate parameters of the algorithm to derive good solutions within a reasonable time, e.g. the selection of the temperature decrement and the function to compute the acceptance probability.

Variants Experience has shown that the computation of the exponential probability function may account for one-third of the computation cost of the optimization [JAMS89]. Different alternatives to remedy the problem have been proposed. Besides the proposal of probability functions which are easier to compute, deterministic acceptance has been proposed. Examples of this are 'threshold acceptance' and the 'sintflut'[9] algorithm [DuSW93]. With threshold acceptance, non-improving solutions are accepted if $\Delta q \leq aT$, where a is some constant. Thus, acceptance is a deterministic function of the temperature. With the sintflut algorithm, acceptance depends on the absolute quality of the solution compared to the relative quality employed with threshold acceptance. Thus, as the temperature is decreasing, the required quality of new solutions to be accepted is increasing.

Another issue is the cooling function. Experience has shown that the middle part of the cooling process is most important while the beginning and the final parts are not. Proposals to take this into account comprise a nonlinear cooling process, or an adaptive cooling function which takes into account the number of accepted solutions at a temperature, or adapts the function to the behavior of the quality of the found solutions [Dows92]. Another alternative is the use of reheating to improve the results.

[9] *Sintflut* is the German term for 'deluge'.

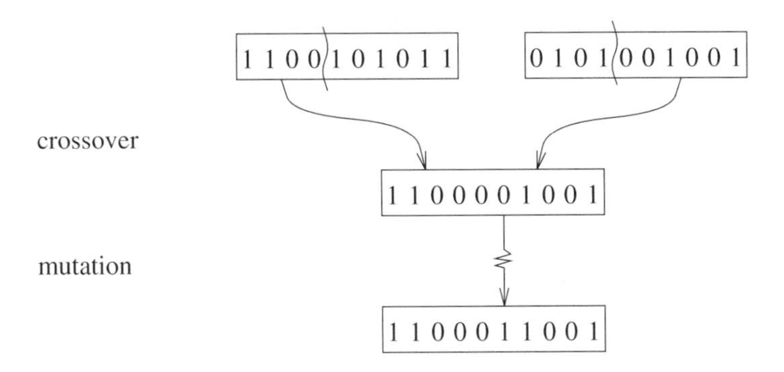

Figure 2.29 Crossover and mutation

Other variants concern the definition of the neighborhood of the search. In [SeBr88], the use of a larger neighborhood in the beginning of the search is proposed, which is reduced during cooling.

Possible approaches to exploit parallelism are discussed in [AaKo89].

2.5.2.6 Genetic Algorithms

Basic idea The idea of genetic algorithms is borrowed from evolution theory. It is based on the observation that populations adapt to their environment over time. The basic principle for this to work is the survival-of-the-fittest principle. It states that individuals that are well adapted to the environment will have more descendants and that their descendants also have a higher chance of being well adapted to their environment.

In order to transform these results to the optimization of technical systems, the genetic operations *crossover* and *mutation* are applied to bit strings (as the representation of individuals or chromosomes). A bit string represents a solution of the optimization problem. The crossover operation is exchanging sections of two bit strings. For example, a new bit string is derived by taking the first k bits of the string from the first bit string and the remaining bits starting from bit $k + 1$ from the second bit string. A mutation is the random switching of a bit of the bit string. Both operations are graphically depicted in figure 2.29.

The initial development of genetic algorithms is mainly due to Holland [Holl75]. A good survey of genetic algorithms and a discussion of its application to a range of problems can be found in [Reev93]. Two textbooks on the topic are [Gold89] and [Davi91].

Classification

Search strategy: probabilistic selection of a set of solutions from the population; higher quality solutions are selected with higher probability; pairs of selected solutions are built and a crossover operation applied to them to derive new

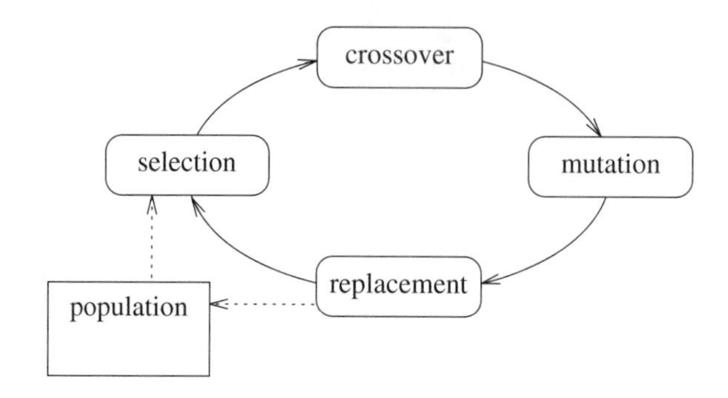

Figure 2.30 Evolution principle

solutions; new solutions may be mutated with a certain (small) probability

Acceptance criterion: new solutions replace older solutions in the population
Termination criterion: static bound on the number of iterations (number of generations) or based on the quality of the population

Algorithm The optimization principle as borrowed from evolution theory is depicted in figure 2.30. It shows the important genetic operations of selection, crossover, mutation and replacement. The four operations are applied to the population, i.e. a set of solutions, to improve its quality over time.

```
(1) population:= initial_population
(2) generation_counter:= 0
(3) REPEAT
(4)   parent_set:= selection(population)
(5)   set_of_pairs:= pairing(parent_set)
(6)   descendant_set:= crossover(set_of_pairs)
(7)   descendant_set:= mutation(descendant_set)
(8)   population:= replacement(population, descendant_set)
(9)   generation_counter:= generation_counter + 1
(10) UNTIL generation_counter = MAX_NUMBER_OF_GENERATIONS
```

Figure 2.31 Genetic algorithm

The structure of genetic algorithms is shown in figure 2.31. First an initial population, i.e. a set of solutions (or bit strings), is generated. In each iteration of the genetic algorithm a subset of the solutions in the population is selected. The selection is guided by the quality of the respective solutions. Thus, a solution of higher quality has a higher chance of being selected. Next, the selected solutions are paired. From the respective pairs, new solutions are generated employing the crossover operator. Crossover denotes the fact that some information from each of the two selected parents is taken to obtain the new solution. Thus, the new solutions are considered the descendants (or children) of the selected parents. The descendants derived in this way may be mutated with a certain (typically very small) probability. Thus, small changes may be applied to them. In the last step, the newly generated descendants replace some of the solutions (parents) in the population.

Typically, the number of solutions that form the population is constant. A typical termination condition is the use of a fixed number of iterations.

Discussion There are many parameters to be tuned in a genetic algorithm. Examples are the replacement function, the selection strategy, the mutation probability, the size of the population, and the termination criterion. Concerning the replacement, the population may be replaced completely by their descendants within a single generation or be replaced gradually by employing some strategy. Concerning the crossover, the intersection may be selected probabilistically or based on some problem-specific information. In addition, applying the crossover operation to a pair of solutions, two descendants result. Thus, one or both may be selected for the next generation. Concerning the selection of the individuals for the crossover, various alternatives exist for the probability function applied by the selection operation.

Parameter tuning is very critical and complicated due to the large number of parameters and also due to the fact that the influence of different parameter settings on the search process is not very intuitive [Duss98]. Another problem is that the optimal parameter set depends on the specific problem class or even the specific problem instance at hand.

Besides tuning of the genetic algorithm, another problem exists. It is a design problem rather than a tuning problem. Since the solutions of the specific optimization problem have to be presented as a bit string, a problem is to find a code that (1) contains a bit string for each possible solution and (2) ensures that each possible bit string represents a valid solution of the optimization problem. If the second condition is violated, an increasing number of constraints on an instance of an optimization problem may lead to a considerable reduction of the number of feasible solutions in the population. This in turn results in a very inefficient search. With many optimization problems dealt with in this book, many constraints on the optimization exist. In order to avoid the usage of a complex encoding scheme that satisfies the above conditions, the use of intelligent or problem-dependent genetic operators has been proposed in [HoAR94]. Thus, the operators for crossover and mutation ensure that the result of their application is again a feasible solution. This avoids the use of a complex encoding scheme for the price of more complex operations.

Altogether it can be stated that the use of genetic algorithms is probably not the easiest approach to solving an optimization problem. This is especially true if complex constraints are present. On the other hand, genetic algorithms are rather generic concerning the kind of problems they can solve. Thus, for an optimization problem to be solved by a genetic algorithm, the major issue is to provide an appropriate encoding of the possible solutions of the problem into a bit string representation.

Variants As can be seen from the above discussions many variants exist. Many variants are derived by using different parameter settings rather than major variations of the basic scheme. The most important variation is probably the application of problem-dependent genetic operators as discussed above. A discussion of variants and extensions of genetic algorithms can be found in [RaPa97].

Due to its parallel nature, several approaches to exploiting parallelism in genetic algorithms exist [Schw97]. One approach is to compute the quality of the solutions in parallel. However, the parallelism may be too fine to be exploited efficiently. Several approaches to exploiting a coarser degree of parallelism have been proposed. Most approaches are based on the partitioning of the population in a set of subpopulations. Each subpopulation is dealt with by a separate processor. From time to time, the subpopulations have to be reshuffled. Various strategies exist to do this. Examples are a complete reshuffle, or the distribution of the best solutions in the subpopulations to the other subpopulations.

2.5.2.7 Tabu Search

Basic idea While the previous approaches rely on nondeterminism, tabu search takes the opposite approach. It is based on the idea that an intelligent search is deterministic rather than probabilistic. The philosophy of tabu search is based on a collection of principles from intelligent problem solving. Tabu search mainly relies on a flexible memory and does not use any nondeterministic features. Thus, the principle of avoiding running in cycles when allowing nonimproving steps in the search is to use the history of the search to detect and avoid this. This is in contrast to genetic algorithms and simulated annealing, which rely on nondeterminism to avoid cycles.

Even though tabu search has many ancestors, it is a relatively new technique. Its modern form is mainly due to Glover [Glov86]. A good survey on tabu search can be found in [GlLa93].

Classification
Search strategy: deterministic local search
Acceptance criterion: acceptance of the best solution in the neighborhood which is not tabu
Termination criterion: various

Algorithm The structure of the tabu search algorithm clearly resembles that of the hill-climbing algorithm. The major differences are the acceptance of nonimproving solutions, the flexible neighborhood, and the use of history. The structure of tabu search is depicted in figure 2.32.[10]

Like other techniques, tabu search starts with an initial solution which is then subsequently improved. In this rather trivial version, we assume that the history consists of a simple list of the n previous solutions. We will see more sophisticated approaches below. The solutions in the tabu list are tabu during the search for a new solution. Thus, if the best solution in the neighborhood is tabu, the solution is not allowed to be selected. In the worst case, all solutions in the neighborhood are tabu. In this case, the neighborhood has to be extended to find a feasible solution.

[10]Note that we have organized the algorithm to optimize readability, not efficient implementation.

```
 (1) solution:= initial_solution
 (2) best_solution:= solution
 (3) tabu_list:= EMPTY
 (4) REPEAT
 (5)   size_of_neighborhood:= INITIAL_SIZE
 (6)   REPEAT
 (7)     neighborhood_set:=
           generate_set_of_neighbors(solution, size_of_neighborhood)
 (8)     neighborhood_set:=
           remove_tabu_solutions(neighborhood_set, tabu_list)
 (9)     size_of_neighborhood:= size_of_neighborhood + INCREMENT
(10)   UNTIL size_of(neighborhood_set) > 0
(11)   new_solution:= select_best_neighbor(neighborhood_set)
(12)   tabu_list:= update_tabu_list(tabu_list, new_solution)
(13)   solution:= new_solution
(14)   IF improvement(best_solution, solution)
(15)     THEN best_solution:= solution
(16) UNTIL termination(...)
```

Figure 2.32 Tabu search

In the algorithm, one new solution is generated in each iteration of the outer loop. The inner loop is used to increase the neighborhood where needed. The neighborhood is subsequently increased until a solution is found which is not tabu.

A typical termination condition is the use of a fixed number of iterations. An alternative is to base the termination on the behavior of the quality of the solutions during the search. In this case, the search is terminated if a given number of iterations has passed without improvement of the best solution.

Organization of the history The intelligence of tabu search lies mainly in the organization of the history information. The tabu list as introduced above is too inflexible to support this. The tabu list above just prohibits cycles in the search of size n or smaller. In addition, it does not support an efficient implementation.

Several other approaches to implement the history have been proposed. The most important are to store the operations applied on the solutions to derive the next solutions, and the use of attributes of solutions or moves rather than the solutions themselves. This adds flexibility and supports an efficient implementation.

In the following, we focus on the use of attributes since it provides a very flexible way to control the search. Examples of attributes for a simple optimization problem with two variables x_1 and x_2 (e.g. a processor assignment problem with two processes) are

(A1) change of the value of variable x_1 from 1 to 2,

(A2) change of the value of variable x_2 from 1 to 0,

(A3) increment of the value of variable x_1,

(A4) decrement of the value of variable x_1,

(A5) the combined change of some attributes, e.g. (A1) and (A2), or

(A6) the improvement of the quality of two subsequent solutions over or below a threshold value.

If a search step (or an accepted solution) exhibits one of the attributes, two major alternatives to use the attributes in the history exist. The attribute may be directly included in the history or its reverse. In the first case, consecutive steps or solutions may not exhibit the same attribute. If the reverse is used, consecutive steps or solutions are not allowed to exhibit the reverse. The effect this has on the search depends on the kind of attributes used.

As can be easily seen from the example attributes, there are many possible tabu restrictions which can be imposed on the search. Thus, they provide an enormous flexibility in controlling the search. The attributes may serve various purposes in the control of the search. Basic examples are

- the prevention of backward steps and cycles in the search, and
- the prevention of the repetition of some search steps, e.g. to ensure that the current neighborhood is searched before the search moves to another neighborhood.

The history information is used to support three important principles which are often employed with tabu search. The *aspiration criteria* allow tabu conditions to be ignored if:

- a tabu solution is the best solution found so far,
- all solutions in a neighborhood are tabu, or
- a tabu solution is better than the solution that triggered the respective tabu condition.

Intensification checks whether good solutions share some common properties and restricts the search based on this information. Conversely, *diversification* searches for solutions that do not share common properties. While intensification supports the exploitation of the current neighborhood, diversification forces the search to move on to another neighborhood.

Another question concerning the history information is how to update it. There are two main approaches, namely the recency-based approach and the frequency-based approach. With the recency-based approach, the tabu attributes are active for a given number of search steps after their activation. Conversely, with the frequency-based approach, the time the attributes are active depends on the frequency of their activation. Thus, the lifetime of an attribute is extended if it is activated several times during a time period, For example, each activation of an attribute may prolong its lifetime by a fixed number of search steps.

Discussion Tabu search is often quoted as being easy to implement. This is certainly true for the search part of the algorithm which is based on a neighborhood search. On the other hand, tabu search algorithms have many parameters. Their tuning is not easy and, like the other advanced optimization techniques, the optimal parameter setting is problem dependent. If a tabu search algorithm is not properly tuned, the effects may be worse than with other approaches. In the worst case, the search may be trapped in a cycle.

An advantage of tabu search – or of local search strategies in general – is that the results of the evaluation of the quality of the current solution can be used to control the search. Thus, a part that is crucial for the quality of the solution, e.g. a part responsible for a performance bottleneck, can be selected and modified [DöMS00]. In addition, parts of the evaluation of the current solution can be reused under certain conditions to quickly determine the quality of a neighboring solution without computing the complete cost function from scratch.

Variants Many variants of tabu search exist. The variants employ different techniques to control the search. A comprehensive survey of variants can be found in [GlLa93]. A newer extension is reactive tabu search where the parameters that control the search are dynamically adapted during the search [Batt96].

2.5.3 Other Techniques

In the following, we survey other techniques not based on the iterative improvement principle. Most other techniques focus on the construction of a single solution. In other words, they try to do it right the first time, rather than explicitly searching through the search space.

2.5.3.1 Single Pass Approaches

Basic idea Several optimization techniques exist that construct just a single solution for the given problem. The idea of these approaches to optimization is to intelligently decide the different steps in the construction of the solution. Thus, the final solution is aggregated from the decisions for a set of subproblems. Often the solution of the complete problem is constructed by the subsequent decision of parts of the problem, i.e. the subproblem being solved is subsequently increased. Probably the most important examples of this approach are list-scheduling algorithms (see below).

Algorithm The framework of algorithms following this approach is depicted in figure 2.33. An important part of this approach is often the derivation of the guidelines for the construction of the solution. Often the guidelines represent the intelligence of the algorithm.

```
(1) derive guidelines for the construction of the solution
(2) REPEAT
(3)   select subproblem
(4)   decide subproblem based on guidelines
(5)   possibly recompute or update the guidelines
(6) UNTIL final solution is constructed
```

Figure 2.33 Framework of single pass techniques

Application to list scheduling List scheduling algorithms are employed to decide scheduling problems for parallel or distributed systems. The outline of this class of algorithms is given in figure 2.34. The problem solved here is to assign a set of precedence-constrained tasks, e.g. modeled by a task graph, on a set of processors.

```
(1) assign priorities to the tasks according to some strategy
(2) REPEAT
(3)   select the executable task with highest priority
(4)   assign the task to a processor according to some
        assignment strategy
(5) UNTIL all tasks are scheduled
```

Figure 2.34 List scheduling

List scheduling algorithms can be considered as algorithms that model the behavior of the scheduler of an operating system. Thus, executable tasks are subsequently selected and assigned to the processors. The selection and assignment follow a temporal order, i.e. time is subsequently advanced during the scheduling process.

The algorithm starts with the assignment of priorities to the task. Then the executable task with the highest priority is subsequently selected and assigned to a processor according to some strategy. A task is executable when all of its predecessors have been completed. The typical assignment strategy is to select the processor which is available at the earliest time. A survey and comparison of various strategies to assign the priorities to the tasks is given in [AdCD74].

List scheduling algorithms typically assume a fully connected system, or at least that each processor in the system is able to communicate with every other processor. Most constraints concerning the assignment of tasks to processors cannot be dealt with by list scheduling due to the greedy nature of the algorithm. The most important constraints that cannot be handled by list-scheduling algorithms are neighborhood constraints, i.e. the requirement that communicating tasks are assigned to connected processors. Two approaches to deal with this are to employ a look-ahead to the complete solution, or to employ backtracking.

Classical list-scheduling algorithms as surveyed in [AdCD74] neglect communication cost. However, several extensions to include communication cost exist, e.g. see [HCAL89, SiLe90, ElLe90].

Application to clustering Clustering algorithms solve the contraction problem, i.e. to cluster the nodes of a graph according to some criteria. Thus, clustering algorithms derive a graph with a smaller number of nodes from the initial input graph. In other words, clustering algorithms partition the set of all nodes into a set of disjunct subsets where each subset contains a number of nodes of the initial set. Clustering algorithms are often applied to solve the process assignment problem, i.e. to assign a number of processes to a smaller number of processors. The general structure of clustering algorithms is depicted in figure 2.35.

```
(1) assign each node to a different cluster
(2) REPEAT
(3)   compute the distance between clusters
(4)   select the pair of clusters with the smallest distance
        for which the constraints hold and merge the two
        clusters
(5) UNTIL some termination criterion holds
```

Figure 2.35 Clustering

Typically, some constraints are imposed on the clustering, e.g. to limit the size of each cluster. An overview of clustering algorithms to solve the assignment problem can be found in [Heis94].

Instead of the subsequent derivation of a single solution, clustering algorithms are sometimes also employed to subsequently improve the solution, i.e. to iteratively search for better solutions. In this case the algorithm follows the structure given in figure 2.25.

Application to partitioning The reverse of clustering is partitioning. While clustering algorithms start with the single nodes that are subsequently merged, partitioning starts with the complete graph and subsequently partitions the graph in subgraphs. The typical approach is the subsequent partitioning of the graph (and subsequently each subgraph) in two subgraphs. This is called the bipartitioning problem. Basic algorithms to solve the bipartitioning problem are described in [FoFu62, KeLi70].

Extensions of the basic idea Many extensions of the basic ideas are possible and have been discussed. The most important are extensions that employ backtracking if the selected solution does not exhibit the required quality, or if the derived configuration does not represent a feasible solution.

Dynamic programming can be applied to problems where an optimal sequence of decisions has the property that whatever the initial state and decision are, the remaining decisions must constitute an optimal decision sequence with regard to the state resulting from the first decision [HoSa78].

Discussion Algorithms based on the single pass approach are rather restricted in the class of problems they can be applied to. Especially critical in this respect are complex constraints on the optimization which excludes certain solutions. This is mainly due to the greedy nature of these algorithms. The more complex the constraints are the more complex becomes the required future knowledge of the consequences of the decisions made in each step. Thus, single-pass algorithms are often not flexible enough to be easily extensible when new constraints emerge. However, where constraints are small, the approach represents a fast optimization technique that is easy to implement.

2.5.3.2 Neural Networks

Basic idea Neural networks model the function of the brain of thinking creatures. Neural networks consist of very many highly interconnected basic cells. Thus, there is a large degree of parallelism. The idea of neural nets is to train the network with the application, i.e. the optimization problem, at hand. After training is completed, the state of the neural network represents the solution of the optimization problem.

Hopfield nets Hopfield nets [HoTa85] are considered appropriate to solve various combinatorial optimization problems. Hopfield nets have been applied to various problems, including scheduling, mapping and partitioning problems (see [PeSö93] for a survey).

For example, [PeSö93] reports on the application of a Hopfield net to decide the graph bisectioning problem, i.e. the partitioning of a graph in two parts with minimal connectivity. They compare the quality of the solution and the runtime of the optimization with a simulated annealing technique. They state that the quality of the solution is comparable to the quality of simulated annealing. However, the neural net computes the solution in a considerably shorter time.

Kohonen nets A relatively straightforward way of using neural networks represents the application of Kohonen nets [Koho89]. However, the use of Kohonen nets is rather limited. Kohonen nets especially support the topological mapping of graphs. Thus, they can be applied to optimize the mapping of processes on processors [Heis94].

Discussion A major advantage of neural networks is their amenability to parallel execution which is due to their highly parallel nature. In general, the application of neural networks to the optimization problems to be dealt with here is rather complicated. However, exceptions exist, e.g. the application of Kohonen nets for the task mapping problem. Usually, the application of a neural network requires a lot of mathematics to map the problem at hand on the selected neural net with its specific requirements. Especially problematic are problems with complex constraints. Typically, the constraints have to be reformulated and integrated into the goal function, which may be very cumbersome. Another problem with neural nets is that the search may be trapped in a local optimum.

2.5.3.3 Branch and Bound

Basic idea Branch-and-bound algorithms represent a combination of constructing a single solution and the iterative search of the search space for an appropriate solution. Thus, they keep intermediate solutions and save them for a later exploration. In addition, branch-and-bound techniques try to exclude parts of the search space which do not look promising with respect to the quality of the expected solution.

A variety of branch-and-bound techniques exist. Some support the computation of the optimal solution, others employ heuristic rules to prune the search space. Intrinsic with branch-and-bound is backtracking. Backtracking is employed in case a track represents solutions of poor quality or if the violation of some constraints is detected. Due to its support for backtracking, branch-and-bound techniques represent a very flexible approach to optimization.

The most important branch-and-bound algorithm is probably the A^*-algorithm [Nils77]. It is an optimal algorithm that employs bounds on the quality of the solutions. The use of bounds allows the search space to be reduced by bounding subsolutions for which the quality of the solution does not seem to be promising at the current state of the search. An application of the A^*-principle to the scheduling and mapping problem can be found in [ChYu90, Mits93, WaTs88].

Discussion The advantage of branch-and-bound techniques is their flexibility in dealing with different constraints. Their major disadvantages are their high processing cost and their memory requirements. An efficient implementation requires tight bounds or estimates to prune the search space. The exploitation of parallelism with branch-and-bound algorithms is possible but often not trivial.

2.5.4 General Issues

Besides the issues specific for the different techniques, some general issues concerning optimization techniques exist.

Mixing of techniques So far we have assumed that for a specific optimization problem, exactly one optimization method is applied. While this is true for many applications, it does not necessarily hold. Instead, a mixture of two techniques may be applied to solve a problem, or ideas from one method are integrated into another.

In addition, the kind of interaction between different methods varies. Most important is the sequential application of two techniques, e.g. to first apply a random search to derive an initial solution or population followed by a tabu search or the application of a genetic algorithm. An alternative is the full mixture or interaction of two methods, e.g. the random generation of solutions and its local exploitation employing a hill-climbing search. Another example is the integration of randomness into tabu search. Still another kind of combination is the use of one optimization method to optimize the parameter setting of another optimization technique.

Different approaches to integrating foreign principles in the different optimization methods and their combination are described in [Reev93a].

Approximation of metrics With model-based optimization virtually all principles for dealing with complexity (refer to section 2.3.1.1) are applied.

Here, we focus on a principle that is important to all optimization approaches that are based on a neighborhood search and to some extent to approaches employing a subsequent construction of a solution. The major part of the computation time of the optimization is typically consumed in evaluating the quality of selected solutions and in checking whether constraints are violated, rather than the search itself. In order to remedy this at least to some extent, the approximation of the metrics used to derive the quality of the solution, or to check whether the constraints are met, is important. The approximation of metrics may be applied for new solutions for which neighborhood solutions with similar properties exist. It often allows a quick estimation of the difference of the new solution from solutions for which the exact values for the metrics are known. However, since the computed differences are estimates in most cases, the exact value has to be computed from time to time to keep the error within limits. This approximation technique has been applied to a range of optimization methods, e.g. clustering and tabu search. However, note that its applicability depends highly on the kind of cost function employed, i.e. whether an approximation is feasible and accurate.

Hierarchization Another important principle applied to reduce the complexity of the optimization process is the hierarchization of the decision process.

Hierarchization of the decision process aims at the partitioning of the problem in such a way that it can be considered as a disjunct set of subproblems. The subproblems are then decided independently of each other. This is graphically depicted in figure 2.36.

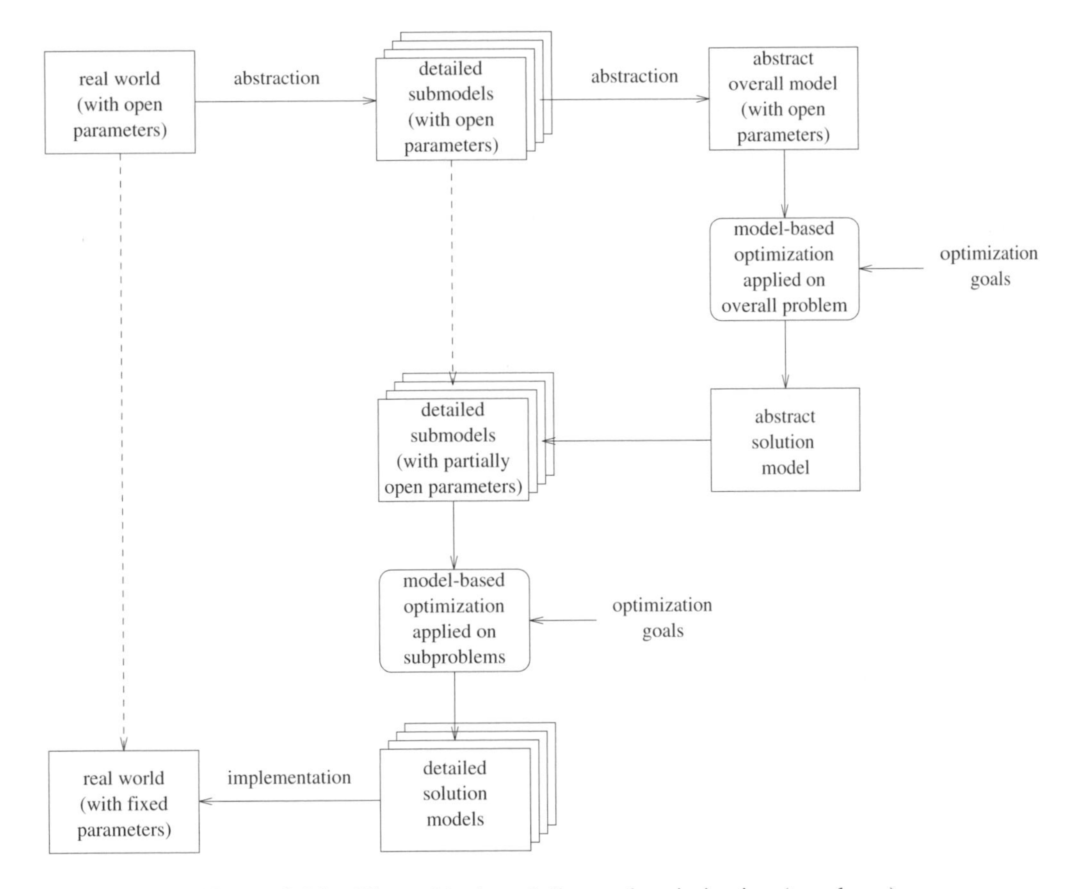

Figure 2.36 Hierarchical modeling and optimization (top-down)

Depending on the specific problem, the optimization may be organized in a top-down or a bottom-up fashion. Alternatively, the optimization may be done iteratively in the different modes. With the top-down optimization, global decisions are decided first followed by local decisions. An example of this is the decision of the spatial mapping of the processes on the processors in the first step, followed by the local optimization of the schedules on each of the different processors. Conversely, the bottom-up approach starts with local decisions and ends up with the global decisions. An example of this is the application of different local optimization techniques on the different processes of the application followed by the global decision of the spatial mapping. Iterative approaches subsequently switch between the different problem domains.

In addition to the hierarchical optimization, the models on which the different optimizations are based may vary. Thus, hierarchization also concerns the models employed. This allows the use of the most appropriate model and granularity for each of the different subproblems.

Note that the merits of the hierarchical optimization depend on the context dependences between the different subproblems. In the simplest case, only the quality of the solution suffers. If context-dependent constraints are present, decisions may have to be revised in another level of the hierarchy. This may happen if there are contradictions due to constraints of the optimization in different levels or between the different subproblems at the same level.

Visualization As we have seen above, tuning the parameters of the search algorithm is an important activity of the development of appropriate heuristics. Important for tuning is to know what exactly is going on during the search. In order to support this, visualization is essential. This is because the optimization process as such is often very complex and hard to follow. Thus, good visualization tools are needed that show how solutions are developing during the search and to indicate why they are developing in a certain way and not differently.

Chapter 3

Design and Implementation of Communicating Systems

Communicating systems, especially communication protocols, number among the most complex systems to develop. In addition, the often stringent performance requirements of communicating systems have a major impact on the development of these systems. Thus, performance engineering plays a very important role in the development of these systems. For these reasons, and also because communicating systems are the prevailing application area for the formal description technique SDL (which is a main focus of the book), we focus on communication protocols and protocol architectures rather than giving a general description of the design and implementation of parallel and distributed systems.

As stated above, communication protocols are highly critical from the performance viewpoint. The performance of parallel and distributed systems is highly influenced by the efficiency of the protocol implementation, its design and the selected protocol mechanisms. Many of the problems and the strategies described here to improve the performance of protocol implementations are also applicable to parallel and distributed systems in general. In order to support this, we concentrate on general issues of protocol design and implementation rather than going into the very details of communication networks.

Experience with practical protocol implementations shows that protocol performance depends at least as much on the design and implementation of the protocol architecture as on the design of the specific protocol mechanisms itself [CJRS89, WaMa87]. In addition, the protocol functionality and mechanisms are typically defined by standard documents provided by ISO, ITU-T, ETSI or other organizations. Thus, the protocol mechanisms are often not negotiable. For these two reasons, we focus on the design and implementation of protocol architectures rather than on their detailed functionality.

Communication protocols are usually implemented by hand. As we will see in this chapter, the techniques employed for hand-coded implementations have been continuously improved. Examples range from the simple elimination of copy operations between protocol layers by data referencing to more sophisticated optimizations such as application level framing, common path optimization and integrated layer processing.

In section 3.1, we start with an overview of the services provided by protocol architectures. In addition, the major structuring principle of protocol architectures is introduced in the section, i.e. protocol layering, and we describe the two most common layering reference models. The basic elements of single protocol layers within the layered architecture are discussed in section 3.2 together with implementation alternatives. This is followed by a discussion of the design and implementation techniques for complete protocol architectures in section 3.3. In the section, various sources for performance limitations are identified and remedies are discussed.

3.1 Basics of Communicating Systems

Before going into the details of protocol implementation, we survey important requirements of protocols, outline the functionality provided by protocols and protocol architectures and survey the major structuring principles. The intention of this section is to provide the ground to understand the problems and issues involved with the efficient implementation of communication protocols, and not to provide a complete discussion of communication protocols.

More details on communication protocols and computer networks can be found in various textbooks on the topic, e.g. [Tane96, PeDa96, Hals96]. [Tane96] provides a comprehensive coverage of computer networks. A very pragmatic book is [PeDa96], which is more oriented towards protocol mechanisms rather than protocol architectures. [Hals96] is more biased towards the OSI reference model.

3.1.1 Requirements

The requirements on protocol architectures, or communicating systems in general, depend on the perspective. Thus, the users of the network, i.e. the application programmer or the end user, the owner of the system and the maintenance staff have different requirements on the system.

Important requirements for protocol architectures and communication networks are

- the provided *functionality*, i.e. the services provided by the system,
- the *reliability* of the provided services, i.e. the probability of failures of the system and of errors,
- the *performance* of the system, i.e. throughput and response times for the different services provided,
- the *cost effectiveness* of the system, i.e. the question of how much performance, reliability, etc. is provided at what cost,
- the *maintainability* of the system, e.g. its flexibility with respect to changes, or the quality of the documentation, and
- the support for *security* it provides.

More requirements may exist depending on the specific application area of the protocol architecture.

3.1.2 Protocol Functionality

Communication architectures provide communication services to the applications which are built on top of the communication services. Thus, the application itself does not have to deal with problems related to communications, e.g. transmission errors, routing, connection management, etc.

In the following, we discuss the basic protocol services. We start with the simple problem of point-to-point connections and subsequently add functionality to provide more advanced features known from computer networks.

3.1.2.1 Encoding and Decoding

Encoding deals with the encoding of bits for the physical connection, i.e. the wire, fiber or radio link, in a way they can be transmitted over the physical link and understood by the receiving communication partner. Thus, encoding deals with translating the bits into physical signals appropriate for the respective physical medium. Conversely, decoding addresses the reverse problem, i.e. deriving the bits from the physical signals received on the link.

The simplest encoding is to map a 0 onto a high signal value and a 1 onto a low signal. The main problem with this simple coding is that a prolonged low signal may correspond either to a sequence of 1's or to the absence of any signal. Thus, there is no difference between a long sequence of 1's and a dead link. In order to solve this and related problems, e.g. the transmission of data over air, a considerable number of coding and modulation schemes have been proposed.

3.1.2.2 Framing

Framing deals with the question of where a portion of transmitted data, i.e. a frame, begins and where it ends. In order to identify the beginning and end of a frame, a special bit string is often used. For example, the HDLC protocol uses the bit sequence 01111110 to identify the beginning and the end of a frame as well as an idle link. As a consequence of this, the bit string including a sequence of six 1's is not allowed within the remaining part of the frame. In order to avoid this, 'bit stuffing' is typically used that inserts a 0 after any five consecutive 1's that are transmitted.

3.1.2.3 Error Detection

Due to electrical interference or other reasons, bit errors may occur during the transmission. Thus, mechanisms are needed to detect these errors. Many approaches for detecting errors in transmission exist. The basic idea is to add some redundant information to the transmitted frame from which the occurrence of bit errors can be derived. An extreme approach would be to transmit the same information in the packet twice. Thus, data of size n would require $2n$ bits in the data field of the frame. In practice, much less redundancy is needed to detect errors with a reasonable probability.

A popular example is the cyclic redundancy check (CRC), which can be implemented efficiently in hardware. Other examples are the use of a parity bit and the checksum algorithm. With the parity bit, an extra bit is simply added to every 7-bit word to balance the number of 1's in the word such that the number of 1's in the word is even (or odd). An extension is the two-dimensional parity scheme, which additionally uses a parity byte for the whole frame. This ensures that the number of 1's in a certain bit position of the bytes of the frame is even (or odd). The checksum algorithm, as employed for example in the Internet, performs a simple addition on the words contained in a frame and adds the computed sum to the transmitted frame. At the receiving side the same is done and compared to the transmitted sum.

3.1.2.4 Reliable Transmission and Flow Control

So far, we have focused on frame processing issues which did not influence the control flow of the frames. However, with the occurrence of errors and losses, correcting actions are required.

In case an error is detected and, in addition, enough redundancy is included in the frame to correct the error, forward error correction (FEC) can be employed to correct the error without retransmission of the frame. However, retransmission is the typical choice to correct errors. This is due to the typically small probability for the occurrence of an error, and the large number of potential error cases that can be introduced on a physical link. Thus, some action which influences the control of the data transmission is needed.

Note that retransmission also requires buffering of the sent packets to ensure their availability in the case of error or loss. Thus, sent packets have to be buffered until their correct transmission is acknowledged.

Stop-and-wait algorithm The simplest approach to supporting reliable transmission is the stop-and-wait algorithm. Stop-and-wait works as follows: after a frame has been transmitted, the sender waits for an acknowledgement before sending the next frame. If the acknowledgement does not arrive within a certain time period, a timeout is raised and the frame is retransmitted by the sender.

The main disadvantage of the stop-and-wait algorithm is that at any time it allows the sender to have a single unacknowledged frame only. Thus, the usage of the link is far below its capacity. In practice, the link usage may be easily limited to one-tenth of its capacity. In this case, the sending of 10 frames before an acknowledgement is expected would allow the full link capacity to be used.

The stop-and-wait algorithm implicitly preserves the order of the frames. In addition, it implicitly supports flow control, i.e. a feedback mechanism which allows the receiver to throttle the sender.

Sliding-window algorithm A popular approach to supporting error correction (as well as flow control) is the sliding-window algorithm. With the sliding-window algorithm, a sequence number (*SEQ*) is assigned to each frame. The sender maintains three variables, the sender window size (*SWS*) which denotes the maximum number of nonacknowledged frames, the sequence number of the last acknowledgement received (*LAR*), and the sequence number of the last frame sent (*LFS*).

Similarly, the receiver also maintains three variables. The receiver window size (*RWS*) denotes the maximum number of out-of-order frames the receiver is willing to accept. The other two variables are the sequence number of the last frame acceptable (*LFA*), and the sequence number of the next frame expected (*NFE*). A received frame is accepted only if its sequence number *SEQ* is within the bounds of the window opened by the variables *LFA* and *NFE*, i.e. if $NFE \leq SEQ \leq LFA$. Otherwise the frame is discarded.

Acknowledgements return the number of the last frame received for which it holds that all previous frames have been received too. With the sending of the acknowledgement, the variables at the receiver are updated. Similarly, the variables of the sender are also updated upon receipt of the acknowledgement.

Besides error correction, the sliding-window algorithm also supports the preservation of the order of the frames (employing the sequence number) and flow control. In addition to restricting the maximum number of unacknowledged frames, flow control allows the window size to be dynamically adapted to the capacity of the sender, e.g. its buffering capacity.

3.1.2.5 Media Access and Mediation

So far, we have assumed point-to-point links for the data transmission, i.e. the case where the sender and receiver are directly connected and where no other communication devices have to be addressed. If this restriction is lifted, i.e. multiple devices are connected to the link, conflicts may arise if more than one device wants to send data at the same time.

Concerning media access and mediation, there are two popular classes, the Carrier Sense Multiple Access (CSMA) approach and token-based approaches. The CSMA approach allows multiple senders to concurrently access the medium. Before sending on the link, the sender ensures that no other device is sending data over the link. However, due to delays in the physical transmission and for the setup of the transmission, conflicts may arise.

A popular approach to deal with conflicts is the scheme taken by the Ethernet. It employs the CSMA/CD approach, where CD stands for Collision Detection. If more than one sender concurrently sends on the link, this is detected by the senders and they back off for some time which is influenced by a random factor. This reduces the probability that a second attempt again results in a collision. The advantage of the approach is the small average response time for low load. The disadvantage is the poor maximum throughput achievable and the large response times at high load, which are both due to the fact that the collision probability increases with high load on the link.

A different approach is taken by token-based schemes employed by the IEEE 802.5 token ring, the IEEE 802.4 token bus and the Fiber Distributed Data Interface (FDDI). With these approaches, arbitration is organized by means of a token. Thus, a sender has to wait until it receives a unique token. When the token is captured by the potential sender, access to the link is granted. Due to the uniqueness of the token, conflicts are prevented right away. However, special care is needed to ensure the reconstruction of the token in case it gets lost or corrupted.

In addition, a wide variety of other approaches exist, e.g. based on polling, reservation of the medium, division of the medium in slots, etc. A comprehensive overview of media access techniques can be found in [As94].

3.1.2.6 Multiplexing and Demultiplexing

The problem of mapping several application data streams onto a single physical link, virtual circuit or connection may arise in various layers of the protocol architecture. This is called multiplexing and demultiplexing.

The mapping of multiple streams on a shared medium has been described above. Besides physical restrictions of the communication medium, other motivations for multiplexing exist. Examples are pricing policies which encourage the reduction of the number of transport connections (upward multiplexing), or the limitation of the bandwidth of a single connection, e.g. with ISDN or GSM, which encourages the use of multiple connections to increase the available bandwidth (downward multiplexing).

3.1.2.7 Segmentation and Concatenation

Segmentation in a protocol layer deals with partitioning of a packet received from the higher layer into a number of smaller packets. This is mostly motivated by limitations of the packet size in lower layers which are imposed in order to support efficient buffering and transmission. Upon reception of the packets, blocking (upward blocking) is employed to reassemble the packets into a single packet.

Conversely, multiple packets received from the higher layer may be concatenated to form a single, larger packet to be passed to the lower layer. This requires blocking at the sending side (downward blocking) until the data are available that justify the transfer of the larger packet to the lower layer. Downward blocking is usually employed to improve the performance.

3.1.2.8 Switching and Forwarding

Networks based on single links or busses are limited with respect to the maximum number of devices that can be hooked on the network, and the geographical area the network can serve. To see this, consider that with a single link, all transmitted data are visible to all devices hooked on the link. In addition, the transmission of data is limited to one sender at a time. Due to these restrictions, an important issue is to support communication between devices that are not directly connected to the same physical link.

This introduces the switching problem, i.e. the problem of forwarding and routing of packets. While forwarding deals with the problem of switching an input signal to the right output, routing deals with the accumulation of data that allows the switch to select an appropriate output. Note that switching also aggravates the flow control problem since each link of the network involved in the communication may now be a potential bottleneck of the data transmission.

There are two important approaches to switching in networks, namely connection-oriented and connectionless switching.

Connection-oriented switching Connection-oriented or virtual circuit switching employs a connection setup phase to install a route between the communicating partners which is then used by all packets transmitted on the connection. Thus, each switch serving as a routing point in the connection is prepared for the packets of the connection and may reserve some resources for it.

This eases flow control and the guarantee of quality of service parameters. In addition, the use of a virtual connection provides better support for the delivery of the packets in the order in which they are sent. This reduces overhead at the receiver to reorder the packets. Another advantage of connection-oriented services is the reduction of the address information transmitted with each frame. This is due to the fact that each switch in the network may store the address of the next hop, which can be identified by a simple circuit identifier.

However, note that the maintenance of state information for each connection in the switches also raises problems. For example, if a switch or a link of a connection fails, a new connection needs to be established. In addition, the old connection needs to be torn down in order to free resources and to update the tables of the switches.

Another drawback of the connection-oriented mode is its delay in setting up a connection before data can be sent.

Connectionless switching Connectionless switching, also often denoted datagram service, sends data right away without explicitly establishing a connection. Thus, no state information is maintained by the intermediate nodes. In addition, each packet is forwarded independently. Thus, subsequent packets may travel on different routes and may arrive at the receiver out of order.

For this scheme to work, it has to be ensured that enough information is contained in the packet itself to enable it to find its destination. Thus, the full address is needed instead of just a single identifier as with the connection-oriented mode.

3.1.2.9 Routing

So far we have ignored the issue of deciding the route on which the packets are traveling. Routing is important for the connection-oriented as well as for the connectionless transmission mode. However, for the connection-oriented mode, the route is fixed during the connection establishment phase, while it has to be decided for each packet in the connectionless transmission mode.

The routing decision is typically based on routing tables. The important issues of routing are how to derive the routing information contained in the routing tables and how to maintain this information.

The basic problem of routing is to find the path with the lowest cost between any two nodes of the network. Note that, in principle, different metrics for the cost of a link may be used, e.g. to denote its bandwidth, latency or current load. In order to deal with dynamic changes to the network, the information has to be updated dynamically. Due to the complexity of networks, routing is typically decided in a distributed manner. Thus, it is possible to maintain a limited amount of information in each switch and also to react quickly to changes in the network topology, caused by failures of links or switches, for example.

3.1.2.10 Internetworking

Using a switching network, we have assumed so far that the different subnets employ the same protocols. However, with the heterogeneity of networks, this is not the case. Thus, provisions are needed to allow the user of one network to communicate with users connected to another network.

Connecting similar networks The simplest internetworking device is a bridge connecting two identical networks. These bridges simply forward frames without making any routing decisions. An extension of simple bridges is so-called learning bridges that only transmit frames destined for the other network. Typically, the bridges learn the location of devices in the network from the return addresses provided by the senders when transmitting frames.

However, note that the scheme only works as long as the extended network does not contain redundant links. In the case of loops in the extended network, redundancy has to be removed to avoid infinitely transmitted frames. This is typically done by employing a distributed spanning tree algorithm that removes redundant links.

Connecting different networks A prerequisite for the application of a simple bridge as described above is that the underlying network technology is similar. For example, these bridges may connect Ethernet networks with each other or connect two FDDI networks. In order to connect networks based on different technologies, e.g. an Ethernet network with an FDDI network, more complex mechanisms are needed.

Important problems to be dealt with in a heterogeneous internetwork are segmentation and reassembly. This is because different networks support packets of different size. Thus, a packet from one network may be segmented in a number of packets for another network and later on be reassembled into a single packet. Other important problems that emerge or are aggravated with the connection of different networks are addressing, routing, flow control and congestion control. While flow control prevents the sender from overrunning the receiver, congestion control deals with congestion or overload within the network itself.

3.1.2.11 Application Services

So far, we have covered the transport-oriented mechanisms, i.e. questions related to the reliable transportation of a portion of data from one point in a network to another point. Besides this, specific application areas are supported by protocols to ease the task of the application programmer. Examples are remote procedure calls, file transfer, terminal emulation, audio and video data transmission and others. Important problems to be dealt with by application-oriented services are the data representation (e.g. big-endian and little-endian representations of computers), data compression and security.

3.1.3 Structuring Principles

Above, we have surveyed important basic services and mechanisms provided by protocols. As can be easily seen from this introduction, the design and implementation of communication networks has a considerable complexity. This is due to a combination of various problems, most notably problems resulting from the spatial distribution which results in the lack of global state information and the presence of errors and failures. In particular, performance and time issues are important and influence the design and implementation of communicating systems.

Before looking into protocol implementation, we take a look at how the different services provided by protocol architectures are organized.

3.1.3.1 Dealing with Complexity

In order to deal with complexity, structuring the system into (mainly) independent building blocks is a major issue. The main structuring principle for communication architectures is hierarchical layering.

Protocol architectures are based on a hierarchical set of layers. Each layer provides some communication services. Lower layers provide simple basic services while higher layers provide complex and more application-oriented services. Higher layer protocols rely on services provided by lower layer protocols, i.e. add some functionality to the services provided to them by lower layers.

For example, consider a data link protocol providing a reliable transport service between two directly connected devices. Thus, a network protocol placed on top of the data link protocol can concentrate on providing the transport service for the end-to-end partners, i.e. concentrate on routing and end-to-end flow control, without having to bother about error detection and correction on the point-to-point connections.

Note that the layering model is a major means of mentally dealing with complexity. In implementations, the layers are often much more integrated into each other to support efficiency.

3.1.3.2 Protocol Architectures

There are two popular models for the layering of protocol architectures, namely the OSI reference model standardized by ISO [ISO84] and the TCP/IP model [LCPM85] employed in the Internet world. Both models are graphically depicted in figure 3.1.

OSI reference model The OSI reference model has for a long time been regarded as the way to structure protocols. It divides the protocol architecture into seven layers. The lower four layers provide transport-oriented services, while the upper three layers deal with application-oriented issues. The transport-oriented layers comprise

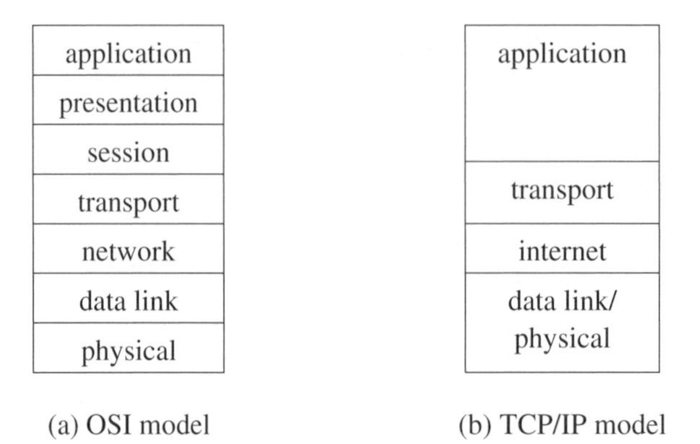

Figure 3.1 The OSI and TCP/IP reference model for structuring protocol architectures

- the *physical layer* dealing with the transmission of raw bits over a physical medium,
- the *data link layer* providing point-to-point connection establishment, error detection and correction, and flow control,
- the *network layer* supporting end-to-end communication, which includes switching and forwarding, segmentation, and end-to-end flow control, and
- the *transport layer* providing (reliable) end-to-end data transfer, containing mechanisms for multiplexing, segmentation and flow control.

The application-oriented layers comprise

- the *session layer* supporting session management,
- the *presentation layer* supporting a common generic representation of data independent of specific hardware architectures, and
- the *application layer* providing specific application support for file transfer, terminal emulation and other functions.

Even though the OSI reference model has been overtaken by the enormous growth of the Internet community, it is still a good model to understand and deal with the complexity of protocol architectures.

TCP/IP reference model The TCP/IP model is a descendant of the ARPANET sponsored by the US Department of Defense. A major design goal of the TCP/IP model was to support communication even in the presence of failures of some links or hardware components. Thus, reliability and fault-tolerance have been major issues. Unlike the OSI reference model, the TCP/IP model is less generic and more specific about the services provided.

The TCP/IP model focuses on the following three layers:

- the *internet layer* with the connectionless Internet Protocol (IP) supporting the detection of erroneous headers, routing and forwarding, and segmentation,

- the *transport layer* comprising two important protocols, namely TCP (Transmission Control Protocol) that supports reliable connection-oriented communication, and UDP (User Datagram Protocol) supporting unreliable connectionless transmissions, and

- the *application layer* comprising application-oriented higher-level protocols, supporting file transfer (FTP), terminal emulation (TELNET), electronic mail (SMTP), domain name services (DNS), transfer of hypertext (HTTP), and other services.

Unlike the OSI reference model, the TCP/IP model does not address the data link and physical layers. In addition, the presentation and session layers of the OSI model are not explicitly present in the TCP/IP model. A comprehensive discussion and comparison of the two models can be found in [Tane96].

3.1.3.3 Functional View of Protocol Processing

Above we have looked at the static structure of protocol architectures. In order to get a better impression of the dynamics of protocol processing, we take an additional look at the functional aspects, or more specifically, at the dynamic flow of data through the protocol layers.

The flow of data can be most visibly described as enveloping. A packet, inserted into the protocol architecture by the application running above it, is subsequently packed into envelopes. Finally, when the information is put on the physical network for transmission it is wrapped in a number of envelopes. On the receiving side, the envelopes are subsequently removed in reverse order until the information can be passed to the receiving application. The envelope is labeled with control information concerning the transfer of the envelope, most notably target and return addresses.

With computer communication, envelopes are implemented by headers (and sometimes trailers) which are added to the data. The addition and removal of headers is depicted in figure 3.2. A header is added in each protocol layer upon sending and removed upon receipt of the packet. The header (or possibly trailer) contains the *Protocol Control Information* (PCI) needed by the peer protocol on the receiving side to process and forward the data on its way to the receiving application.

The data contained in the envelope are often denoted as the payload. Upon sending, each protocol layer takes the payload provided by its upper layer and adds its protocol control information to it. In addition, the payload given by the upper layer protocol may be transformed into another representation for various reasons. The protocol control information is used to provide the different protocol functionalities as described above, e.g. error detection and correction, forwarding to the receiver, etc.

In addition, segmentation or concatenation may be applied to adapt the size of the packets. As described in section 3.1.2.7, segmentation divides the payload into smaller pieces and packs

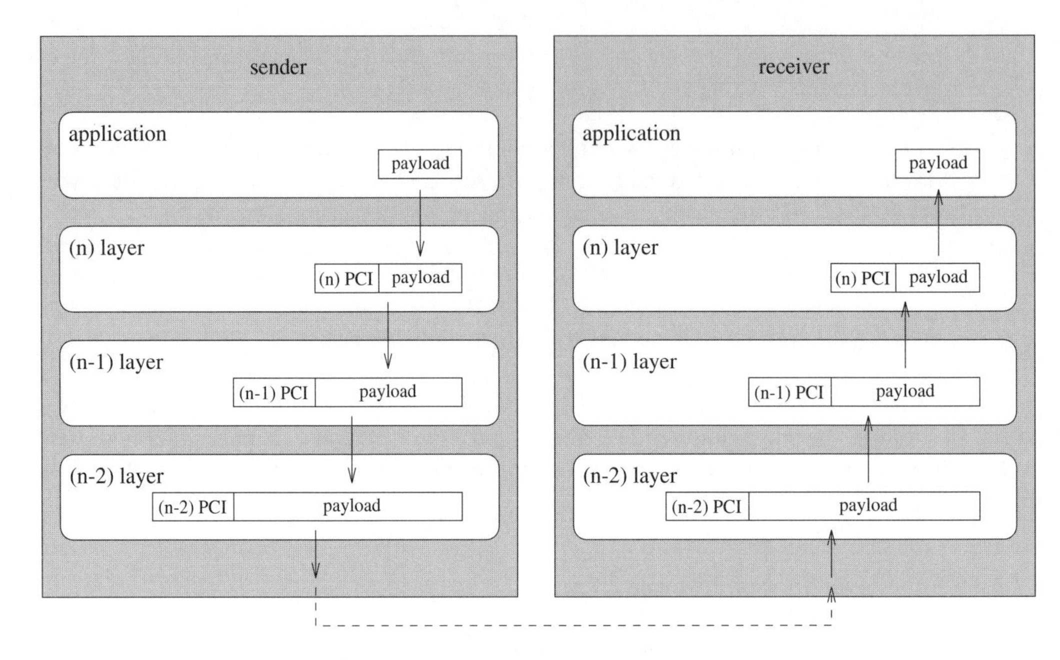

Figure 3.2 Outline of the enveloping principle employed by protocol architectures

them in separate packets before they are passed down. With concatenation, several payload units are packed into a single envelope (packet).

3.2 Elements of Protocol Design and Implementation

After identifying the basic functionality of communication protocols and the principal organization of protocol architectures, we discuss the alternatives for the design and implementation of communication protocols and protocol architectures. The focus in this section is on the functions implemented by single protocol layers, while the focus of section 3.3 is on issues concerning several layers and the protocol architecture as a whole.

Concerning the protocol design as discussed here, we assume that the principal protocol mechanisms are given. This is due to the fact that protocol mechanisms are typically specified by standards. For this reason, we focus on implementation-oriented design issues, rather than the protocol mechanisms itself. Thus, we deal with the question of how to structure and implement the system such that given performance goals can be met.

Besides vertical layering of the protocol architecture, the manner in which a single protocol layer is organized is of interest. Each layer of a protocol architecture provides some protocol processing functionality. Classifying protocol processing functionality, we can identify two main elements of a protocol layer:

- *data processing* denoting the processing of the payload, i.e. transforming data into another representation, possibly including segmentation and concatenation, and
- *control processing* denoting the processing of the protocol control information in conjunction with the protocol state, i.e. the analysis (receive) or synthesis (send) of the header, taking alternative actions depending on the contents of the header and the payload.

In order to provide the protocol processing functionality, an additional element can be identified. This comprises

- *support functions* typically provided by the runtime environment, e.g. to support buffer, process and timer management.

In the following, we look at the three basic elements in more detail.

3.2.1 Data Processing

3.2.1.1 Functionality

In protocol processing, three basic classes of data processing activities can be identified (compare also [ClTe90]). The operations are employed in different layers of the protocol.

The motivation for the first two classes of data processing activities is mainly to move data from one location to another. Thus, data movement is a major concern, rather than changing the representation of the data itself, which is the focus of the third class of operations.

Network interfacing The data to be transmitted have to be moved onto the physical link. Similarly, the data received on the physical link have to be retrieved. Since this involves a transformation of parallel data into a serial stream of bits (and vice versa), this is typically performed by special hardware.

User interfacing Typical computer systems are divided in at least two protection domains for security reasons. In order to support this, data are typically copied when passing a security border. Upon reception from the network, data have to be copied at some point from the system domain to the user domain to make them available to the application. The same applies in the sending direction, where the data are copied from the user domain to the system domain before they can be transmitted on the network.

Data manipulations Besides the copy operations described above, the data have to be manipulated for various reasons. These include

- *Error detection:* Typically, the transmitted data are attributed with some redundant information to detect errors. The derivation of the redundant information on the sending side, e.g. a checksum, requires processing of the data. The same holds for the receiving side to detect errors.

- *Buffering for retransmission and reordering:* In order to recover from errors or lost packets, a copy of the transmitted data is kept at the sending side. This allows the data to be retransmitted upon request. Similarly, buffering may be employed at the receiving side in the case that the retransmission of data is selective, i.e. only the lost or erroneous packet is transmitted rather than the whole sequence of packets following the corrupted packet. Another reason for buffering at the receiving side may be packets arriving out of order and having to be reordered before they can be passed to the application.

- *Buffering due to segmentation and concatenation:* Packets may be segmented and transmitted in smaller pieces. In order to support this, buffering of the received packets may be required until the data to be passed to the higher layer as a single packet are available. Conversely, buffering at the sending side is needed if concatenation is employed.

- *Encryption/decryption:* Encryption may be employed for security reasons. This involves processing of the data at the sending and receiving side.

- *Presentation formatting:* Different computer systems may use different internal data representations. Thus, conversion of the data presentation may be needed. Typically, the presentation is changed into a (standardized) transfer syntax at the sending side and transformed into the syntactical representation of the receiver at the receiving side. The machine-independent definition of data structures is supported by ASN.1.

- *Data compression:* In order to reduce the size of the transmitted data, data compression techniques may be employed.

3.2.1.2 Design and Implementation

The above operations are typically implemented by a loop sequentially processing the data words in a linear order. Within each iteration of the loop, a word is retrieved from memory, manipulated and written back to memory. This is depicted in figure 3.3. For simplicity of the figure, the details of the address manipulation are omitted.

```
(1) FOR EACH word OF packet
(2)   FETCH (word, addr)
(3)   MANIPULATE (word)
(4)   STORE (word, addr)
```

Figure 3.3 Framework for the implementation of data manipulations

Data flow operations are a main source of overhead and limit the maximum throughput of the protocol architecture. Under some circumstances, two or more of the operations may be merged or combined to reduce the number of transfers of data between the processor and the main memory (see below for details).

3.2.2 Control Processing

3.2.2.1 Functionality

In addition to the operations applied to the data itself, additional control functions are applied. The control functions control the data processing activities. The result of a data processing activity may be an input to control processing, e.g. the checksum computation may prompt the disposal of the packet. Conversely, the data manipulation may be controlled by some control operations.

The major control actions performed by communication protocols are

- *connection control* comprising the functions to set up and release a communication, and
- *transfer control* denoting the control functions needed to control the data transfer.

In the following, we focus on the second class of functions. This is because they represent a major source of overhead during data transfer. Conversely, connection control deals with functions to initialize and finalize a communication. Even though this has an impact on the setup latency, it typically does not have a major influence on the performance of the data transmission.

The following important transfer control operations can be identified (compare also [ClTe90]).

- *Flow and congestion control*

  In order to prevent flooding the receiver or the network with more packets than they can handle, flow control and congestion control are employed. In order to support this, some functions are needed that throttle the sender and monitor the arrivals at the receiver. This is complicated with real-time and multimedia applications where the timing of packet arrivals underlies rather strict requirements.

- *Detecting network transmission problems*

  Data may get lost or corrupted during the transmission. Thus, functions are needed to detect these problems and to remedy them, e.g. to prompt a retransmission of lost or corrupted packets.

- *Acknowledgement*

  Acknowledgements are a fundamental concept of communication. Acknowledgements are transmitted by the receiver to notify the sender of the correct reception of a packet.

- *Multiplexing*

  Since several data streams may be interleaved at a communication end system, multiplexing the data to the right application (or lower-layer connection) is an important issue.

- *Segmentation and concatenation*

  Mainly due to network limitations, data may be segmented in smaller units for transmission. Thus, functions are needed to allow for the units to be correctly reassembled at the receiving side. Conversely, a number of packets may be concatenated to a single packet before the transmission. Again, the respective functions to control the concatenation and the separation are needed.

The discussed functions require additional support functions typically provided by the operating system, e.g. to provide time stamps or to support buffering.

3.2.2.2 Design and Implementation

The control part of communication protocols is typically specified by a finite state machine. The automaton decides on the actions to be selected based on

- the current state of the automaton,
- the protocol control information contained in the packet, and
- the result of the data processing.

Typically, the current state and the present event, e.g. information contained in the PCI, determine the entry point into a protocol processing function. In typical implementations many of the details, i.e. the detailed actions taken on behalf of the packet, are decided within the selected function. In addition, the next state of the automaton is decided based on the processing. In OSI protocol specifications, often additional predicates are used to decide on the selected transition.

The state automaton is triggered by events. Triggering events are arriving packets (from the application or the network), and expired timeouts.

The state of a communication (or a connection) is defined by the state of the finite state automaton and the values of supplementary variables, e.g. to implement the sliding window algorithm.

There are two basic approaches to implement the protocol automaton of a communication protocol, namely the table- and the code-based approach. With the *table-based approach*, a table is employed that maps the input and the current state of the automaton on the actions to be performed. Typically, the table contains the pointers to the functions that perform the respective actions. In this case, a generic interpreter is employed to process the state automaton, i.e. to select the appropriate actions based on the input and the state of the automaton.

An alternative technique is the *code-based approach*. Here, the state automaton is directly implemented in code using `IF` or `CASE` statements. The code-based approach has been quoted for being more time efficient [Svob89] while the table-based approach may require less memory. However, there is no general agreement on this in the literature. Especially for implementations automatically derived from formal descriptions, time and space efficient table-based implementations have been reported [KöKr96] while others advocate a code-based implementation where runtime efficiency is an important requirement [BrHa93].

3.2.3 Support Functions

In order to process packets by a communication protocol, a set of supplementary services is required. These functions typically require support by the operating system or are supported by special hardware.

The protocol support functions are services used by the respective protocols. In addition, some of the services are also important for the protocol architecture as a whole. Examples are buffer management, which is often implemented in a distributed fashion, i.e. buffer allocation and deallocation are performed in various protocol layers, and process management. As we will see below, this is important for reasons of efficiency.

In the following, we concentrate on the basic functionality of the support functions and some issues concerning their implementation. Where the application of the support functions comprises multiple protocol layers, i.e. where the functionality is used in an integrated way over a set of protocol layers, the discussion is deferred to section 3.3.

3.2.3.1 Timer Management

Many protocols rely on timers, most notably to detect and react to packets lost during transmission, for connection management, and to support flow control.

Different alternatives for the use of timers exist depending on the specific protocol mechanism. A timer may be used for each packet that is sent. Alternatively, just one timer may be used for each connection.

There are different approaches to implement timers. The minimum requirement for the use of timers is a command that provides some notion of time. In subsystems solely dedicated to protocol processing, a feasible approach is to (periodically) poll the actual time and to decide on the further actions based on the current time. The remaining functions can be implemented in the software processing the protocol stack or some additional support layer.

In more general systems, often the timer management facilities provided by the operating system are used. Typical functions supported by operating systems are `SET`, `RESET` and `CANCEL` operations and a mechanism to trigger a process due to a timeout. Alternatively, special timer hardware may be employed to provide an efficient implementation of the timer management functions.

A problem with timer management by the operating system is that these timer routines are optimized for `SET–EXPIRE` sequences, while communication protocols use `SET–RESET` sequences most of the time.

A discussion of the design and implementation of timer management functions can be found in [Svob89]. Approaches to hardware support for timers are discussed in [Hedd95].

3.2.3.2 Buffer Management

Protocol processing manipulates large amounts of data. Packets are of variable size, are manipulated possibly changing their size, segmented and concatenated. Memory has to be allocated for the packets upon arrival and freed upon their departure. In addition, buffer management is often required during the protocol processing, e.g. to add header information or to store the packet in another representation.

The usage of memory and the extent to which copying is employed are considered the major sources of overhead of protocol processing and, thus, are very important in deriving efficient implementations. This is especially significant as the control processing functions of the protocols are often not very computationally intensive.

A discussion of buffer management schemes can be found in [Svob89]. A performance evaluation of three buffering strategies is given in [WoMo89]. As can be easily seen, buffer management is a topic that goes beyond single protocol layers. This is especially true where optimizations to minimize copy operations are employed. In this case, buffers are passed between the protocol layers wherever possible instead of allocating a new buffer in each protocol layer. We will take a closer look at strategies for the minimization of copying in section 3.3.3.

3.2.3.3 Process Management

The protocol functions implemented in code have to be executed by some hardware units. Thus, some active resources are needed, i.e. some kind of processes provided by the runtime environment, or alternatively, directly implemented in hardware.

Process management deals with the creation, scheduling, and destruction of processes on behalf of the application. Process creation and destruction are especially important during connection setup and release.

Conversely, scheduling is important during data transfer and may highly influence the performance of the system, especially the packet delay. An important question concerning scheduling is the scheduling strategy, i.e. the strategy to handle contention, and the dynamic influence of the protocol mechanism on this. The problem is even aggravated when the communications underlie real-time constraints, e.g. as is the case with multimedia systems. Scheduling can be supported by hardware, e.g. as is the case with Transputers [INMO88].

Highly related to scheduling is also the support of the operating system for synchronization between processes, e.g. to enforce mutually exclusive access to shared data.

By its very nature, process management is an issue that goes beyond a single protocol layer. It is an issue that defines the interleaving between the processing of different layers and different packets. We will discuss different process models for protocol processing in section 3.3.2.

3.2.3.4 Process Communication

Processes communicate with each other. Thus, support for interprocess communication is an important issue. Alternatives are synchronous and asynchronous communication. Obviously, process communication is highly interrelated with buffer management and process management.

Besides communication between the processes implemented in software, communication with hardware devices connecting to the physical link is an important issue. This is because of the limited capacity of the network interface, which requires immediate action upon the arrival of data.

3.3 Design and Implementation of Protocol Architectures

Above, we have introduced the basic elements of single protocol layers and have discussed some design issues. The focus in this section is on how to map a protocol architecture comprising a set of protocol specifications into an implementation. Even though protocol specifications are typically provided by standardization bodies, the protocol standards leave many design and implementation decisions to the developer. In this section, we especially focus on how to glue the layers together in an implementation and how to map the protocol layers on executable units for the available resources, i.e. processes, threads or special hardware. Thus, the focus here is the design and implementation of the protocol architecture as a whole rather than the specific protocols.

3.3.1 General Issues

3.3.1.1 Layered Implementation of Protocol Functions

The mapping of the conceptual (or design) units of the protocol architecture on the underlying system with its physical resources is an important topic. It is highly related to scheduling and process management issues.

The question is how to distribute or map the functionality needed for protocol processing onto the layers of the underlying system. In other words, in which layer of the system architecture is a specific functionality provided? For example, which system layer provides timer management, process management and communication? In principle, functionality or services may be implemented in

- the application-specific part,
- the runtime support system,
- the operating system kernel, or
- special hardware.

The discussion is often subtle and has a major influence on the performance of the derived implementation. However, note that due to other constraints not all mappings are feasible or useful.

A related question concerns the granularity of the units to be mapped on the active entities. This is especially important where the process management overhead and possibly the communication overhead are high compared to the execution cost of the protocol mechanisms itself. We will see different alternatives for the mapping in this section.

3.3.1.2 Implementation of Protocol Layering

As discussed in section 3.1.3.2, the top-level design of protocol architectures is clearly oriented towards a hierarchically layered approach. Thus, higher protocol layers provide more advanced services based on more basic services provided by lower layer protocols. In other words, each higher-layer service request has to be mapped on some simpler service requests provided by the layer below. As a result, each layer is of limited complexity, compared to the complexity of the services provided by the complete protocol architecture.

The direct implementation of the layered model, i.e. mapping each layer on an independent implementational unit, is very appealing from the software engineering viewpoint. It allows one to derive a highly structured system with clearly defined interfaces and separated functionality. However, as we will see below, the major drawback of a puristic implementation of the layered approach is its inefficiency.

On the other hand, even highly optimized implementations do not completely avoid layering. Thus, the layered approach is suitable for the implementation if applied with care. In addition, as layering is the most important structuring principle in communications, an implementation based on layering can be derived much faster than an unstructured implementation. Moreover, layering eases the derivation and maintenance of high-quality software. As we will see in chapter 5, optimized implementations can also be derived automatically from (layered) formal descriptions. In this case, the extent to which the layering principle is adhered to in the derived implementation is a minor issue provided the implementation is correct and does not have to be maintained at the code level.

3.3.2 Process Models

The process model defines to a large extent the implementation of the protocol architecture. It defines the way the protocol layers are implemented and interact. We introduce the two basic process models, namely the server model and the activity model, and discuss variants and design issues.

3.3.2.1 The Server Model

The server model provides (at the first glance) a straightforward mapping of the layered design on an implementation.

Basic principles Protocol processing within a protocol architecture can be viewed as follows: there are a number of active entities, in the following called servers, each entity implementing a protocol layer or a sublayer. The servers are asynchronously executing units, communicating with each other through a clearly defined interface. Communication is also asynchronous, employing buffering to decouple sending and receiving entities. This implementation principle is called the server model. The principle is graphically displayed in figure 3.4.

The generic tasks of each of the servers are as follows:

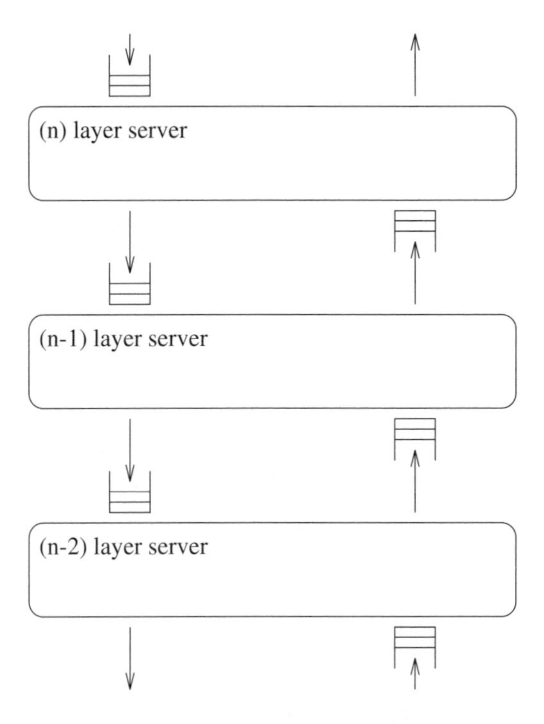

Figure 3.4 The basic server model

- A server implementing a protocol layer maps a single input, or possibly a set of inputs (events), on a number of outputs (i.e. zero, one or several). This way, segmentation, concatenation, erroneous packets, timeouts, or other issues can be handled.
- Each server has a single interface where it blocks (or waits) for triggering events, e.g. packets from the upper or lower layer or triggers due to timeouts.[1]
- Each server executes one event at a time, e.g. a packet or a timeout. After completion, it again blocks on its interface, waiting for new events. This simple principle is depicted in figure 3.5.

The discussion implies that the server model is an ideal approach from the software engineering standpoint. It clearly supports good engineering principles such as modularity, encapsulation, and clear interfaces. This is especially important in systems where some protocol layers are replaced over time, and for conformance testing of protocols. The server model allows one to directly transfer the protocol design principles described above to the implementation.

[1]Note that this is different from the simplified view given in figure 3.4, where two input queues per server are displayed to simplify the figure. In fact, multiple queues can be employed in the case that the concurrent blocking on more than one interface is supported by the underlying system.

```
(1) FOREVER
(2)   WAIT_FOR (event)
(3)   protocol_processing (event)
```

Figure 3.5 Framework for the implementation of a server

Implementation of servers as processes of the operating system The generic view given above represents the major principles of server model implementations. It supports the fully asynchronous processing of packets and other events.

However, so far we have not discussed how the servers and the support functions needed to manage the servers are implemented in a real system. Several approaches to this exist. A straightforward approach is to implement each server by a process handled by the operating system. Thus, the operating system has full control of the protocol processing activities.

Multiple connections can be implemented by creating a new process for each new connection. This approach is common for protocol design and, as such, can be easily mapped into the implementation. Conversely, multiple connections can be handled by a single process, too. In this case, only new data structures have to be allocated upon connection establishment, to store and maintain connection-specific data. This approach is taken by most implementations. In either case, an additional process is typically used to accept and handle new connection requests.

With this simple implementation strategy, process management (e.g. scheduling), timer management, and interprocess communication are fully handled by the underlying operating system.

Besides the principal problems encountered with the server model – which are described below – there are some specific drawbacks of the implementation of servers as operating system processes. Most of them relate to performance issues. Most notably, the implementation of different protocol layers by different processes typically does not allow memory to be shared between processes. Thus, a major source of overhead results from the movement of data between processes and for buffer management. Other disadvantages are overhead for process management (context switching), and the use of other support functions provided by the operating system which are typically not optimized for protocol processing.

Implementation of servers as lightweight processes In order to remedy the problems described above, lightweight processes have been proposed. Unlike traditional (heavyweight) processes, lightweight processes operate on a common memory area. Thus, they support shared data and fast context switching. Lightweight processes are contained within a traditional process and share the resources available to this process. Lightweight processes are often called threads of control, or just threads, for short.

Threads are either handled by support functions provided by a thread library, i.e. providing user-level support, or implemented by the operating system, i.e. system-level threads. The main difference between user-level and system-level threads concerns the handling of possibly blocking operating system calls.

With *user threads*, blocking system calls have to be replaced by nonblocking calls. This is necessary due to the fact that a user-level thread blocked on a system call would block the whole process. Thus, no other thread of the process would be able to proceed with its execution until the blocking is released. In addition, user threads employ cooperative multitasking, i.e. the threads of a process rely on the currently executing thread to release the processor in order to get a fair share of the processor. Thus, single threads are not preemptable; only the whole process can be preempted.

User threads introduce an additional layer in the system architecture, i.e. a layer between the application and the operating system. The new layer provides its own primitives for thread management, e.g. thread creation, scheduling and synchronization. For this reason, user-thread packages are often also denoted as runtime support systems. The big advantage of these packages is the fact that they are built on top of the operating system. Thus, the thread package can be quickly adapted to the specific needs of the application or the underlying operating system. With user threads, there are two levels where scheduling is dealt with, i.e. the operating system supporting preemptive process scheduling, and the thread scheduler within each process.

Conversely, with *system threads*, thread and process management are both supported by the operating system. Thus, a much tighter integration is achieved, and a more homogeneous handling of scheduling can be supported.

An approach that supports the use of shared memory is vital for most of the other approaches discussed below to improve performance. Threads are an important approach in meeting this prerequisite.

General problems of the server model Even though the server model is based on a clear concept and, thus, very intuitive, there are some subtleties. One problem is interlayer flow control, i.e. flow control within the protocol stack. Interlayer flow control is needed to prevent an (n)-layer server from producing data faster than the consumer, e.g. the underlying (n-1)-layer server, can consume them. In order to prevent this, some feedback mechanism is needed. Approaches to support interlayer flow control are sketched in [Svob89] and discussed in more detail in [GaHM88].

A related drawback of the asynchronous execution implemented by the server model is that a large number of buffers is needed to support interlayer communication. As pointed out in [Svob89], additional complications may arise due to the need for the atomic execution of events. For example, this may be needed to handle connection releases, where synchronous interaction between the layers may be required in addition to an asynchronous mechanism.

An alternative to asynchronous communication between processes is synchronous communication. This may alleviate some of the problems. However, it is not suitable for the case where several connections are handled by a common process.

Another problem of the server model relates to the scheduling strategy. Typical operating systems require priorities to be assigned to processes. In order to minimize the response time of protocol processing, the shortest-job-first (SJF) scheduling strategy is an important approach.

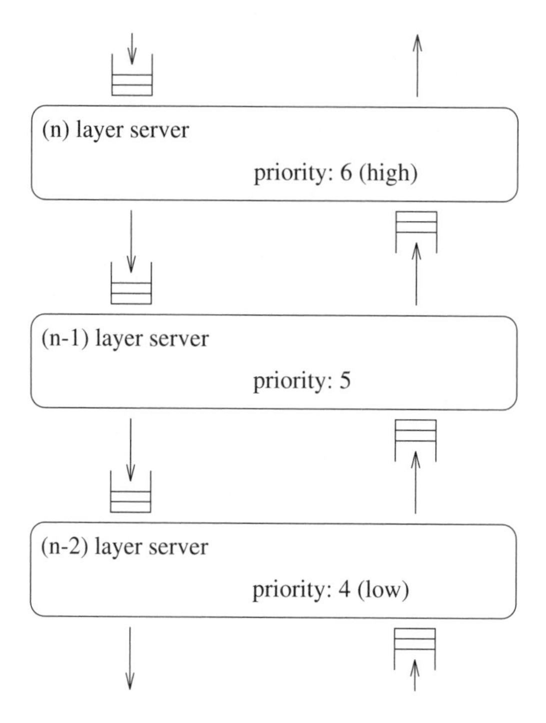

Figure 3.6 Process priorities for the server model

It assigns priorities to jobs (in our case events such as packets or timeouts) to be processed according to their remaining processing demands. Jobs with a high processing demand are assigned low priorities, while jobs with small demands are assigned high priorities. A direct implementation of SJF would require one to continuously increase the priority of jobs as they approach their completion. Concerning inbound traffic, this would mean increasing the priority of a packet as it approaches the application layer. Since the assignment of priorities to packets is often impractical, an approximate implementation of SJF is typically to assign priorities to processes according to their position in the processing pipe. In figure 3.6, this is graphically depicted for a three-layer protocol stack.

In the figure, the priorities assigned to the layers are in increasing order of the level of hierarchy of the protocol layers. Thus, higher layers are assigned higher priorities, while lower layers are assigned lower priorities. As a result, packets entering the protocol stack for inbound processing at the lower interface are first processed with low priority and later – as processing through the layers proceeds – with higher priorities. This ensures that the processing of an inbound packet can proceed even if a subsequent packet has arrived at the lower interface. In fact, processing of a subsequent inbound packet is not commenced before the first packet has been fully processed by the stack.

A major problem with this priority assignment scheme is that it only works for one direction.

This is because typical servers process inbound and outbound traffic. The alternative, to employ separate processes for inbound and outbound traffic, i.e. to have a dedicated server for each direction, is often not practical due to the fact that inbound and outbound traffic require common data, e.g. for connection management.

A feasible alternative that is supported by some specialized operating systems is to assign priorities to packets rather than to processes. In this case, scheduling of processes is based on the priority of the packets they hold in their input queue. Obviously, this supports different priorities for inbound and outbound traffic at a server.

As pointed out in [Svob89], the complete, non-interleaved processing of a single packet by the whole protocol stack may not necessarily be advisable. Especially with the implementation of servers by operating system processes, the overhead for process switching may be considerable. Thus, it may be advantageous for a server to process a number of packets, e.g. to empty its input queue, before control is switched to another process.

3.3.2.2 The Activity Thread Model

With the server model, each layer is represented by a set of servers that process packets or other events. Thus, the protocol instances or layers are the active entities of the system.

Basic principles With the activity thread model, the packets that are processed by the protocol stack represent the active entities, rather than the protocol instances. The protocol instances themselves are passive entities. The main ideas underlying the activity thread model are due to Clark [Clar85], who introduced multi-task modules and upcalls. Before going into the details of this proposal, we discuss the general ideas underlying the activity thread model (see also [Svob89]).

With the activity thread model, each protocol layer is implemented by a set of procedures, typically a procedure per event. An implementation based on the activity thread model is outlined in figure 3.7. The example comprises three protocol layers. The figure shows the basic execution principle of the activity thread model. Every incoming packet arriving at the interface triggers a procedure call. The communication between the protocol instances is implemented by (synchronous) procedure calls rather than asynchronous communication. Thus, each incoming packet results in a hierarchy of procedure calls – called the activity thread. The activity thread terminates (i.e. the procedure initially called at the interface returns) when the packet has been completely processed by the protocols that constitute the activity thread implementation.

The simplest form of the activity thread model does not support concurrent execution. Thus, the packets or other events arriving at the interface are executed sequentially, one packet at a time. The advantage of this approach is that no locks are needed within the activity thread implementation itself, which reduces runtime overhead. Blocking may occur only at the external interface of the activity thread implementation.

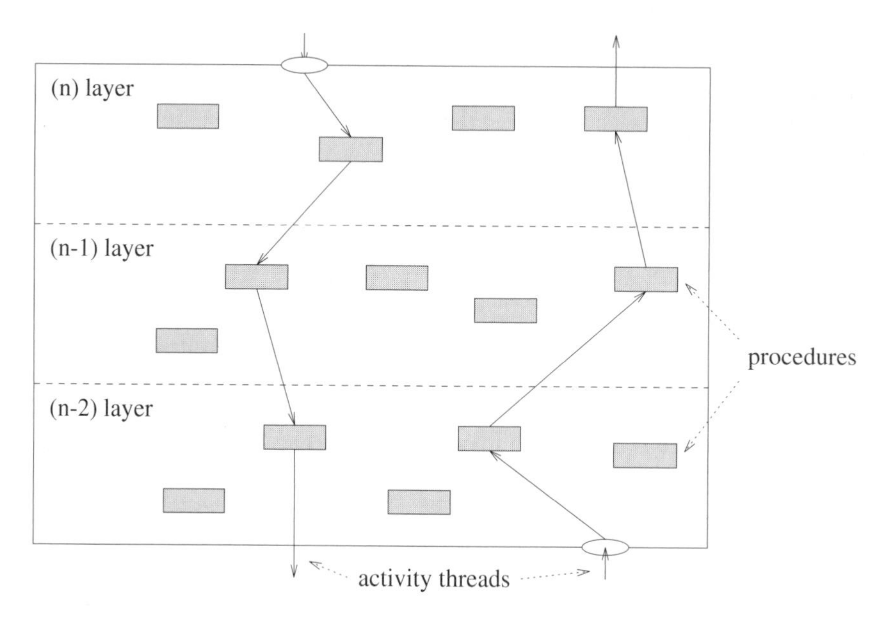

Figure 3.7 Packet processing according to the activity thread model

Multi-task modules and upcalls Clark [Clar85] proposed multi-task modules and upcalls to implement protocol architectures or other layered systems. Multi-task modules denote a set of procedures that share some common data. Since multiple concurrent threads of control are supported, shared data have to be protected by locks.

Clark proposed to implement upward and downward communication in layered architectures by procedure calls. An upcall is a procedure call in the upward direction in a layered architecture, i.e. a call of a higher-layer procedure by a procedure of a lower layer. Conversely, sidewards communication, i.e. communication between the procedures within the same layer, is proposed to be implemented by asynchronous communication.

The support for concurrency requires some care in implementing multi-task modules and upcalls. Special care is needed to avoid deadlocks where upcall–downcall sequences are employed and to recover from failures. Clark proposed a methodology to deal with these problems. For example, before an upcall is issued, shared data in a multi-task module have to be put in a consistent state and locks acquired in the layer have to be released. Another rule is to prohibit recursive downcalls that are issued by an upcalled procedure.

As noted in the original paper, the approach supports the implementation of flow control by an intermediate layer. In this case, so-called arming calls are used by an upper layer to inform the lower layer, i.e. the layer that is in charge of flow control, that data for transmission are available. The arming call immediately returns after setting a flag in the lower layer. This prompts the lower layer to issue an upcall to get the data whenever flow control permits the sending of the packet.

Problems and limitations In order to apply the activity thread model rather than the server model to derive code, the following differences between the two models have to be taken into account:

- The active elements in the server model are the protocol entities. Thus, protocol entities are implemented by active servers. The execution sequence of the events is determined by the scheduling strategy of the operating system or the runtime support system. Communication is asynchronous via buffers.
- The activity thread model is event-driven. It can support an optimal execution sequence of the events. Communication is mostly synchronous, i.e. no buffers are employed. Thus, the model is well suited for the layer-integrating implementation of protocols.

As a result of the semantic differences between layered protocol architectures and the activity thread model, i.e. synchronous computation and communication, the implementation of a protocol stack according to the activity thread model is less straightforward than with the server model. On the other hand, the server model exhibits extra overhead which is not present with the activity thread model, e.g. overhead for asynchronous communication including queuing operations and overhead for process management. However, with the activity thread model, special care has to be taken to correctly map the semantics of the protocol specification on the implementation. Potential problems may arise with timer handling, interrupt handling, and protocol mechanisms that require feedback to the sender.

Since the direct implementation of upcalls poses many problems, the management of activity threads by a runtime support system has been proposed [FICE87]. Instead of direct up- and downcalls by the calling procedure, the procedures are called and managed by the runtime support system. This allows clean software engineering principles to be employed.

Also due to the potential problems with upcalls, often a mixing of the server and activity thread model is implemented. Thus, traffic in one direction is handled by procedure calls while traffic in the opposite direction is implemented by asynchronous communication employing buffers.

3.3.3 Minimizing Data Movement

The copying of data is the most published don't in implementing communication systems. The problem is continuously aggravated by the fact that processor performance is increasing faster than the performance of memory.

Especially with the basic server model implementation described above, numerous copy operations are employed within a protocol stack. Copying is employed at each process interface to move data from the producing process to the buffer and subsequently from the buffer into the consuming process. Additional copying may be employed within the servers to manipulate the data.

Compared to the overhead for copying as introduced with the basic server model implementation described above, the time for protocol processing itself, i.e. the implementation of the protocol mechanism, may even be negligible. Often the processing employed by a protocol layer is rather small, i.e. a lookup in a table, adding or removing a header or similar things.

A simple performance evaluation based on the memory bandwidth of the computing system, together with the number of copy operations and the respective size of the packets being copied, allows the derivation of an upper bound on the throughput of the communication architecture.

Due to their enormous influence on performance, the minimization of data movements is a major issue in protocol implementation. In the following, we discuss the basic techniques. A detailed discussion of advanced approaches to minimizing copying in the context of communication subsystem design can be found in [DAPP93]. Approaches to supporting buffer management in hardware are described in [Hedd95].

3.3.3.1 Common Buffer Management

As noted above, often only a small part of the packet received by a layer is changed, i.e. the addition or removal of a header. Thus, there is no real need to copy the whole packet just to change a small portion of it.

In order to support this, a common buffer pool can be used. A requirement for this to work is the existence of memory that can be shared between the protocol entities, as with the thread-based approach introduced above.

In addition to the use of a common buffer pool that is used by several protocol entities, the data are not copied into the manipulating server (or procedure) itself. Instead, the packet is maintained and manipulated outside. Thus, the manipulated packets are not physically owned by the protocol instance manipulating the data. Rather, ownership is logical, i.e. a buffer is only temporarily owned by a protocol entity. Ownership is passed to the protocol entity consuming the packet after the initial, producing protocol entity has completed its protocol processing on the packet. The application of common buffer management with inbound traffic is depicted in figure 3.8.

Employing common buffer management, copy operations can be eliminated wherever the protocol processing only has a minor influence on the packet. However, copying is still needed where new data are added to the packet, i.e. a larger buffer is required. For example, this is the case with outbound processing, where headers are subsequently added, and where concatenation and reassembly are employed. It may also happen with the application of presentation formatting functions.

The following three techniques deal with the further reduction of data movement. The first two approaches, i.e. the offset and the gather/scatter technique, are based on a common buffer management scheme. The third approach, namely integrated layer processing, proposes an even tighter integration of the protocol layers.

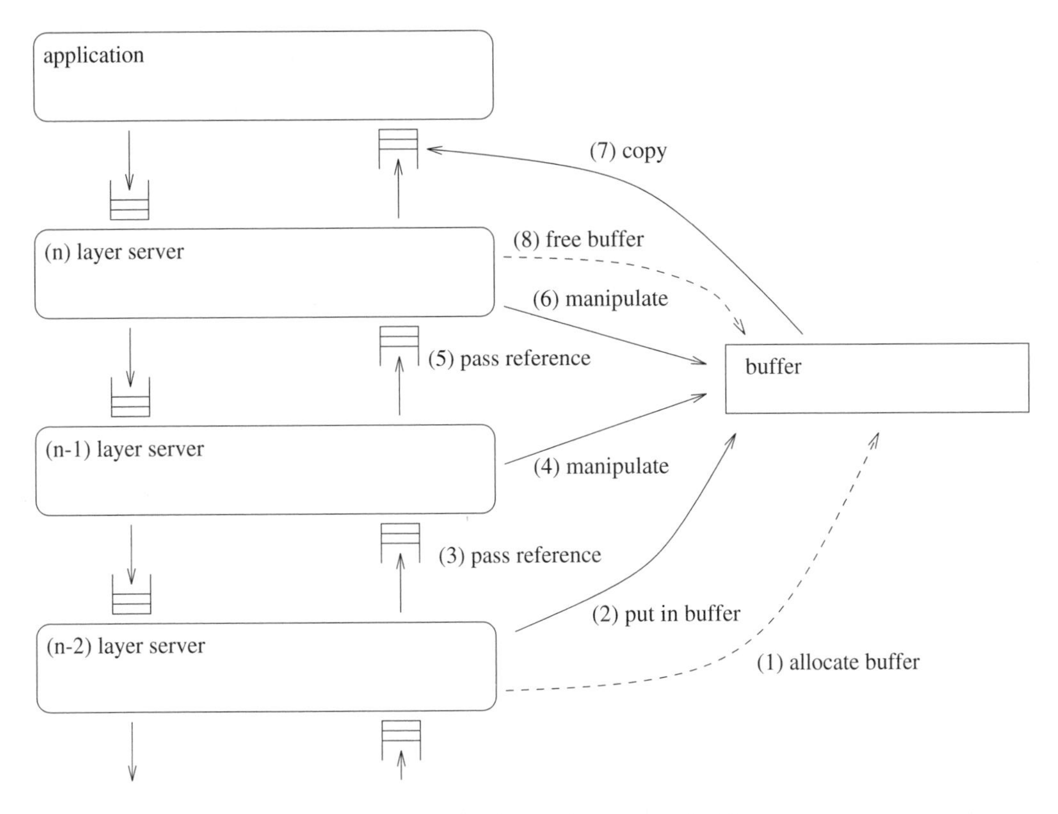

Figure 3.8 Inbound processing with common buffer management

3.3.3.2 Offset Technique

While the above principle works fine for inbound traffic, i.e. the case where headers are removed rather than added, the scheme is not useful at all for outbound processing. In addition, it does not work for other more complicated protocol mechanisms, e.g. such as packet reassembly.

The problem arises when more memory is needed during the protocol processing, e.g. to accommodate protocol control information in a header.

The basic idea of the offset technique is to allocate enough memory with the buffer allocation such that the headers subsequently added during protocol processing can be directly accommodated in contiguous memory. This is graphically displayed in figure 3.9.

The offset technique is useful where the protocol layers employ a one-to-one mapping between input and output, i.e. the case where a single outbound packet is directly mapped onto a frame to be transmitted on the network. The offset technique does not help where segmentation or concatenation is applied. In addition, the offset technique is not useful when the data have to be aligned to certain word or page boundaries, which requires copying of the data such that they

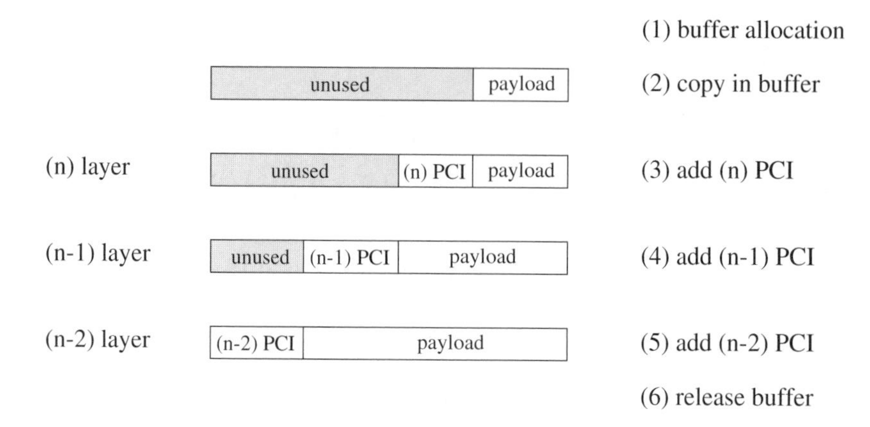

Figure 3.9 The application of the offset technique to outbound processing

fit into a certain addressing scheme.

3.3.3.3 Gather/Scatter

While the offset technique deals with a contiguous buffer, the gather/scatter technique supports managing a non-contiguous buffer consisting of a set of segments.

Gathering is employed on outbound traffic. With gathering, headers added during protocol processing are stored in separate buffers, one header per buffer. Finally, when the packet is transmitted on the network or another physical interface, the headers and data are gathered to form the final packet for transmission. Thus, the packet is physically distributed in memory up to the point where the data pass a physical interface.

The reverse of gathering is scattering. Scattering is employed on inbound traffic. Thus, upon receipt of a packet from the network or another physical interface, the logical parts of the packet are identified and stored in different buffer segments. However, scattering is much more complicated than gathering. This is due to the fact that the logical boundaries of the different parts of the packets are not known in advance. In other words, layer (n) does not know what exactly constitutes the protocol control information (headers) of the layers above it, i.e. layer (n+1) and higher. The identification of these higher layer headers may require some processing in the lower layers first. Depending on the specific protocol mechanisms employed, this may make scattering very complicated, if at all possible.

Gather/scatter may also be employed where segmentation and concatenation is applied. With segmentation, gathering can be used to extract the segments to be transmitted directly from the buffer that holds the initial (unsegmented) packet. Upon transmission of the segment, the remaining headers are gathered as described above. A similar approach can be employed for the reassembly of the packet at the receiving side, and to support concatenation and separation.

A problem with the gather/scatter approach, especially when applied to segmentation or concatenation, is to properly manage the buffer segments. Keeping track of the buffers and properly deallocating the buffer space may be a complicated issue. This is especially true where buffers cannot be deallocated as long as the acknowledgements confirming the correct transmission of all segments in the packet buffer have not been received.

Another problem may be the overhead to manage a large number of buffers, each holding a small chunk of data only.

3.3.3.4 Integrated Layer Processing

So far, we have looked at techniques to minimize the movement of data between different areas in memory. Thus, the focus was on the elimination of copying of the packets or parts of the packets.

The approach taken by Integrated Layer Processing (ILP) [ClTe90, AbPe93] has a much wider scope. With ILP, the goal is to integrate data copying with data manipulations or to integrate several data manipulations into each other. In order to exhibit a maximum increase in performance, the tight integration of several protocol layers is often employed. Thus, the application of ILP may have a major influence on the architectural design of the system.

The basic idea of ILP is as follows. Copying data from one area of memory to another involves the execution of a sequential loop of fetch and store operations subsequently applied on the words that constitute the packet. This is depicted in figure 3.10(a). For simplicity of the figure, the details of the address manipulation are omitted. The idea of ILP is to integrate a data manipulation operation (shown in figure 3.10(b)) with the copy operation. The motivation is that when a data word has been fetched by the processor anyhow (i.e. for copying), then the additional application of a data manipulation on the word just requires an additional processor cycle. Thus, the two operations, i.e. the copy operation and the data manipulation, can be done in a single fetch–store cycle instead of two fetch–store cycles. In other words, two sequential loops can be merged to a single loop. The integration of these two operations is depicted in figure 3.10(c). As a result, four memory accesses can be replaced by two, i.e. a single read from memory and a single write to memory.

A speedup of 50% compared to a sequential implementation has been achieved with the simple integration of the copy operation with a checksum computation [ClTe90]. Typically, several data manipulation operations executed on a packet can be integrated within a single fetch–store loop.

However, problems with the integration may arise [AbPe93, Brau96]. For example, data manipulations do not necessarily preserve the word boundaries or operate on differently sized units of data, which complicates the integration. In addition, data dependences that influence the data manipulation operations may be present that prohibit an interleaving of transfer control and data manipulation operations. For example, the size of headers should be known to effectively apply ILP. Otherwise, a data manipulation operation may have to be applied on a part of the packet first before a second data manipulation can be applied. Another problem is that ILP conflicts with a modular implementation. Thus, the layers in the protocol architecture may vanish in the implementation, which increases effort to design, implement, debug and maintain the implementation.

```
(1) FOR EACH word OF packet
(2)   FETCH (word, source_addr)
(3)   STORE (word, target_addr)
```

(a) Packet copying

```
(1) FOR EACH word OF packet
(2)   FETCH (word, addr)
(3)   MANIPULATE (word)
(4)   STORE (word, addr)
```

(b) Data manipulation

```
(1) FOR EACH word OF packet
(2)   FETCH (word, source_addr)
(3)   MANIPULATE (word)
(4)   STORE (word, target_addr)
```

(c) Integrated layer processing

Figure 3.10 Integrated layer processing applied to a copy and data manipulation operation

Newer research has shown that the speedup achievable with ILP is often much smaller and depends on many different factors. A throughput improvement of 10–20% has been measured for a file transfer application with encryption run on top of TCP [Brau96]. This is due to the complications described above and the fact that the size of caches is continuously increasing. Thus, the whole packet may be kept in the cache, which reduces the positive impact of ILP. The influence of caching on the performance of ILP implementations has been studied in [Brau96, AhBG96]. In [AhBG96], examples have been given where ILP even degrades performance. Note that the relative performance gain achievable by ILP is increased with a decrease of the time complexity of the data manipulations integrated into the loop [Brau96].

3.3.4 Other Sequential Optimizations

Besides the selection of an optimized process model and the minimization of copying, other ideas to improve the performance of protocol architectures exist. In the following, we take a closer look at three important approaches, namely at application level framing which is a protocol structuring principle that has a positive impact on performance, at the common path optimization, and at the use of specialized operating systems.

Additional techniques have been proposed to improve the performance of protocol implementations. These are the use of header templates to speed up header processing [CJRS89], the simplification of layering in order to remove unnecessary functionality which is no longer needed in modern communication systems [Gree91], the use of a function-based communication architecture rather than a layered architecture [Haas91], and the development of new, so-called lightweight protocols that are specially suited for an efficient implementation [DDKM+90]. Lightweight protocols apply a set of principles to support an efficient implementation, including the use of fixed header formats, word-aligned parameters, implicit connection setup, and the minimization of timer usage.

3.3.4.1 Application Level Framing

An important functionality of protocol architectures is to retransmit lost packets and to reorder packets that arrive at the receiver out of order. The traditional approach to handle out-of-order packets is to buffer the packets at the receiving side and defer passing the data to the application until all missing packets have been arrived. This is typically implemented in the transport layer. A result of this is that the application is blocked until the data are contiguously available. In other words, a packet is passed to the application-oriented layers only if all packets with a smaller sequence number are present and have been passed to the application before. In addition to causing delay in the application-oriented layers, this causes overhead for buffering and buffer management. Since the application-oriented layers, especially presentation formatting, are often a bottleneck in protocol processing, blocking and reordering by the transport layer has a detrimental influence on the performance.

In order to tackle the problem, Application Level Framing (ALF) has been proposed in [ClTe90]. ALF is an optimization technique that aims at the structuring of the protocol architecture, i.e. deals with the question of how to distribute functionality or services between layers. Thus, ALF is an architecture design principle, i.e. comparable in scope to the OSI reference model, that eases the derivation of efficient implementations, rather than an implementation technique for a given protocol architecture. ALF is employed with several newer Internet protocols, e.g. with the real-time transport protocol RTP.

The basic principles of ALF are to move functionality into the application layer, and to frame packets in the application layer in a way that out-of-order processing is possible in the application layer at the receiving side. Thus, presentation formatting can be applied to packets even if preceding packets have not arrived yet. In order to support out-of-order processing of packets, the units of information passed between the application and the transport layer have to be meaningful for the application. Thus, packet boundaries have to be oriented along the natural boundaries of the data as present in the application. Otherwise, out-of-order presentation formatting would not be possible due to missing context information.

Since ALF is based on packet reordering in the application layer, any protocol function that may result in reordering of packets, e.g. packet segmentation, has to be avoided in the lower layers. Where applicable, ALF allows the received data to be continuously processed by the complete protocol hierarchy. Thus, ALF supports the application of efficient implementation strategies, e.g. the application of integrated layer framing techniques. A more detailed discussion of the merits of the application of the ALF principle can be found in [AhGM96, Brau97].

3.3.4.2 Common Path Optimization

With most systems, the 80/20 rule applies, i.e. 80% of the load is caused by less than 20% of the code. In general, this also holds for protocol processing. In fact, with most systems only a small number of execution scenarios, or in other words execution paths through the protocol stack, are in common use. The common cases are usually represented by the data transfer modes where packets are sent and received in order, possibly on an established connection. These cases are addressed by the common path approach as proposed in [CJRS89].

The common path approach as suggested exploits this knowledge about the typical usage of the protocol stack. In [CJRS89], a separate implementation of the common path is proposed for the Transport Control Protocol (TCP). Upon arrival of a packet, the implementation just performs a couple of checks to see whether a common case holds. In case the common path is recognized, the specialized optimized implementation is selected for execution.

The research reported in [LeOe96] extends the basic idea to several layers of a protocol architecture. The approach makes some assumptions with respect to the packets, i.e. assumes that a packet to be processed fulfills the common path assumptions. Thus, the implementation employs the processing steps in the common path in a speculative way.

Concerning the common path implementation, this results in the fixing of data-dependent decisions in the protocol processing path. The advantage of the restriction of the degree of freedom in the implementation is that certain optimizations can be applied which would otherwise not be possible or would at least complicate things. An example of this is the application of integrated layer processing.

The drawback of this approach is that complex recovery mechanisms are needed in case the common path assumption does not hold for a packet, e.g. in the case of transmission errors or out-of-order arrivals.

3.3.4.3 Specialized Operating Systems

As noted above, protocol implementations depend on underlying services provided by operating systems. Since typical operating systems are not specially optimized for communications, some services can be provide more efficiently with specialized operating systems. Examples of such performance-critical services are buffer management and support for data transfers across protection domains, as well as process and timer management.

Of special importance is the support for data transfer between different protection domains. On its processing path through a protocol architecture, a packet may have to pass several protection domains. The traditional approach taken with the crossing of a protection domain is to copy the data. As data copying is the major performance bottleneck with the implementation of protocol architectures, the avoidance of these copy operations is vital. In order to support this, various techniques have been proposed. They are all based on some scheme that supports sharing or remapping of virtual memory pages between protection domains. An overview of the different techniques can be found in [DAPP93].

Examples of operating systems specialized for the implementation of protocol architectures are the Scout operating system [MMOP+94], the x-kernel [HuPe91], and the V system [Cher88]. Even though these specialized systems are not applicable to general computer systems, they provide an important alternative for application-specific systems as communication subsystems, and special network devices.

3.3.5 Exploitation of Parallelism

A last resort for the optimization of protocol implementations is to improve hardware or alternatively to employ more hardware, i.e. to do things in parallel. A major focus of studies concerning performance improvements of protocols and protocol architectures in general has been the exploitation of parallelism. Most studies have focused on the exploitation of parallelism on multiprocessor systems.

In addition, the use of specialized hardware devices to implement complete layers in hardware or just special functions has been a research area. Hardware support typically exploits layer parallelism or functional parallelism (see below). A survey of the implementation of various protocol functions in hardware is given in [Hedd95]. An important disadvantage of hardware implementations is their inflexibility regarding changes.

An important question when exploiting parallelism is to identify the exact goal of the optimization, e.g. to improve the throughput for a single connection, the overall throughput over all connections, or the response time of the system.

There are a number of approaches to exploiting parallelism with protocol processing, and we will discuss each of them in turn (see also [ScSu93, Walc94]). We focus our discussion on the exploitation of parallelism rather than a discussion of parallelism within specific protocol architectures. Besides the pure application of one form of parallelism as discussed below, also the combined application of two or more of the principles is possible.

As noted above, possible parallel resources employed to exploit parallelism may be parallel processors or special hardware, especially VLSI circuits. For simplicity of the discussion, we denote the resources as processors since this is the major focus of most studies. However, in principle a direct hardware implementation is often possible as well, and we indicate this where appropriate.

3.3.5.1 Connectional Parallelism

The exploitation of connectional parallelism is based on the use of a separate set of processes for each connection that is supported by the system. Thus, each connection can be handled by a separate dedicated processor. Alternatively, a set of connections may share a common processor. In any case, each packet is dealt with by a single processor only. Thus, no interprocessor communication is needed within the protocol stack.

Since there are typically no dependences between different connections, the implementation of connectional parallelism is straightforward. Another advantage of the absence of dependences between connections is that there is only very little additional overhead. Thus, the additional hardware units provide a maximum of improvement.

Connectional parallelism can improve the overall throughput of the communicating system. Due to the removal of bottlenecks, the approach may also improve the response time at least at high load. However, the approach does not improve the response time in the case of low load, i.e. in the case that only one connection is active.

3.3.5.2 Layer Parallelism

Layer parallelism denotes the case where different layers of the protocol architectures are implemented by different processors. For example, one processor for each layer, or a processor for the transport-oriented layers and one processor for the application-oriented layers, respectively. Thus, a processor implements the functionality of one (or more) layer(s) for all connections. In other words, each packet passing the layered architecture passes through all processors of the system.

Alternatively to the use of a processor, dedicated hardware, i.e. VLSI circuits can be employed. This approach is often applied for the lower protocol layers where bit and byte level data manipulations are prevailing rather than complex computations.

Due to its closeness to layered protocol architectures, the implementation of layer parallelism is rather simple. The model is especially suited for server model implementations. However, the problem is the high degree of interprocessor communication. Thus, special care is needed to keep the communication overhead down. For this, multiprocessor systems that efficiently support shared memory are essential to achieve a speedup.

Concerning the achievable performance improvement, there is one important thing to note: the throughput of the system is limited by the bottleneck processor, i.e. the processor that has the highest load. Thus, an even distribution of the layers, or more accurately of the load they cause, is essential. Even though the exploitation of layer parallelism may decrease the response time at high load, this definitely does not hold for low load. This is because of the fact that the critical path through the protocol processing is not decreased by the exploitation of layer parallelism.

3.3.5.3 Packet Parallelism

Packet parallelism denotes the approach where the packets to be processed by the protocol architecture are assigned to different processors. For example, two subsequent packets are processed by two different processors. The selected processor is solely responsible for the handling of the respective packet from its reception to its departure. Thus, each processor provides the full functionality of the protocol stack.

Since the processing of subsequent packets belonging to the same connection is not quite independent, interactions between the processors are often needed. This especially holds for the connection data that have to be maintained. Additional problems emerge where packet segmentation or concatenation is applied. Thus, the availability of efficient access to shared data between processors is essential for this approach to work efficiently. In order to ensure the consistency of the connection data, synchronization and locking are important.

As the packets of a single connection may be processed in parallel, the approach may provide a performance improvement even for the case that only one connection is active. Performance improvements may be achieved with respect to the response time as well as for the throughput. However, note that the speedup is reduced by the extra synchronization overhead needed to keep the shared data consistent.

3.3.5.4 Directional Parallelism

Directional parallelism denotes implementations where the directions of the packet flow are separated. Thus, a separate processor is dedicated for each of the two directions, i.e. for inbound and outbound traffic. This requires a separation of the protocol functionality into a part that implements inbound traffic and a part that implements outbound traffic.

The implementation of this approach is also not quite easy. Especially with connection-oriented protocols, synchronization between the inbound and the outbound path is needed. Thus, some prerequisites for the design of the protocol mechanisms are important to efficiently support this kind of parallelism. An example of a protocol that has been designed with directional parallelism in mind is the Xpress Transfer Protocol (XTP) [Prot92].

The possible speedup of the approach is limited by interdependences between the inbound and outbound path.

3.3.5.5 Functional Parallelism

Functional parallelism denotes the finest degree of parallelism. This approach identifies the operations in the protocols that can be computed in parallel. In other words, the goal is to process a single packet in parallel, e.g. to perform a protocol control operation in parallel with a data manipulation, or even to perform a single data manipulation on the words of the packets in parallel.

Due to the fine granularity of the parallelism exploited here, efficient implementation is rather complicated and relies on very efficient interprocess communication. A natural approach to exploiting functional parallelism is by specialized hardware, e.g. a VLSI circuit that performs a cyclic redundancy check. In order to support efficient implementation, the processor and the specialized hardware device typically operate on a common memory.

Besides throughput optimization, the approach may decrease the response time for packet processing.

3.4 Summary

In this chapter, we have discussed the basics of protocol processing, covering the basic services provided by communication protocols, structuring principles for protocol architectures, and important principles and techniques for the design and implementation of efficient communication systems.

This chapter builds the ground for the engineering of communicating systems with SDL. As we will see in chapter 5, the described techniques are important when describing communicating systems with SDL as well as for the derivation of efficient implementations from given SDL descriptions.

Chapter 4

System Development with SDL

As already discussed in section 1.1, the use of formal description techniques is motivated by the need to detect functional errors early in the development cycle. This reduces the cost and time for redesign and reimplementation activities.

This also applies to the formal description technique SDL and its companions, most notably MSC and TTCN, which form a coherent set of techniques that support different phases of the development process.

The chapter is structured as follows. Section 4.1 provides an introduction to SDL covering structural and behavioral concepts of the language. Section 4.2 covers the most important companion of SDL, namely MSC. In addition, it briefly discusses TTCN and the use of object-oriented analysis and design techniques in the context. An overview of methodological concepts for the use of SDL and its companions is provided in section 4.3.

4.1 Specification and Description Language (SDL)

4.1.1 Introduction

In the development cycle, SDL is employed for the formal specification and design of the system. SDL supports the specification and description of structural and behavioral aspects of the application under development. SDL comes in two syntactic forms, the textual representation SDL/PR (SDL Phrase Representation) and the graphical representation SDL/GR.

Purpose of SDL SDL is a specification and description language. Thus, it supports a specification of systems at an abstract level as well as the description of a system design. SDL is a formal description technique (FDT). Thus, its dynamic semantic is formally defined. In the case of SDL, a combination of Meta-IV [BjJo82] and CSP [Hoar85] is employed for this.

SDL may serve a number of purposes from reasoning about systems at an abstract level to the automatic derivation of implementations. We will come back to this in section 4.3.

SDL abstracts from the final implementation to a considerable extent. SDL implicitly assumes unlimited resources, e.g. unbounded queues and infinite processing capacity. Related to this, SDL currently has a rather vague time semantics.

Application domain There are a number of potential application areas for SDL. SDL has been designed to support reactive, concurrent, real-time, distributed and heterogeneous systems. Today, the major application area of SDL is in telecommunications. SDL is used by standardization bodies such as ITU and ETSI to formally specify communication protocols and distributed applications. Currently, the usage of SDL in standardization is in transition from just providing an additional non-normative supplement to the establishment of SDL as language for the normative specification of standards [HoEl98].

Especially in the telecommunication industry, there is a large community of SDL users. Most of the major telecommunication companies, especially in Europe, employ SDL and respective tool support during system development.

Evolution of SDL SDL has evolved over more than two decades now (see [Reed97]). SDL has been standardized by the ITU-T [ITU93]. In 1992, a major revision of SDL took place to include support for object orientation. SDL 96 differs from the 1992 version only in minor points. The changes are documented in [ITU97b]. In this book, we refer to SDL 96 unless otherwise noted. In the new version of SDL, SDL 2000 [ITU99a], several modifications and adaptations have been made in order to suit the language for newer developments.

Tool support for SDL Today several commercial tools are available that support the development of systems with SDL. Tool support comprises graphical editing, validation, verification, simulation, animation, code generation and testing. The best known tools are ObjectGEODE [Veri98] and TAU [Tele98]. In addition, many tools from academia have been developed to support various aspects of the development cycle. In section 6.3.2 we will take a closer look at tools that support performance-related activities in the context of SDL.

Focus of this section In our introduction to SDL, we focus on the aspects that are important for the design and implementation of time-critical and efficient distributed systems. Thus, we are more interested in the semantics of SDL than in the exact syntax of the SDL constructs. This also implies that behavioral concepts are more important here than structural concepts, which may or may not be reflected by the structure of the final implementation derived from the SDL description.

Further reading A more detailed general description of SDL can be found in [BrHa93, OFMR+94, ElHS97]. [OFMR+94] is an excellent reference book for SDL providing an in-depth coverage of all aspects of SDL 96 including object orientation and data. Conversely, [BrHa93] is more oriented towards the engineering of SDL systems with an emphasis on the implementation of SDL descriptions. Both [BrHa93] and [ElHS97] are suited for novice users of SDL.

4.1.2 Structural Concepts

SDL allows for the structured, hierarchical specification of large systems. It provides a set of structuring concepts including support for the object-oriented specification of systems. Structuring concepts are an important means of mentally dealing with complexity and supporting reuse.

However, the structuring concepts of SDL do not have a major influence on the implementation to be derived from an SDL description. Thus, we will only sketch the important structuring concepts, rather than giving a detailed description.

We start with a description of the most important structuring concepts of SDL from top to bottom. These are the system, blocks, processes, services and procedures. Since processes, services and procedures are also main behavioral concepts in SDL, their detailed discussion is deferred to section 4.1.3. In addition to these basic building blocks, channels and signal routes are important concepts for communication.

In the following, we start with these concepts, neglecting support for object orientation. Thus, we first focus on the specification of instances (or objects in object-oriented terminology) and defer the discussion of object orientation to the end of the section.

4.1.2.1 System

The highest level in the hierarchy represents the *system*. The system is the entry point into the SDL specification. A system diagram comprises, among other entities, a set of *blocks* and *channels*. Blocks are connected with each other and with the environment by channels.

A system is a static entity in SDL. Besides the specification of a system instance, SDL also supports the specification of system types.

An example of a system definition of a remote sensoring system is depicted in figure 4.1. The system contains two blocks, the block *sensor_device* that implements the sensoring device and the block *client* that uses the sensor. The two blocks are connected by the channel *CH_device*, and connected to the environment of the SDL system by the channel *CH_external*.

The text area in the upper right corner of the figure specifies the types of signals transmitted in the system. The signal types transmitted on the channels are specified within brackets right next to the arrows which indicate the direction in which the respective signals are sent.

Note that the hierarchy in SDL is mainly a static structuring concept. It has only very little influence on the dynamic behavior of the system. Thus, the hierarchy is of little importance for the implementation and execution of an SDL specification. In this respect, SDL is very different from other specification languages such as, for example, Estelle [ISO89], where the hierarchical structure of the system has a major influence on the system behavior.

4.1.2.2 Channels

Communication between blocks itself and between blocks and the environment is only possible along the defined channels. Channels can be unidirectional or bidirectional communication devices.

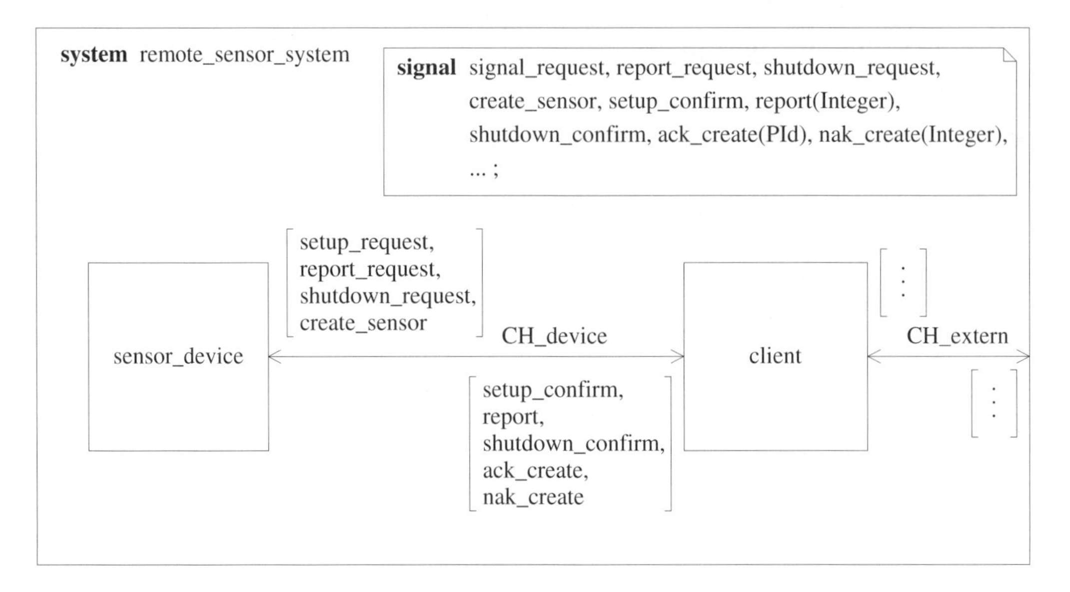

Figure 4.1 Top level description (system diagram) of a remote sensoring system with two blocks

The communication structure between blocks is static. Channels may be specified as delaying or non-delaying communication devices. Communication on channels is free of errors and preserves the order of the transmitted signals. In order to refine the properties of channels, i.e. to model errors or reordering of signals, channels may be refined by channel substructures.

4.1.2.3 Blocks

A block can be refined either by a set of *processes* or by a set of block substructures. However, there may not be a mixture of blocks and processes within a single block. Blocks are static entities in SDL. Thus, they are created during the initialization of the system. Besides the direct specification of blocks, SDL also supports block types.

The block *sensor_device* referred to in figure 4.1 is detailed in the block diagram given in figure 4.2. The octagons in the figure specify the processes contained in the block. The block contains two processes, the process *sensor* that implements the sensoring device and the process *sensor_manager* that creates and controls the operation of the sensor.

The two processes are connected by the signal route *SR_int*, and connected to the block interface by the signal routes *SR_manager* and *SR_sensor*, respectively.

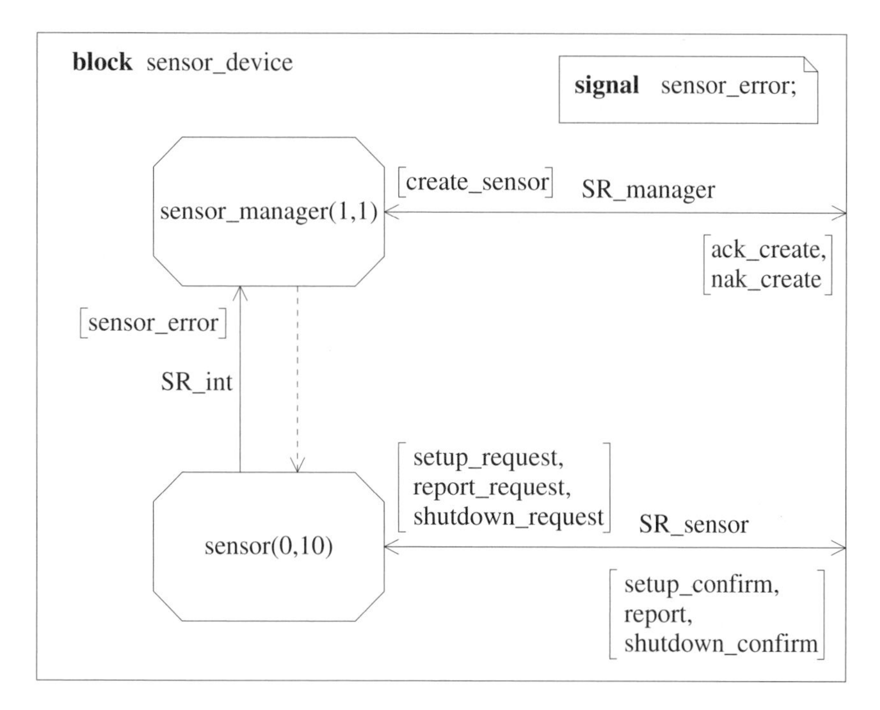

Figure 4.2 Refinement of the block *sensor_device* (block diagram) of the remote sensoring system

4.1.2.4 Signal Routes

Communication between different processes and between processes and the block interface is done via signal routes. Signal routes are non-delaying and may or may not be explicitly given in the block diagram. However, either all or none of the signal routes in a block diagram have to be defined.

The most important difference between channels and signal routes is their usage. Signal routes are used within blocks to connect processes, while channels connect blocks. In case a signal is sent to a process within a different block, the signal travels along the signal route in the same block, the channel(s) connecting the blocks, and finally the signal route defined in the block where the receiving process is located. Another difference is that signal routes are always non-delaying, while channels may delay signals.

4.1.2.5 Processes

Blocks are finally refined by processes. Processes are the most important entities of an SDL specification. They are the major component to specify the behavior of the system. Similar to blocks, SDL also supports process types. We will discuss the process behavior in detail in section 4.1.3.

In SDL, several process instances may exist that represent an SDL process. Process instances can be created either at initialization time or dynamically during runtime by other process instances belonging to the same block (see below).

Two parameters concerning the number of process instances of a process set are supported by SDL. With the process declaration, the initial number of process instances of a process upon system initialization, and the maximal number of process instances active during system execution, can be specified. Note that it is not necessary to bound the maximum number of process instances.

The two parameters are given in parentheses following the process names. For example in figure 4.2, *sensor(0,10)* defines that no process instance of the process is generated during system initialization, and that at most 10 process instances of this type may exist at runtime.

4.1.2.6 Object Orientation

So far, we have focused on single entities or instances in the SDL specification. In addition to the direct specification of the system, blocks and processes, etc., SDL supports type definitions of structural entities and to some extent of data objects. Type definitions allow one to generate several entities supporting instantiation, generalization and specialization. Thus, new subtypes may be derived by specialization of other types. The type concept of SDL is very expressive. It allows a very flexible refinement of types or subtypes.

Types in SDL support the object-oriented definition and the reuse of structural entities. However, note that the concept does not support the object-oriented description of data structures very well.

Types are supported for systems, blocks, processes, procedures, services, signals and data entities. We will not go into the details of the use of types as it does not have a major impact on the implementation derived from the SDL description.

SDL also supports a concept to group a set of type definitions together and to use them in different system specifications. This is supported by *packages* which allow the separation of type definitions from the system specification.

4.1.3 Behavioral Concepts

4.1.3.1 SDL Process as Finite State Machine

The specification and description of behavioral aspects in SDL is based on the concept of Finite State Machines (FSM).

Finite state machine A finite state machine – or more specifically an input–output automaton – is defined by

(1) a set of states S,

(2) an initial state s_0,

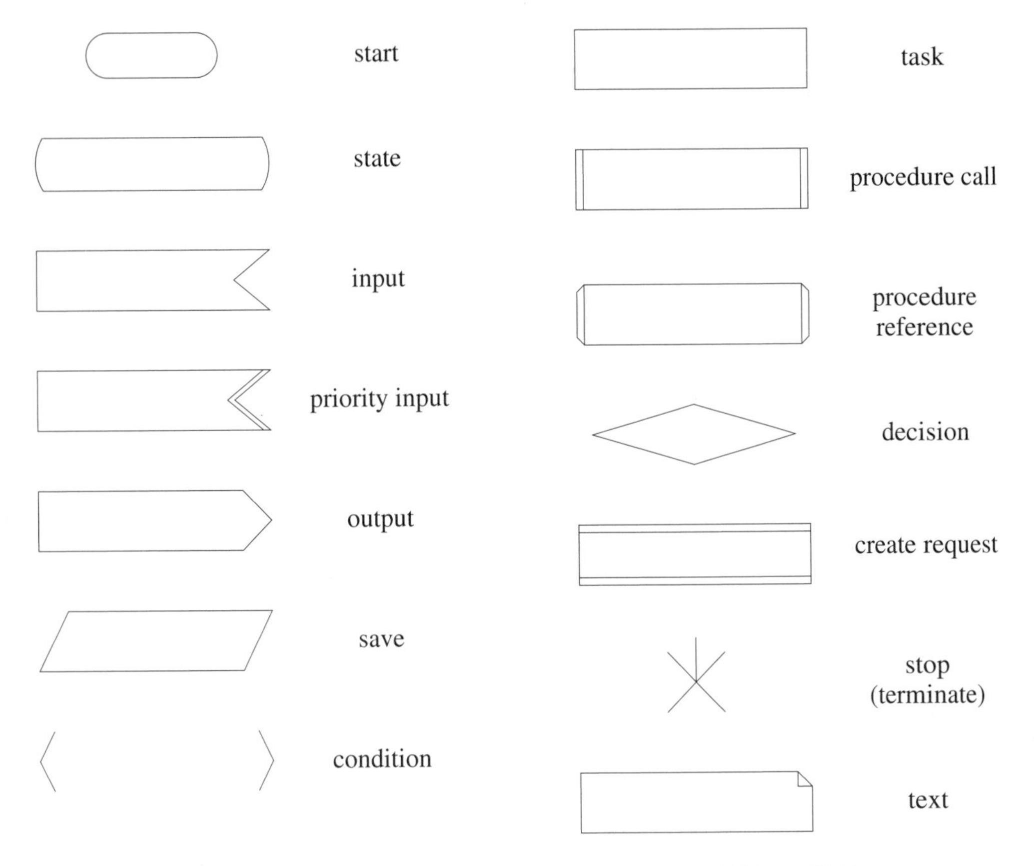

Figure 4.3 The most important graphical symbols used within an SDL process

(3) an alphabet A, and

(4) a function $f(S, A) \rightarrow (S, A)$ that describes the possible transitions, i.e. the mapping of a state and an input to the successor state and to the respective output.

The finite state machines are the active entities in an SDL system and are called *processes* in SDL. Processes are the main concept used in SDL to define the dynamic behavior of the system. The most important graphical symbols used within an SDL process are summarized in figure 4.3.

We will describe these symbols as we move along. In addition to these basic symbols, many other constructs are supported by SDL. However, most of these other language constructs deal with the description of algorithmic features, of control structures and data manipulations, or simply add syntactic sugar. Thus, they are not of particular importance in our context.

An example of the specification of a simple SDL process is outlined in figure 4.4. The figure depicts the *sensor* process of the remote sensor system introduced in the previous section. The

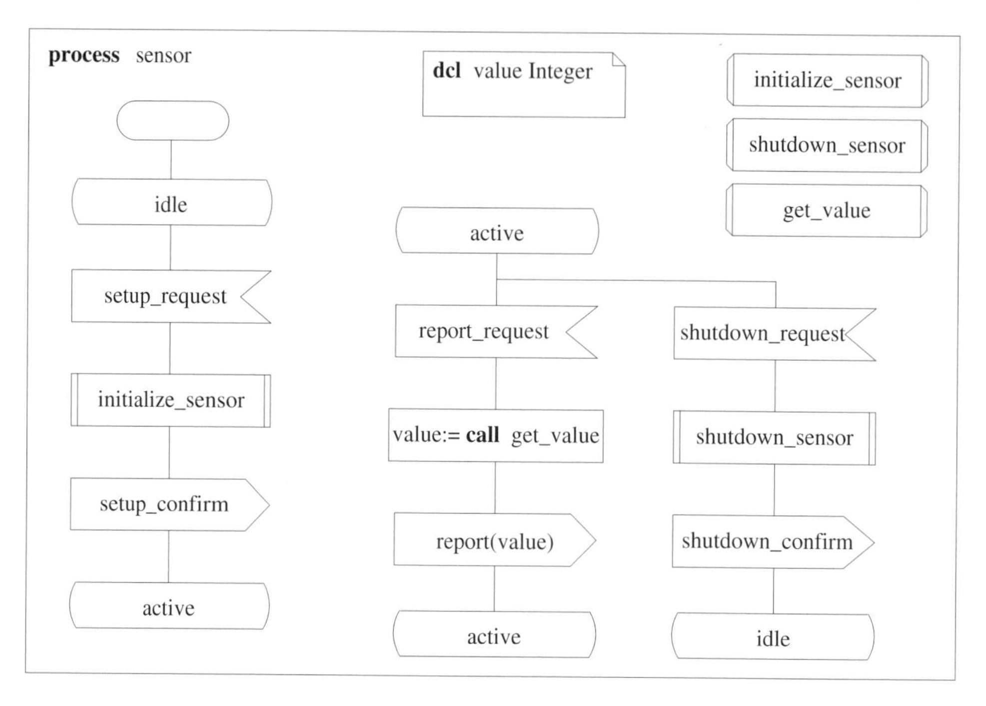

Figure 4.4 The behavioral specification of a simple remote sensor process with SDL

process is used to remotely check the state of some device.

The SDL process can be in one of the two states, *idle* and *active*. Its set of input signals comprises the three signals *setup_request*, *shutdown_request*, and *report_request*. As response to the input, the sensor outputs three different signals, namely *setup_confirm*, *shutdown_confirm* and *report*. The output signal *report* holds a parameter called *value* that indicates the current state of the sensored device.

The respective FSM diagram for the SDL process is given in figure 4.5. It shows the two states *idle* and *active*, and the transitions with their respective input (above the line) and output signals (below the line).

Extended FSM SDL extends the basic FSM concept by associating data with the state machine. This is typically called Extended FSM (EFSM). This denotes the fact that the state of an SDL process is defined not only by the explicit states as described above, but additionally by the state of its data objects, i.e. the values of its variables. An example of local data is the *value* variable used in figure 4.4.

In SDL, no data objects exist above the level of a process. Thus, data cannot be declared directly in an SDL block or system. All data objects are local to some entity at the level of a process or below. However, data types (or classes in object-oriented methodology) can be described at the

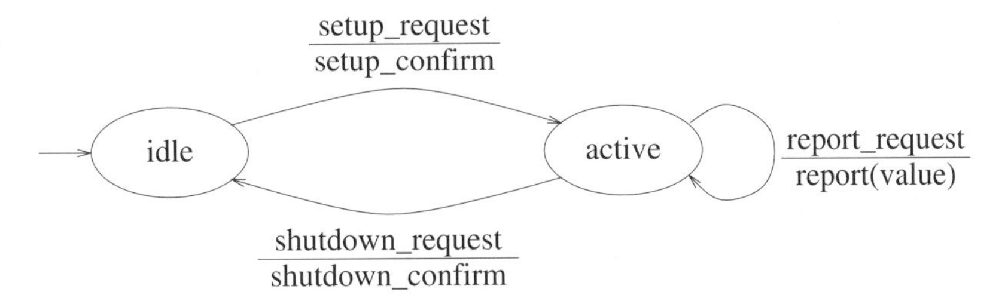

Figure 4.5 Representation of the SDL process as a finite state machine

levels above. Data are declared in an SDL process, service or procedure. With the exception of the reveal/view concept (see below for details), the direct sharing of data objects between SDL processes is not supported. However, mechanisms to remotely access the data of another process in a controlled manner exist, e.g. by remote procedure call or by using the export/import concept (see below for details).

Note that the local data associated with an SDL process may influence the behavior of the state machine. In fact, SDL supports various trigger conditions for the execution of a transition (see below for details).

Each SDL process maintains a set of intrinsic variables which are of the predefined data type PId (Process Identity). The following variables of this type are defined for each process instance:

- *self* denotes the PId of the process instance itself,
- *sender* denotes the PId of the process instance from which the most recent signal has been consumed,
- *parent* denotes the PId of the process instance that has created the instance,
- *offspring* denotes the PId of the most recent process instance created by this process instance.

4.1.3.2 Communicating Extended Finite State Machines

An SDL specification typically comprises more than one process. In general, it comprises a number of cooperating and communicating processes. Concerning communication, SDL is based on the concept of Communicating EFSMs (CEFSM). SDL processes typically cooperate and communicate with each other and with the environment by exchanging signals.

Signals Signals are the primary communication mechanism in SDL. With one exception, all advanced communication mechanisms provided by SDL can be modeled by the exchange of signals.

All signals are identified by a signal type identifier. In addition, signals may carry data by means of parameters. Due to the powerful data concept of SDL, the data transmitted as parameters may be of variable size. A simple example of the use of a parameter with a signal is the *report* signal in figure 4.4.

All signal types used in the SDL systems have to be defined. In addition, the channels and signal routes have to be attributed with the signal types allowed to travel on them.

Input port Each process instance in an SDL system owns a dedicated input port. The port allows received signals (and timeouts) to be queued until they are consumed (or discarded) by the owning process instance. The queues have an infinite capacity. Thus, no blocking of the sender due to a full queue can occur.

The queue is mainly organized according to the FIFO principle (first-in first-out). However, since there exist two priorities for inputs and the possibility of saving signals for later consumption (see below), the FIFO order in processing the queue is not always preserved.

Concurrency and time Each process instance in an SDL system represents an independent asynchronously executing CEFSM. Conceptually, there is no interdependence between the executing process instances other than explicitly defined by the various communication mechanisms.

The definition of time in SDL is very vague. A process instance may be executed as soon as one of its trigger conditions holds. However, the execution may be deferred as well. Similarly, the execution of a transition may or may not consume time. However, the transitions of a single process instance are executed sequentially. Thus, the next transition of the same process may not be triggered before the previous transition has been completed.

Transitions are not strictly atomic. Thus, there may be an arbitrary interleaving of different actions concurrently executed by different transitions in different process instances. However, since the externally visible behavior is limited to a small subset of the constructs supported by SDL, i.e. to the constructs that support communication, process and timer management, the number of relevant, externally visible interleavings is much smaller.

4.1.3.3 Transition Triggers

For each state of the CEFSM, a set of trigger conditions can be specified. Besides the presence of a signal in the input queue, other triggering conditions may be present. Examples are the expiration of a timeout (which is also reflected by a signal in the input queue) or a certain state of the local data of the process instance. If a trigger condition holds, a set of actions is performed, typically including the (asynchronous) sending of signals to other SDL processes. As a result of these actions, a subsequent (possibly the same) state is entered.

Time and timers SDL supports two concepts to deal with time, i.e. to directly access the time and the use of timers.

To support time, two predefined data types are used in SDL, namely the data types *Time* and *Duration*.

The **now** construct allows one to access the current time, e.g. 10.45 pm, and to store the time in a variable of type *Time*.

In addition, SDL supports timers. Timers operate on the predefined data types *Time* and *Duration*. Unlike the data type *Time*, the data type *Duration* specifies a time span rather than an absolute time.

Timers can be **set**, **reset** and inspected (**active** construct). The **active** construct allows one to check whether a timer has expired or not. When a timer expires, a special timeout signal is generated. The timeout signal is considered as input to the process instance that owns the specific timer, and is handled in a very similar way to the input of a regular signal. Thus, a timeout can be consumed by the process instance at its convenience. Note that a timer is a local mechanism. Thus, each timer is specific to a single process instance. This relates to the setting, resetting, inspection and expiration of the timer.

When a timer is reset, three cases can be distinguished. When the timer has not expired yet, it is simply deleted. The same holds when the timer has expired but the timeout has not been consumed yet. In case the timer has expired and consumed, i.e. is not active, the **reset** construct has no effect. In case a timer is **set** again before it has expired or consumed, it is set to the new value.

Besides simple timers, SDL also supports the use of parameters with timers. The usage of parameters with timers is similar to the use of parameters with signals. Thus, the parameters of an expired or consumed timer can be used within the SDL process. With timer parameters, arrays of timers can be supported.

An example of the use of timers is given in figure 4.6. It shows the extension of the sensor process. In the example, a timer is used to deactivate the sensor (return to the *idle* state) in case there is no request for a report for more than 10 units of time during normal operation. The timer is reset when the sensor returns to the *idle* state. As we will see later, the **reset** operation is not strictly needed since a signal or timeout which is not expected in a state is simply discarded. Thus, a timeout issued while the process is in the *idle* state is discarded and has no effect.

Prioritizing triggers In systems, often different events are of different importance. In SDL, this is supported by the priority input construct. The signal or timer specified in a priority input symbol has priority over all other signals or timeouts given in a regular input symbol. When a priority input is present in a state, all other inputs are implicitly saved.

A priority input for the sensor process is given in figure 4.7. In the example, a priority input is employed for the *shutdown_request* signal. Thus, in the *active* state, the processing of the *shutdown_request* signal has priority over the processing of the *report_request* signal. Figure 4.7 shows the fragments of the sensor process where the changes are made. All other parts are unchanged from figure 4.4.

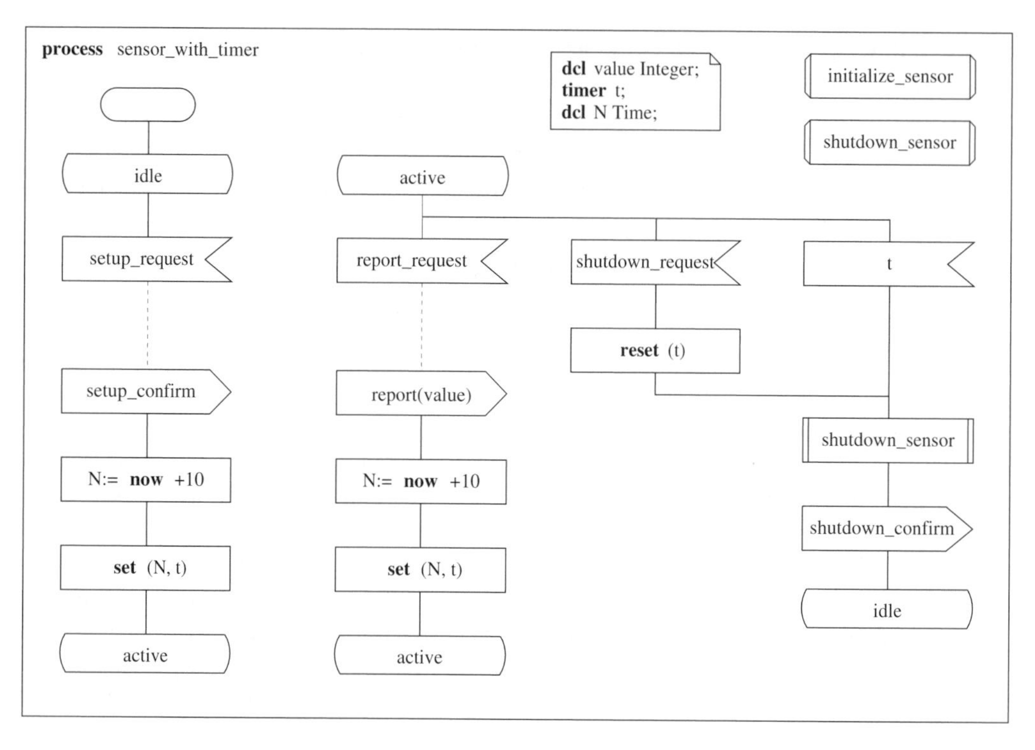

Figure 4.6 An example (incomplete) of the use of a timer in the sensor process

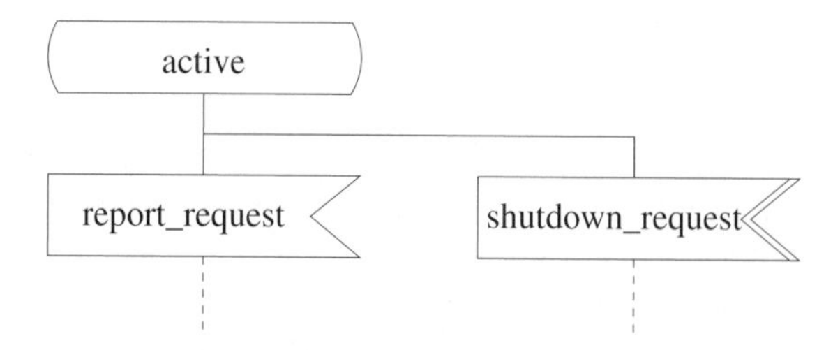

Figure 4.7 An example of a priority input in the sensor process

In case several events (signal or timeout) which are specified in a priority input are present, the order of the high priority inputs in the queue decides on the selection of the first event. Note that the higher priority holds only for the specific state in which the priority input is specified. In any other state, the signal may be processed with regular priority or even discarded if not explicitly saved (see below).

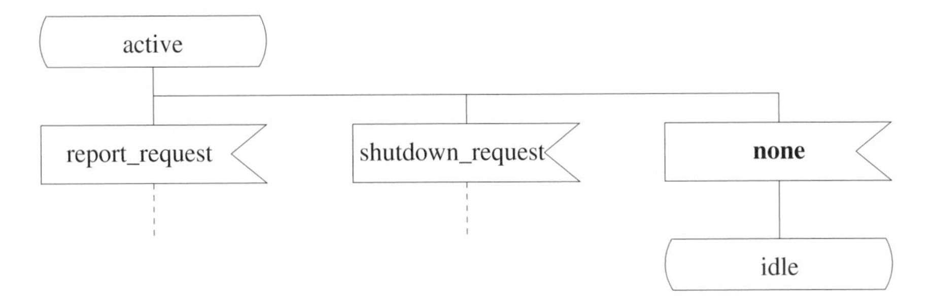

Figure 4.8 An example of a spontaneous transition (**none**) in the sensor process

Spontaneous transitions – *none* construct SDL supports the modeling of nondeterminism, i.e. to describe some behavior of a system without precisely defining what exactly is causing this behavior. The most important application for this is to model the failure of system parts. This is supported in SDL by means of the **none** construct. The keyword **none** within an input symbol specifies that the respective transition may be triggered nondeterministically, i.e. at any time when the process is in the specified state.

An example of the usage of the **none** construct with the sensor process is given in figure 4.8. The figure shows the added fragment with the spontaneous transition. In the example, the spontaneous transition triggered by the **none** construct is intended to model some failure of the sensor that causes a reset.

Postponing triggers – *save* construct In case a signal or timeout is encountered by processing the queue, which is not specified in the current state of the process, the signal (or timeout) is simply discarded. For example, let's again consider the remote sensor example given in figure 4.4. If a *report_request* signal is encountered in the *idle* state of the process, it is discarded. The same happens to the *setup_request* signal in the *active* state.

In order to prevent this, i.e. to save the signal instead of discarding it, the **save** construct can be employed. In this case, the respective signal is saved and reconsidered in the next state the process is entering.

An example of this for the remote sensor is given in figure 4.9. In the example, the **save** construct is used to save a *report_request* signal in case the sensor process is in the *idle* state.

Conditions – continuous signal So far, we have seen three classes of triggers for a transition, namely the presence of input signals, timeouts and spontaneous transitions. Here, we discuss a fourth trigger condition which differs considerably from the above classes. All of the above triggers depend on some kind of explicit event. Conversely, the last class of triggers depends on the state of the system, typically on the (implicit) state of the process.

This last class of trigger conditions can be considered as preconditions on the state of the data objects accessible by the process. In SDL, this mechanism is called a continuous signal. Typ-

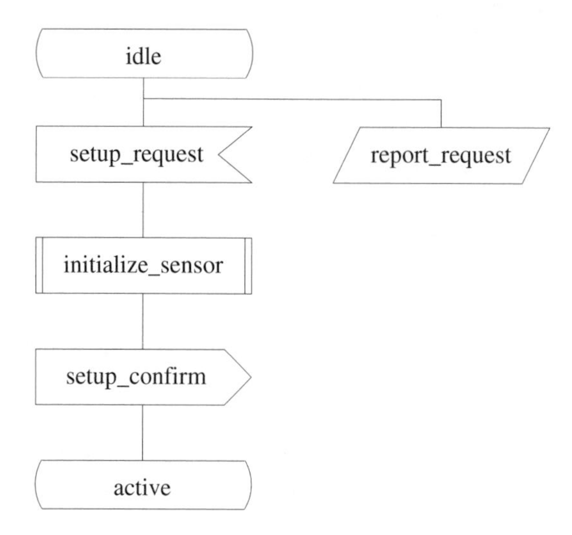

Figure 4.9 An example of the usage of the **save** construct in the sensor process

ically, the continuous signal refers to data local to the process. However, conditions may also depend on remote data imported or viewed by the process. A popular example is the use of the **view** construct in the condition (see below for details), i.e. to base the trigger condition on the state of some data object not local to the executing process instance.

The application of a continuous signal in the sensor process is graphically displayed in figure 4.10. In the figure, the boolean variable *error_status* triggers a transition in the *active* state if no other event (signal or timeout) is present. In this case, an *error_report* signal is sent.

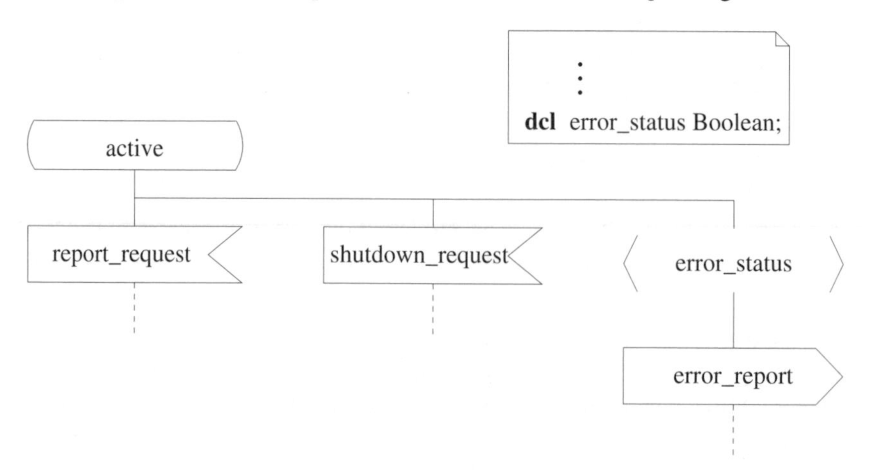

Figure 4.10 An example of the usage of a continuous signal in the sensor process

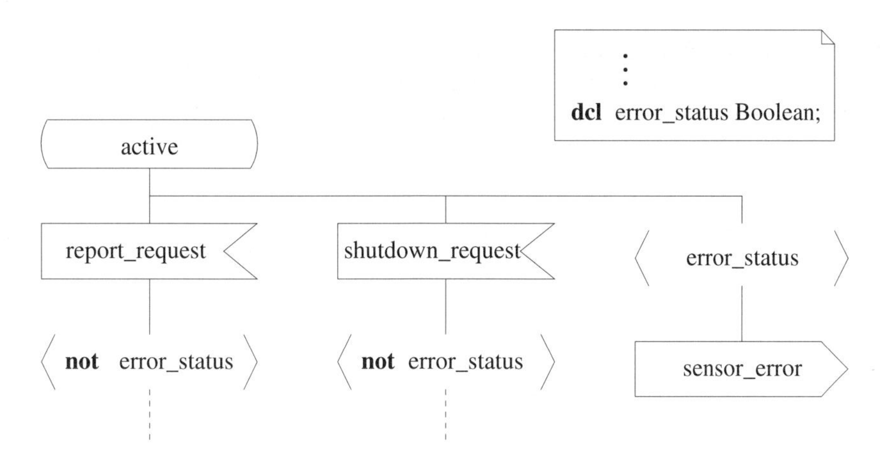

Figure 4.11 An example of the use of an enabling condition in the sensor process

A major difference of triggers described by continuous signals is their priority. Continuous signals have lower priority over all the trigger conditions described above. Thus, a continuous signal may only cause the trigger of a transition if none of the above trigger conditions hold, i.e. no input signal and no timeout is present that may trigger a transition in the current state. Thus, the use of the continuous signal, as specified in figure 4.10, to deal with errors may not be appropriate. In the example, errors are reported only after all input signals, i.e. all *report_requests*, have been processed.

In case several continuous signals are specified and evaluate to true, one of the transitions is selected nondeterministically. Alternatively, a priority may be assigned to each continuous signal.

Combining conditions with input – enabling condition In addition to the separate use of input and conditions to trigger a transition, conditions and inputs can be combined. Thus, the consumption of an input event, i.e. a signal or timeout, can be restricted by an additional condition. This is called an enabling condition in SDL.

An example of the use of an enabling condition in the sensor process is given in figure 4.11. In the figure, conditions are imposed on the consumption of the *report_request* and the *shutdown_request* signal. This ensures that error handling is commenced immediately after an error has occurred. In the example, the use of enabling conditions ensures that the sending of the *sensor_error* signal is not delayed by other input signals, i.e. the *report_request* or *shutdown_request* signal.

Note that the use of an enabling condition implies the implicit use of a **save** on the signal. Thus, if the condition evaluates to false, the respective input signal is saved instead of being discarded.

Summary of transition triggers Of vital importance for the efficient implementation of process instances is the exact understanding of the trigger conditions of the SDL processes. Above we have identified four possible classes of triggers for a transition:

- input signals,
- timeouts,
- spontaneous transitions, and
- conditions.

The first two classes of trigger conditions are organized by means of the input queue of the process instance. Thus, signals and timeouts are queued and typically executed in the order in which they are received (first-come first-served). In case a signal or timeout is encountered which is not specified in the current state of the process, the signal (or timeout) is discarded unless explicitly saved. Input events may be preferred over others with the priority input construct. In addition, trigger conditions may arise nondeterministically (**none** construct) or may depend on the state of the data objects accessible to the process.

The efficient implementation of conditions requires special care. For the implementation of SDL processes, the discrimination between data that are maintained and changed only during the execution of a transition in the same process and data that are changed asynchronously, i.e. even if the process does not execute a transition, is important. Thus, the data objects accessed in a condition may be divided into two classes, namely data that can only be changed by the process instance that executes the condition and data that may be changed outside, independently of the process instance. Examples of variables belonging to the second class are viewed variables (see below) and variables or constructs dealing with time.

4.1.3.4 Structuring the Behavioral Description

Procedures The concept of procedures in SDL is far beyond the procedure concept in typical programming languages. With procedures in SDL, the behavioral description of an SDL process, i.e. its states and transitions, may be partitioned. Thus, a procedure defines behavior with the same elements as a process, i.e. employing states and transitions.

A procedure operates on the input queue of the calling process instance and may consume the same set of input signals as the calling process. However, a procedure uses a set of states separate from the states of the calling process. A procedure may be called by a process, a service (see below) or a procedure. When a procedure is called, the calling entity is suspended until the procedure returns.

Services A process may be described by a set of services. Services are intended to structure the behavioral description of a process in a set of rather independent partitions. For example, different services may be used to describe the processing employed in different main states of a communicating system. Often, services can reduce the complexity of a process specification.

Each service description represents a separate independent finite state machine. All services of a process share the same input queue and the data objects of the process. Additional local data may be declared by each service.

Each of the services defines a set of states and transitions. The sets of states employed by the different services within a process are disjunct. In addition, different services in a process consume disjunct sets of input signals.

Unlike procedures, services support conceptional concurrency. However, the transitions specified by the different services belonging to a process are all executed in mutual exclusion. Thus, the process instance still consumes one signal at a time, and executes one transition after the other.

4.1.3.5 Additional Communication Concepts

Besides signal communication, other communication mechanisms exist. These are mechanisms for remote procedure invocations and two concepts to support the sharing of data between processes.

Remote procedure call Unlike signal communication, remote procedure calls provide a synchronous communication mechanism. Thus, the calling process instance is blocked until the remote procedure has been completed by the executing (remote) process. Remote procedures are the prevailing communication mechanism in client/server systems.

For a procedure to be available remotely, it has to be explicitly exported by the process that executes (or implements) it. The default execution mode is that the remote procedure call is handled by the called process independently of the state the process is in. Alternatively, an arriving call of the remote procedure may be postponed by applying the **save** construct to the procedure. This specifies that the remote procedure call is not processed in the current state of the called process. In addition to this, the call of the remote procedure may trigger the execution of a transition in the process that provides the remote procedure. However, the transition is not started before the remote procedure call has been completed (and has returned control to the caller). Thus, the execution of the transition does not have an immediate influence on the caller.

Remote procedure calls can be implemented on top of the basic signal communication mechanism provided by SDL. Thus, in implementations remote procedure calls are often transformed to basic signal communications used in conjunction with newly introduced states.

Export/import The export/import concept supports the controlled usage of (virtually) shared variables. It is especially suited where the sole purpose of the communication is to read the value of a variable of another process.

Using the export/import concept, variables of a process can be made visible outside the process in which they are declared, i.e. exported. The exported variables are made visible to another process by explicitly importing them.

However, unlike truly shared variables, the export/import concept is based on a value semantics. Thus, the variable explicitly defined as imported (and exported by another process) is actually a copy of the (previously) exported variable. The exported copy of a variable is explicitly updated by issuing the **export** construct in a transition. In order to make the value of the exported variable visible to the importing process, the process has to issue the **import** construct. This ensures that consistency problems like those present with truly shared variables in a concurrent system are avoided.

The **import** construct is especially useful when employed in continuous signals or enabling conditions. This allows a transition to be triggered based on the state of remote data objects.

Similar to remote procedure calls, **export** and **import** constructs can be implemented based on the basic signal communication mechanism.

Reveal/view Unlike the export/import concept, the reveal/view concept supports the physical sharing of common variables. The **view** construct allows a process in a block to read the value of a variable defined in another process of the same block. In case more than one instance of the revealing process exists, the specific PId of the revealing process instance has to be specified with the **view** construct.

In order to make a variable accessible, the revealing process has to declare the variable as revealed, and the viewing process has to declare the variable as viewed.

Similar to the **import** construct described above, the **view** construct is especially useful when used in continuous signals or enabling conditions. However, the use of the **view** construct in a continuous signal may be very costly to implement. Thus, the use of the **view** construct within a continuous signal is not always supported by code generators.

The reveal/view concept has been included in SDL 92 and 96 for compatibility reasons only, and will probably not be supported by the standard beyond the year 2000.

Addressing Addressing is an important subproblem of communication, i.e. identifying the receiver of signals and remote procedure calls. In SDL, there are two basic addressing mechanisms:

- *Explicit addressing* directly specifies the address of the receiving or called process instance. In SDL, this is supported by using the **to** clause in an output construct or a remote procedure call followed by the identifier of the receiving process instance, i.e. its PId. Note that in case only one instance of a process exists, the **to** clause followed by a process name also explicitly identifies the receiver.

- *Implicit addressing* does not explicitly specify the receiving process instance. Implicit addressing is supported by the **to** clause followed by the name of a process (instead of a process instance), or by the **via** clause followed by the name of a signal route or a channel.

Depending on the circumstances, both the **to** and the **via** clauses may or may not uniquely identify the receiving instance.

Alternatively, the receiver may not be given at all. Instead it may be implicitly defined by the signal sent.

If the receiving instance cannot be identified uniquely, i.e. several process instances are potential receivers, one of them is selected nondeterministically. If no process instance exists that may receive the signal, the signal is discarded.

4.1.3.6 Process Management

Dynamic creation and termination of process instances So far we have focused on the cases in which exactly one instance or a statically fixed number of process instances exists for an SDL process. However, SDL provides mechanisms for dynamic process management, i.e. to dynamically create new instances of an SDL process and to dynamically terminate itself.

The creation of process instance is supported by the *create request* symbol. A process instance may create a new process instance defined in the same block provided the maximum number of active process instances of the created process is not exceeded. A process instance may terminate itself using the *terminate* symbol. The direct remote termination of other process instances is not supported.

With the creation of a new process instance, parameters may be passed from the creating to the created process instance. After successful creation of the new process instance, the creator holds the PId of the created process instance in its *offspring* variable. Conversely, the newly created process instance keeps the PId of its parent process instance in its *parent* variable.

Information maintained with each process instance The most important information that is specific for each process instance is as follows:

- a unique identifier of the process instance and some related process instances (as described in section 4.1.3.1),
- a unique input port that stores (queues) received signals and timeouts until they are consumed (or discarded) by the process instance,
- data objects (variables) to store (local) data, and
- timers.

4.1.4 Data Concepts

The data concept of SDL is based on abstract data types. An abstract data type defines a type of data object by its functional properties, i.e. by a set of operations applied to it. Abstract data types focus on functional properties of data objects. Thus, an abstract data type defines the results of the operations applied to a data object, rather than defining how the result is exactly obtained.

In the section, we will only sketch the basic ideas of data in SDL. Detailed discussions of data in SDL can be found in [ElHS97, Olse93].

In SDL, data types are called *sorts* and can be considered as a set of values. SDL supports a number of predefined sorts, i.e. data types, well known from programming languages, e.g. integers, reals, booleans, characters, character strings, etc. Besides these simple data types, SDL supports structures of different sorts and so-called *generators* to define more complex sorts from simpler ones. SDL supports predefined generators that can be used to define arrays, strings and sets. Important to note is that the size of these data types is not necessarily finite.

For the case that the above constructs are not sufficient to define data types, SDL also supports the axiomatic specification of data types based on the ACT-ONE model [EhMa85]. Thus, data types are defined by means of a set of axioms, i.e. equations that define the behavior of operations applied to the data type.

Another approach to specifying data is the Abstract Syntax Notation One (ASN.1) [CCIT88]. ASN.1 represents a standardized mechanism to specify data types. ASN.1 is especially used for the specification of communication protocols and services. SDL supports the specification of data types according to ASN.1. Z.105 [ITU95a] defines how ASN.1 data types can be integrated in SDL.

4.1.5 SDL 2000

In the SDL 2000 standard [ITU99a], several modifications and extensions to SDL 96 have been made in order to suit the language to new developments, especially to harmonize its use with UML, CORBA and ASN.1. In addition, the definition of the formal semantics of the language has been revised and is now based on ASM [GlGP99], i.e. based on operational semantics.

In the following, we give an overview on the most important changes.

4.1.5.1 Structural Concepts

In SDL 2000, the structural concept has been harmonized with respect to the use of the system, block and process entities. All three entities are now based on a common concept, which is called an agent. Several instances of an agent may exist. Unlike SDL 96, where behavior is confined to processes, SDL 2000 supports the specification of behavior by means of state machines in all agents, i.e. also at the system and block level (see below). The state machines are addressed by the name of the enclosing agent. They are connected to other entities defined within or outside the enclosing agent by means of channels. Besides a state machine, an agent may contain the declaration of shared data, and the definition of other agents or agent types.

In SDL 2000, blocks may contain blocks, e.g. defining behavior within state machines, and processes. In addition, processes may themselves contain processes. However, as before, processes may not contain blocks or systems.

Unlike SDL 96, block instances may be created (and terminated) dynamically.

4.1.5.2 Behavioral Concepts

Agents Each agent, i.e. a system, block or process, may define a state machine of its own. The state machine is similar to a process in SDL 96, comprising an input queue, states, transitions, as well as data. Other entities, i.e. other blocks or processes, may communicate with the state machine as previously known from SDL 96, using signals transferred on channels or remote procedure calls.

Defining state machines in these entities results in a hierarchy of state machines (EFSMs), not known from SDL 96. For example, there may be an SDL system including its own state machine and some blocks, that themselves define state machines and additionally contain processes that also define state machines.

Data sharing Different from SDL 96, SDL 2000 supports the direct access to shared data. The data declared in an agent may be accessed by all enclosed entities. Thus, each process or state machine may access all variables declared in the containing agents, i.e. declared in the containing processes, blocks and system. For example, the data declared in a process A may be accessed by its state machine, as well as all processes directly or indirectly enclosed by process A.

Concurrency and data coherence The use of shared data raises the question of concurrency and data coherence. This is solved in SDL 2000 differently for the different kinds of agents. In fact, with the harmonization of the concepts for processes and blocks (and systems), the major difference between the two concepts lies in the manner in which concurrency is dealt with. The underlying idea is that blocks represent truly concurrent entities, while processes describe entities with alternating semantics, i.e. controlled concurrency.

Process instances contained in a process, as well as the state machine of the containing process are interpreted in mutual exclusion at the transition level. Thus, within a hierarchy of processes, only one transition is active at a time. As a result, shared data within processes can be safely accessed within transitions. However, note that due to the application of transformation rules, e.g. to implement remote procedure calls, explicit transitions may be split in a set of implicit transitions. Thus, unexpected interleavings may arise.

State machines defined within blocks or the system are interpreted concurrently and asynchronously. Concerning the underlying semantics, communication among these entities is exclusively by means of signals, e.g. explicit signals or implicit signals resulting from remote procedure calls. Semantically, access to a variable declared in either a block or the system is implicitly implemented by two remote procedures, one to implement a read access and one to implement a write access.

Data model The axiomatic definition of data types is no longer supported in SDL 2000. Generators are also no longer supported. The elements of a sort may be values, objects, i.e. references to values, and PIds, i.e. references to agents. SDL 2000 discriminates between operations and methods. Operations are defined in a similar way to procedures containing a single transition. Methods are similar to operators with additional support for the specialization of data types.

Communication Communication is supported by means of signal communication, remote procedure calls and the export/import constructs. Unlike SDL 96, communication may additionally be by means of shared data (as described above). Shared data replace the reveal/view concept known from SDL 96, which is no longer supported.

The concept of signal routes and channels is merged. Communication is now always along channels, independent of the kind of the connected entities, e.g. blocks, processes or state machines.

Composite state SDL 2000 introduces a concept to define the states of a state machine in a hierarchical manner similar to the respective concept known from Statecharts. Thus, a state may consist of a set of substates.

Exception handling In order to handle exceptional cases during the interpretation of an SDL specification, exceptions and exception handlers have been introduced in SDL 2000. SDL 2000 supports user-defined exceptions as well as a set of predefined exceptions, e.g. to denote division by zero, access to an invalid reference or an undefined variable.

Handling an exception results in a transition. Exception handlers may be associated with different scopes, e.g. an entire process, a specific state, a transition or a specific action within a transition.

An exception breaks the normal flow of control and may be raised implicitly by the underlying system, or explicitly by a **raise** symbol. Exceptions may have parameters. Exceptions raised in a procedure or an operation are handled locally, or if no exception handler is present propagated to the caller of the procedure or operation. Exceptions raised in remote procedures are raised locally as well as propagated to the calling entity.

Interface descriptions SDL 2000 supports the specification of interfaces. An interface definition may define signals, remote procedures, remote variables and exceptions. Note that each agent and agent type has an implicitly defined interface.

4.2 Complementary Languages

In this section, we focus on languages and techniques complementary to SDL. These languages represent important constituents of the system development process in the context of SDL. We will survey MSC, TTCN and relevant object-oriented techniques.

4.2.1 Message Sequence Chart (MSC)

4.2.1.1 Introduction

MSC (ITU-T Z.120 [ITU97]) supports the specification and description of functional aspects of the application under development. In the development cycle, MSC is employed for the formal specification and description of use cases of the system. An MSC describes a specific execution sequence of the system or of parts thereof. Thus, an MSC mainly represents a kind of sequence diagram.

Like SDL, MSC comes in two syntactic forms, a textual and a graphical representation. MSC is often employed in conjunction with SDL. However, MSC is much more general and may be combined with other specification or design techniques as well. For example, UML employs a variant of MSC called a sequence diagram.

Also as in SDL, a formal definition of the semantics of MSC exists (Annex B to Z.120). Due to the incompleteness of MSC concerning a functional specification of the system under development, there are subtle differences between the semantics of SDL and MSC.

Purpose of MSC MSC is a specification language for use cases. It supports the specification of use cases at an abstract level as well as at a very detailed level. MSC is not intended to provide a complete functional specification of the system under development. Rather, its purpose is to support the engineer by providing an additional view of the system.

MSC may serve a number of purposes from reasoning about systems at an abstract level to the automatic derivation of test case descriptions for the implementation. In the development process, MSCs are typically created during the requirements analysis, i.e. before the SDL specification, and subsequently refined during the design process. So far, MSCs have been mainly used to

- formally specify the functional requirements,
- serve as a basis for the generation of SDL skeletons, and
- serve as a basis for testing.

In addition, MSCs can be employed to verify the conformance of the system implementation with its specification, and to verify consistency of the system specification with its requirements. Moreover, MSCs provide a valuable means of visualizing traces which result from various kinds of analyses of the system.

Like SDL, MSC abstracts from the implementation to a certain extent. MSC does not support issues related to the physics of the underlying system, e.g. dealing with bounded queues and finite processing capacity. In addition, MSC does not support the specification of data types and operations on data. Like SDL, MSC currently has a rather vague time semantics. In particular, the kind of interleaving of a set of execution scenarios specified with MSC is not defined.

Application domain There are a number of application areas for MSC. Due to the generality of MSC these are far beyond the application domain of SDL. Similar to SDL, the major current application area of MSC is in telecommunications. MSC is also used by standardization bodies such as ETSI to supplement the SDL specification.

Evolution of MSC MSC is a relatively new specification language. The first standard for MSC appeared in 1992. Subsequently, several extensions of MSC to include constructs to structure use case descriptions and also to better support the specification of algorithmic features have been proposed and integrated into the standard.

Tool support for MSC Today the commercial SDL tools support the integrated use of SDL and MSC. MSC-related tool support comprises graphical editing of MSCs, consistency checking with SDL, the generation of SDL frames from an MSC description, the use of MSCs to control the simulation of the SDL description, the visualization of SDL simulations with MSC, and the derivation of test cases. In addition, tools from academia are being developed to support other aspects of the development cycle, e.g. the MSC-driven instrumentation of SDL descriptions and MSC-based performance evaluation. In section 6.3.3, we will take a closer look at tools that support performance-related activities in the context of MSC.

Focus of this section In our introduction to MSC, we focus on basic MSCs, i.e. on constructs to describe execution sequences, rather than on structural concepts.

Further reading A detailed introduction to the MSC language and its usage can be found in [RuGG96]. Also a good introduction to MSC is given by the example section of the Z.120 standard [ITU97]. Methodological issues of the usage of MSCs are discussed in [Haug95].

4.2.1.2 Basic MSC

An MSC describes an example (or instance) of a possible execution of the system. In the terminology of MSC, it specifies how messages are passed through the instances of the system (see below for details). In the context of SDL, an MSC describes a partial order on the execution of the SDL system.

An example of a simple MSC for the sensor example introduced in section 4.1.2 is depicted in figure 4.12. The given MSC specifies two instances which correspond to the two SDL blocks of the sensor application outlined in figure 4.1. The given MSC describes the handling of a single user request provided by the environment.

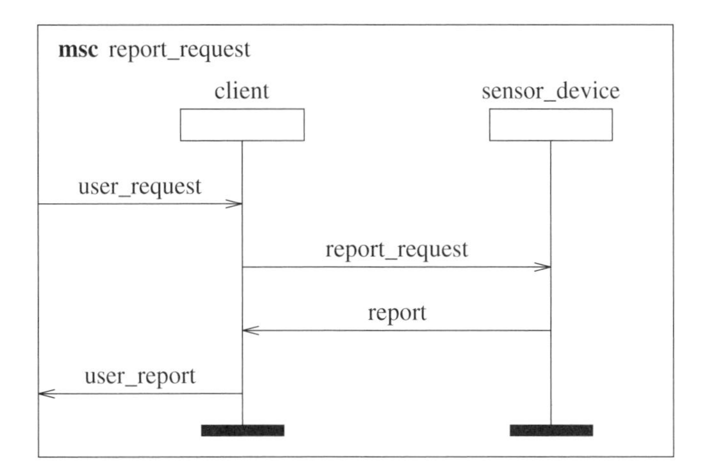

Figure 4.12 An example of an MSC to describe the handling of a simple user request in the sensor system

Instance The basic elements of an MSC are (interacting) instances of entities of the system. Instances are described by vertical lines that define the temporal ordering of the events relevant to the instance. In addition, an MSC instance comprises a header that defines its name and a trailer.

In the context of SDL, an MSC instance represents an instance of an SDL entity as a block, process, or service. In the example given in figure 4.12, the two MSC instances represent instances of two SDL blocks.

Message Messages describe the interaction between instances or between an instance and the environment. A message in MSC is a relation between an output and an input. In the graphical representation, a message is described by a (horizontal) arrow. The arrow is labeled by the message name identification. In the context of SDL, the message name may represent an SDL signal type.

A message comprises two events, an output and an input event. An output may be generated by the environment, or by any instance. Similarly, an input may be consumed by the environment or any instance. In addition, MSC supports messages that emerge from nowhere (found messages), and messages that disappear (lost messages).

Messages define a temporal order of the execution. An additional construct exists to describe temporal orders where no explicit message exchange takes place.

Condition Conditions in MSC represent a notion of state. A condition is graphically specified using a hexagon and defines the state of a single instance or of a set of instances.

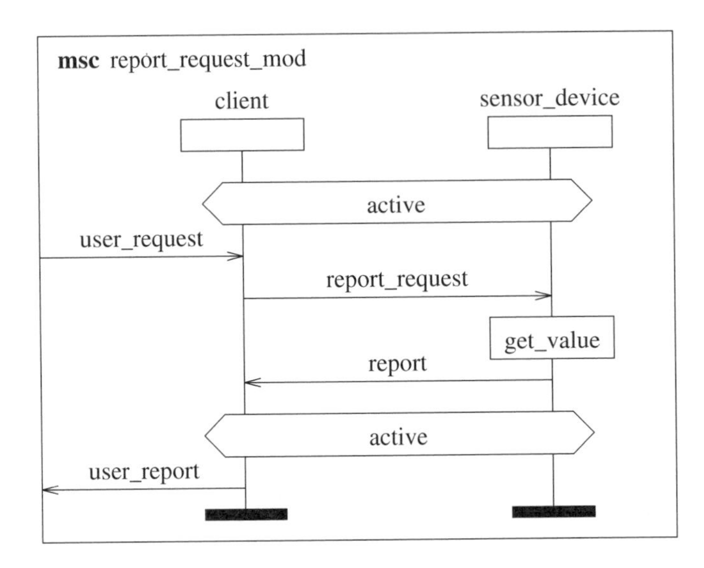

Figure 4.13 Usage of conditions and actions in the sensor example

An example of the use of conditions in the sensor example is depicted in figure 4.13. The MSC specifies the processing sequence for the user request for the case in which the system, i.e. the *client* and the *sensor_device*, are in the state *active*.

Action An action describes an internal activity of an instance. In MSC, an action comprises informal text and is graphically specified by means of a rectangle.

An example of an action (*get_value*) in the sensor example is depicted in figure 4.13.

Instance creation and termination Similar to SDL, MSC supports the creation and termination of instances. Instance creation is described in MSC by a dashed arrow. An example of the creation of an instance for the sensor example is given in figure 4.14. In the example, the instances *sensor_manager* and *sensor* represent SDL process instances rather than SDL blocks.

Termination of an instance is described by a cross in the graphical description, analogous to SDL.

Timer The role of timers is crucial in communicating systems. MSC supports constructs to set and reset timers as well as to react to expired timeouts.

Two possible MSCs for the extended sensor example introduced in figure 4.6 are depicted in figures 4.15 and 4.16. Figure 4.15 shows regular processing without the expiration of the timeout, i.e. a set operation followed by a later reset operation. Conversely, figure 4.16 shows the execution

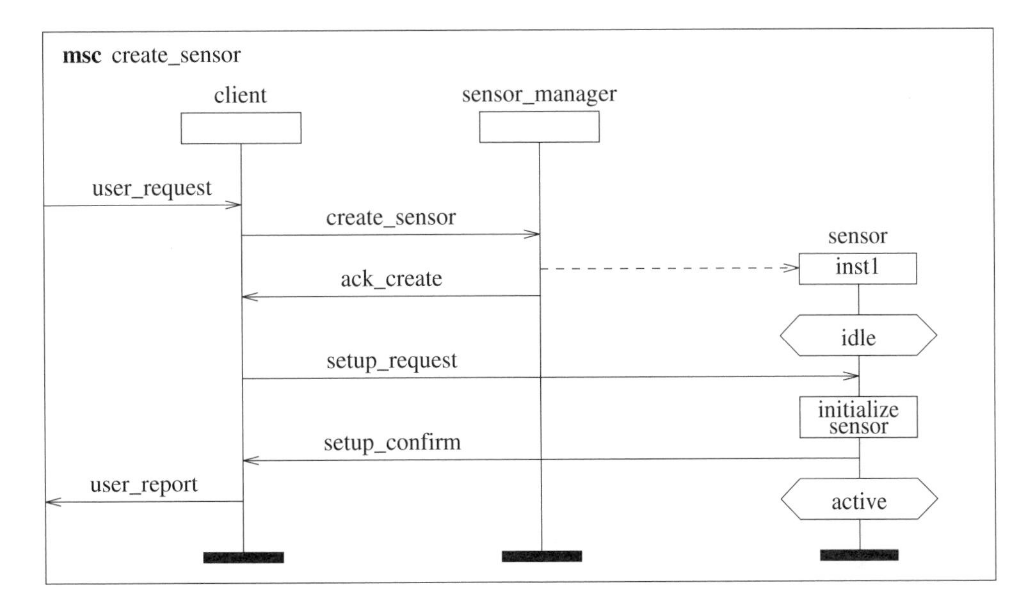

Figure 4.14 Instance creation

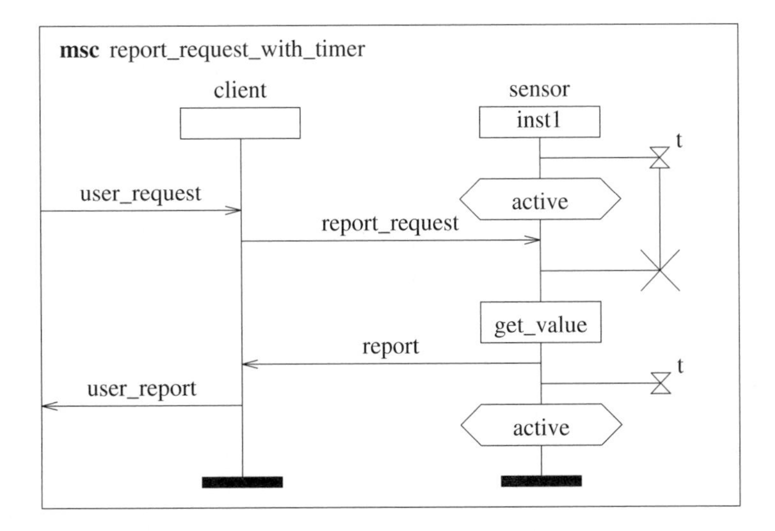

Figure 4.15 Modified sensor example with timer (set and reset)

sequence for the case in which the timer expires before a *report_request* message arrives at the sensor.

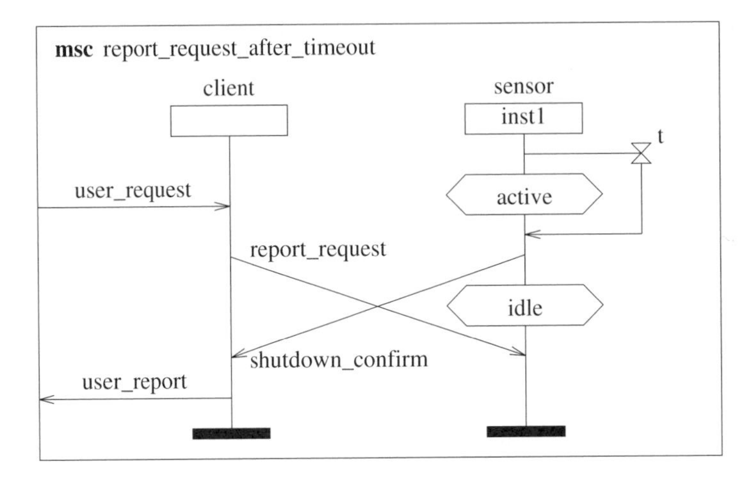

Figure 4.16 Modified sensor example with timeout

4.2.1.3 Structural Concepts

Besides basic MSCs, Z.120 also comprises a set of higher-level constructs that allow one to structure a set of basic MSCs.

Coregion The total ordering of the events on an instance axis may not always be appropriate. The coregion allows two or more events of an instance, e.g. the arrival of messages, to occur in arbitrary order.

Instance decomposition An instance of an MSC can be refined by means of instance composition. This supports hierarchical specification, as refined MSCs can be subsequently integrated into the initial MSC description.

Inline expression MSC supports a set of inline operations to express alternatives, parallel composition, iterations, exceptions and optional regions. Thus, more complex execution sequences can be described.

MSC reference In addition to the direct specification of MSCs, references to other MSCs are supported.

High-level MSC A means of graphically combining a set of MSCs is high-level MSCs. A high-level MSC is a directed graph that comprises, among other constructs, a start and end symbols, MSC references, conditions and connection points.

4.2.2 Tree and Tabular Combined Notation (TTCN)

TTCN [ISO96] is a language to define test suites for protocol specifications. Test suites are used to ensure that different implementations of the same protocol specification conform to the requirements. Test suites are collections of a set of test cases. Each test case is defined with a specific test purpose in mind. A test case defines how to drive an implementation under test in order to reach the test purpose. Test cases are defined by sequences of test events. The most important test events are input and output events.

4.2.3 Object-Oriented Analysis and Design Techniques

The focus of SDL is the specification and description of structural and behavioral aspects of communicating systems, in other words, the description of active objects that interact by means of messages. On the contrary, object-oriented approaches, i.e. class diagrams, focus on the relationship between various kinds of objects. In particular, passive objects, i.e. data objects, can be modeled more appropriately with object-oriented approaches.

There are several ways to use object-oriented approaches in conjunction with SDL. Below, we outline the two principal concepts. The major difference represents the phase and activities for which object-oriented approaches are employed instead of the use of SDL. The prevailing object-oriented technique used in conjunction with SDL is OMT, which is currently being replaced by UML.

Object-oriented analysis With this approach, the object-oriented analysis is mainly based on the object modeling approach supported by OMT (class and instance diagrams). In addition, state diagrams and MSCs are often used to describe the dynamic model. This is followed by a manual – possibly tool-supported – translation of the analysis model to an SDL specification. SDL is employed for all design activities.

This approach is supported by major SDL tools, e.g. ObjectGEODE [Veri98] and TAU (SOMT) [Tele98]. In order to support consistency between the analysis model (OMT) and the design (SDL), links between the entities of the two models can be established.

SDL 2000 recommends the joint use of SDL with UML (see section 2.1.3.5). The new ITU standard Z.109 [ITU99] defines a specialization and restriction of UML for combined use with SDL, i.e. in UML terms, an SDL UML profile. The SDL UML profile focuses on the description of structural aspects of the system. It mainly defines the link between elements of UML class diagrams and their corresponding SDL entities. The SDL UML profile defines a one-to-one mapping which allows each UML entity supported by the profile to be transformed smoothly to a corresponding SDL concept.

Object-oriented analysis and design With this approach, object-oriented techniques, e.g. OMT, are used not only for the analysis but also for a considerable part of the design. Approaches for the use of object-oriented techniques for the design range from

- the use of OMT for the system design, followed by an automated translation of the design model to an SDL description which forms the basis for detailed design (e.g. INSYDE [HWWL+96]), to

- the use of OMT for major parts of the design except for the description of behavioral aspects, for which SDL (without object orientation) is used (e.g. SCREEN [LSIV97]).

For passive objects of the system, the direct translation of the object model to code has been proposed [PeDe97, IRSK96].

4.3 Methodology

As defined in section 2.1.1.2, a methodology defines methods employed in the development process, and procedures and rules when and how to apply the methods.

4.3.1 Z.100 Methodology Guidelines

Some methodological guidelines are provided as part of the SDL (Z.100) standard. The two documents described in the following form a complementary set of guidelines for system development in the context of SDL. The SDL+ methodology focuses on the early design phases while the SDL methodology guidelines given in the appendix of Z.100 focus on implementation design. A new methodology document is expected to be completed in 2001.

4.3.1.1 SDL+ Methodology

In order to support the development of systems based on SDL, the SDL+ methodology [ITU97a] has been devised (see [Reed96] for an overview). The SDL+ methodology focuses on the analysis and the design phases (draft design and formalization).

The methodology suggests the use of an object-oriented method to classify the information analyzed and derived during the analysis. For draft design, the use of a set of concepts is recommended. This comprises SDL, MSC, ASN.1 and an object-oriented analysis technique. MSCs are recommended to specify important use cases. A high-level SDL specification can be derived from the MSCs. In this early stage, SDL is mainly used to describe the structural aspects of the application and the data and control flow in the system. Behavioral aspects described with SDL at this point focus on essential features of the system and are given in a rather informal and incomplete manner. Up to this point, the SDL specification focuses on the functional aspects of the system from the user's viewpoint. During the formalization phase, the behavioral aspects of the system are formalized and added to the SDL specification such that a complete and consistent SDL description results.

4.3.1.2 SDL Methodology Guidelines

The SDL+ methodology as described in [ITU97a] focuses on functional aspects of the system under development. However, it does not cover implementation design and the implementation process in much detail. The methodology is based on the assumption that computing resources are abundant and that performance aspects do not play a major role in the development process.

The implementation design phase and the implementation itself, including a concept for the integration of non-functional aspects in the development cycle, are discussed in the SDL methodology guidelines (Z.100, Appendix I, [ITU93a]). This proposes the integration of non-functional aspects during the implementation design. In the implementation design phase, the given formal SDL specification is transformed to an SDL description. Unlike an SDL specification, the SDL description deals with some non-functional aspects and nonideal features of the target system. Obviously, not all aspects can be expressed with standard SDL. The aspects not covered by the SDL description are informally given in a separate implementation description.

Unlike the SDL+ methodology, the SDL methodology guidelines assume the use of two kinds of SDL documents in the development process. A formal SDL specification should be produced first, which is later transformed to an SDL description which takes into account implementation constraints. This approach is quite natural where the product development is based on standards provided as SDL specifications.

4.3.2 Other Methodologies

Besides the official Z.100 methodology documents, a number of additional documents exist that focus on methodological aspects of the development of communicating and real-time systems in the context of SDL and companions. Most approaches are based on an integrated use of SDL, MSC and OMT/UML or a subset thereof. We will outline the most important projects.

4.3.2.1 The SISU Project

SISU is a Norwegian technology program aiming at the improvement of the productivity and quality of Norwegian companies that develop systems in the real-time domain. A number of companies and research institutes participated in the project, including Alcatel, Siemens, and Ericsson. SISU-I, which ran from 1988 to 1992, produced [Reev93] and had a major impact on the SDL methodology guidelines [ITU93a] as well as on the SDL and MSC standards.

SISU-II, which ran from 1993 to 1996, resulted in the TIMe methodology (The Integrated Method) [BHMM+97]. TIMe aims at supporting the development of complex, reactive, concurrent, distributed and heterogeneous systems. As pointed out in [BHMM+97], TIMe is centered around a set of models and descriptions capable of expressing domain knowledge, system specifications in terms of external properties, system designs in terms of structure and behavior, implementation mappings and system instantiation. TIMe is an object-oriented approach based on OMT/UML, SDL and MSC. The design with TIMe is based on a separation of

- application design, focusing on the functionality of the system,

- architecture design, dealing with non-functional properties, and
- framework design, supporting the design of product families based on reuse.

TIMe is commercially available through SINTEF.

4.3.2.2 The INSYDE Project

The INSYDE project (Integrated Methods for Evolving System Design – Esprit project P8641) aims at defining, implementing and demonstrating a comprehensive methodology for the design and validation of hybrid HW/SW systems, based on OMT, SDL and VHDL.

INSYDE discriminates between the three development stages of analysis, system design and detailed design. OMT is used for the analysis phase, employing object models, dynamic models, and functional models. For system design, OMT* is employed, a subset of OMT to support the automatic translation to SDL. INSYDE propagates the early partitioning of the system in parts to be implemented in hardware and software. SDL is used for the detailed design of the software parts while VHDL is employed for the hardware parts.

4.3.2.3 Others

Methodological guidelines have been proposed by other projects as well, e.g. the SPECS project (RACE) [Reed93, OFMR+94], and ETSI MTS PT60 [ETSI95].

Methodologies have also been proposed by the commercial tool providers, especially ObjectGEODE [Veri98] and TAU (SOMT methodology) [Tele98]. These approaches are based on OMT, SDL, MSC, and TTCN. The advantage of these methodologies is support by the respective tools. Other proprietary methodologies have been developed by telecommunication companies to suit their specific needs and adapted to their specific tool chain.

4.3.3 A Note on Performance and Time

Note that currently only the verification and validation of functional aspects of the SDL specification (or description) and the functional check for consistency between the SDL specification and the MSCs is supported by commercial tools. Due to the limitations of SDL and MSC to express non-functional aspects, an integrated methodology to deal with non-functional aspects more formally is missing. As a consequence, the commercial SDL tools do not support the verification/validation of non-functional aspects. An exception is ObjectGEODE, which supports performance simulation to some extent. We will return to the issue in chapter 6.

Chapter 5

Implementing SDL

SDL is a specification and description technique, not a programming language. As a consequence of this, SDL abstracts from implementational aspects, i.e. neither intends nor supports the specification of the details of how to implement the SDL description. Especially, SDL does not specify how to implement the underlying communication between different SDL processes and communication with the environment. In addition, SDL does not deal with the implementational details concerning process and timer management.

The development phases concerned with the derivation of an implementation from a given SDL description are the design and the implementation phases.[1] In this chapter, we describe the major alternatives and decisions related to these two phases. However, we will focus on design decisions rather than details of the implementation in a specific programming language. We also focus on issues that are specific to the implementation of SDL. General implementation issues such as the implementation of data types, arithmetic operations, the application of data-flow analysis techniques, etc. are not considered.

The chapter is structured as follows. General problems and issues with the derivation of code from SDL are discussed in section 5.1. The basic elements of implementations automatically derived from SDL are introduced in section 5.2. The integration of code derived from SDL with the underlying computing system is discussed in section 5.3. Section 5.4 describes different approaches to organizing the execution of the process instances.

Sections 5.5 to 5.10 discuss the design alternatives for the different aspects to be dealt with when implementing SDL descriptions, namely the implementation of the finite state machines embedded in SDL processes, the management of the process instances, the implementation of process communication, interfacing of the SDL system with the environment, the handling of timers, and SDL data. In order to reduce the complexity, the respective sections are restricted to the discussion for a subset of SDL, mainly the features of basic SDL as defined in Z.100. The issues involved with supporting advanced features of SDL are discussed in section 5.11.

[1]Note that we focus on the generation of product code rather than the derivation of code for simulation.

Section 5.12 discusses optimizations for a specific application area, namely the implementation of protocol architectures. The section is based on earlier work conducted with Ralf Henke [HeKM97, HeKM98, HeMK97]. Section 5.12 includes text and figures from [HeKM97].[2] Approaches to dealing with physical and organizational limitations of real systems when implementing SDL descriptions are discussed in section 5.13. The side effects of applying the performance-optimizing code generation are discussed in section 5.14. Section 5.15 discusses the impact of SDL 2000 on the manner communication systems are described with SDL, and the impact on code generation. Section 5.16 summarizes the optimization concepts discussed in the chapter with respect to prerequisites, scope and merits. In addition, the section provides guidelines on how to derive efficient implementations.

5.1 Introduction

We start with a discussion of the problems and issues involved with the derivation of efficient code from SDL in section 5.1.1. The principal limitations of SDL concerning the derivation of an implementation, and the differences between SDL and typical implementations, are discussed in section 5.1.2. Section 5.1.3 discusses additional constraints which have to be dealt with when code is derived from SDL descriptions.

5.1.1 Problems and Issues with the Efficient Implementation of SDL

In principle, there are many paths to derive an implementation from a formal specification. This especially holds where the semantics of the formal description technique considerably abstract from implementational details as is the case with SDL.

5.1.1.1 Deriving an Implementation from an SDL Specification

The paths, steps and alternatives to derive an implementation from an SDL specification, e.g. as provided by a standardization organization, are outlined in figure 5.1. In each step of the process, decisions are made. The respective decision alternatives, or the potential design space, is defined – or restricted – by the semantics of SDL and the subset of SDL used by the given SDL description.

SDL design The first step shown in figure 5.1 is concerned with the SDL design, i.e. the derivation of an SDL description from an SDL specification.[3] A general introduction to the activities of the design phase has been given in section 2.1.2.2. The design is to a large extent a manual activity. The derivation of the SDL description is very important for the performance of the system.

[2]Reprinted from [HeKM97], Copyright 1997, Pages 397 to 414, with permission from Elsevier Science.

[3]Refer to section 4.3.1.2 for a discussion of the difference between an SDL specification and an SDL description.

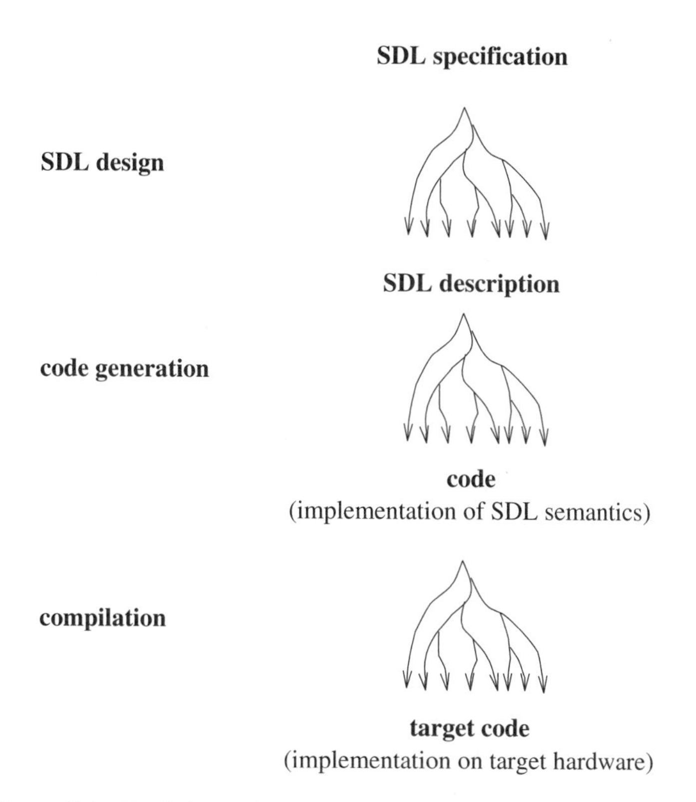

Figure 5.1 Deriving an implementation from an SDL specification

There are two major issues:

- The selected subset of SDL used in the derived SDL description has implications for the code generation strategies that can be applied. There are many issues that have an influence on code generation. Especially important is the degree of dynamics that is used, e.g. dynamically created processes, dynamic size of signal parameters, etc. The reader should have a better understanding of these issues after reading the chapter.

- The architecture of the system is important, especially the module coherence and interaction between modules or entities, i.e. which functionality is provided locally and which part depends on (costly) interactions in the system. Approaches to identifying problems belonging to this class will be discussed in chapter 6.

Code generation The second step is the (automatic) code generation, which is the main focus of this chapter. In this step, the specific way the SDL semantics are implemented is decided. As

SDL leaves implementational details open, different behavior may be in accordance with the semantics of SDL. For example, consider the order in which SDL process instances are scheduled, which may even result in differences of the externally visible behavior. An issue related to code generation is the adaptation of the runtime support system and the selection of the underlying operating system.

Compilation The last step, i.e. the compilation of the derived code, is a general compiler issue. Since the compiler does not know the semantics of SDL, the scope of the optimizations it applies is limited and independent from the SDL semantics. Thus, compilation will not be discussed here.

5.1.1.2 Problems with the Automatic Code Generation

Approaches to the computer-aided derivation of implementations from formal descriptions such as SDL, Estelle or LOTOS have been investigated for many years. So far, the use of automatic code derivation is limited due to its lack of efficiency. Thus, automatically derived implementations are mainly used for prototyping or for applications where optimal performance is not crucial.

The insufficient efficiency of the generated code is the main obstacle to the successful application of computer-aided implementation techniques (and probably of formal description techniques in general). There are several reasons for this [HeKö95]:

(1) The implementation model prescribed by the transformation tool is elaborated during tool design. It does not take into account the context in which the implementation is used.

(2) Formal description techniques are based on formal semantics. The semantics highly influence the design of the implementation model. Optimizations can only be introduced by ensuring correctness and conformance criteria.

(3) So far, automated protocol implementations are mainly focused on one-layer implementations (due to their orientation on prototyping). Thus, optimizations that involve more than one protocol layer are rarely supported.

Current transformation tools mostly follow a straightforward implementation of the semantics of the formal description technique and do not exploit potential alternatives. The process model typically employed is the server model as introduced in section 3.3.2.1 (e.g. [Fisc93, StJG93, Tele98, Veri98]). However, proposals to increase the efficiency of automatically generated implementations exist, and will be discussed in this chapter.

A distinguishing feature of formal description techniques is their foundation on formal semantics. Automatically deriving implementations from formal specifications requires transformation rules which preserve their semantics, i.e. conformance of the implementation with the specification. This may add overhead to the implementation which is not present with implementations manually derived from informal specifications.

On the other hand, formal description techniques such as SDL considerably abstract from implementational details. Thus, numerous alternatives for implementing the underlying semantics of SDL exist. This opens up significant possibilities for performance optimization. Important examples of this are the implementation of the process management of the SDL entities, the communication between the SDL entities and with the environment, and the management of timers.

5.1.2 Differences Between an SDL Description and an Implementation

SDL is a specification and description language. Thus, SDL provides the means to describe structural and behavioral properties of the system in abstract terms. However, SDL abstracts from many implementational aspects of real systems.

5.1.2.1 Physical and Organizational Limitations

The most important abstraction SDL makes is the abstraction from physical and organizational limitations caused by the passage of time and the consumption of memory. Especially, SDL has a very vague notion of time.

On the other hand, all real physical systems are limited in some way. Everything consumes time and space. Thus, some of the assumptions (or abstractions) made by SDL do not hold in practice. This may have serious consequences if not dealt with carefully.

As described before, abstraction is a very appropriate means of handling complexity. Abstraction eases the task of the designer. It allows the designer to deal with one problem at a time without being overwhelmed by the whole complexity of the system.

However, when it eventually comes to the implementation of the system, physics brings the designer back to reality. Thus, a neat design may finally turn out to require resources that are beyond practical limits.

Other consequences that may result when limitations are ignored are deadlock, e.g. due to blocking on full queues, or system failure due to the overriding of a full queue or due to an overloaded processor.

An additional complication is often present in communication systems. In these systems, high load often results in additional load. Thus, the detrimental effect is enhanced. For example, the delay caused by high load may trigger the retransmission of packets, which in turn results in more traffic and, thus, in even more load at the already overloaded instance.

5.1.2.2 Kinds of Limitations of Implementations

Several kinds of limitations exist in real systems. Most of them cannot be expressed in SDL.

Limited number of process instances In SDL, an upper limit of the number of process instances per SDL process can be optionally specified. This allows one to deal with some organizational limitations of real systems.

Limited queues SDL queues are unlimited in size and capacity. Limited queues can only be explicitly modeled by introducing additional SDL (queuing) processes. However, the merit of introducing a limited queue, i.e. by means of an explicit SDL process, surrounded by two unlimited queues is questionable.

Limited processing capacity This relates to the semantics of time in SDL. In SDL, computational actions may or may not advance the time. Conversely, in real systems any action consumes time.

Approaches to explicitly modeling the consumption of time in SDL are the use of timers or of the time construct (**now**). However, as in dealing with limited queues, this is not a very appropriate way to describe physical aspects of the system. This is especially problematic if code is derived automatically from SDL. In this case, the usage of timers as part of the behavioral specification and as a concept to model time consumption have to be kept separate. Or in other words, the constructs that have been introduced in the SDL description solely to model time consumption have to be removed before the code is derived. This is because the developer does not want to derive code for SDL constructs which have the sole purpose of waiting till some time period has passed.

In addition to the sole consumption of time, limited resources cause contention. Thus, the strategies to handle concurrent requests for limited resources have to be decided and implemented.

Since computation takes time, actions and transitions in SDL take time and are not atomic by definition as, for example, is assumed by the formal semantic definition of SDL and by some validation and simulation tools.

Delayed communication Like computation, in real systems any communication takes time. Thus, any communication is delayed, even communication on nondelaying channels and signal routes.

Limitations on abstract data types On computers, data have to fit into some word boundaries. Thus, all data representations are of finite size. As a result of this, overflow and rounding errors may occur.

Errors and failures Physical systems can work erroneously or may fail in part or completely. In general, this cannot be appropriately expressed with SDL. As we do not focus on fail-safe systems, this is not addressed here.

Different approaches to dealing with implementational limitations are discussed in section 5.13.

5.1.3 Implementation Constraints

In implementing SDL descriptions, a large set of alternatives exists. Some of the alternatives are typically fixed by company or departmental policies or constraints; others may be fixed for a specific product or product line, or by the context in which the derived implementation is used.

Here, we discuss the typical requirements and constraints encountered with the implementation of SDL descriptions, which is denoted as the implementation process in the following.

5.1.3.1 General Requirements on the Derived Implementation and the Implementation Process

A general introduction to quality issues has been given in section 2.1.1.5. Important quality requirements of the products derived from SDL descriptions are typically:

- completeness and correctness,
- robustness,
- reliability,
- maintainability,
- dynamic replacement of components,
- reusability, and
- requirements related to performance, time and memory space.

Typically, the development is additionally constrained by requirements imposed on the implementation process. Important issues are:

- the time to implement the system,
- the cost of deriving the implementation, and
- the use of certain methods, concepts, languages and tools to derive the implementation.

Our focus here is especially on performance and time issues and the reduction of time and cost in meeting these constraints. The influence of performance-optimized code generation on other quality issues will be discussed in section 5.14.1.

5.1.3.2 Status of the SDL Description

The effort to derive an implementation depends highly on the status of the SDL description that serves as base. Thus, the question is whether the given SDL document represents a rather abstract specification of the system behavior or a detailed description which considers implementational issues to some extent. For example, an abstract specification of a protocol typically does not specify how to deal with multiple connections to be handled concurrently by the implementation, or how to deal with overload. Thus, the effort depends on the degree of completeness of the given SDL document.

Other issues related to the degree of detail are the implementation of abstract data types, and the appropriateness of the constructs and structure used in the SDL document. Typically there is a trade-off between the suitability of an SDL description for an automatic derivation of code and other goals. For example, constructs that support readability may not be very appropriate when an efficient implementation has to be derived.

5.1.3.3 Integration with the Technical Environment

SDL descriptions – or implementations derived from them – are typically integrated in a larger technical system. Thus, interfacing with the environment is very important. Especially with time-critical systems, the interface of the system with the environment often has a major impact on the implementation.

Especially important are the mechanisms employed at the interfaces and the time requirements on these mechanisms imposed by the environment. For example, consider an I/O interface at which a new set of data is provided to the SDL system every 10 ms. Thus, the processor serving the interface has to be able to read the data set once every 10 ms in order to ensure that no data are lost. Besides these periodic arrivals, nonperiodic arrivals may exist with similar time constraints.

5.1.3.4 Integration with the Computing System

Besides integrating the SDL system with the surrounding technical environment, the SDL description has to be mapped on the underlying services provided by the computing system on which the SDL description is implemented and executed. Thus, the services required by the SDL system, such as support for communication, process and timer management, have to be provided by some entities in the implementation. As a result of this, some adaptation between the underlying system and the SDL description is typically needed to match the requirements and constraints.

From the performance perspective, especially the runtime behavior and service times of the different services provided by the underlying system are important.

Available resources The performance of a system depends on the available resources that execute the application. The most relevant resources from the performance perspective are the processors and the communication links. The available memory to hold code and data, most notably the size of buffers, may also be important. Besides the capacity of the resources, their service strategies are of importance.

Underlying support system Systems are usually implemented by a set of layers to reduce complexity. This technique is also applied where code is automatically derived from SDL descriptions. Thus, the question arises in which layer specific services are provided, i.e. the runtime support system, the operating system or by specific hardware, and which of the services are directly implemented in the application-specific part.

Typical implementations of SDL descriptions in software which are based on the automatic derivation of code employ a four-layer model consisting of the application-specific code, the runtime support system, the operating system and the hardware. We will discuss the details of the functionality provided by the different layers below.

Mapping on the resources In larger systems often a set of identical or similar resources exist. Thus, the question of how to map or distribute the load caused by the application on the available resources is important. A similar question arises where specific hardware is employed, e.g. FPGAs or ASICs, to directly implement specific parts of the application by customized hardware.

Dealing with limited resources Above, we have discussed the limitations of SDL concerning the available resources. An important issue in this context is how to deal with feedback from the physical or organizational resources. Especially critical are questions of how to deal with limitations that are not dealt with in the context of SDL. Examples are overloaded processors, overloaded communication links and SDL queues. In typical system implementations, these problems have to be dealt with.

A related issue is how to dynamically change parameters of the underlying system, e.g. priorities of an instance of an SDL process. This is important in order to react to dynamically changing requirements during system operation. These issues are highly related to performance and time aspects of the system and will be discussed in detail in section 5.13.

5.2 Basic Elements of an Implementation Derived from an SDL Description

In the section, we take a look at important basic elements of implementations automatically derived from SDL.

In order to set up the basics for the efficient implementation of SDL, we employ a somewhat different view on the semantics of SDL, i.e. different from the formal semantic model documented in Annex F of Z.100. Annex F employs a rather distributed programming paradigm based on the CSP model. Conversely, our semantic model is more oriented towards a centralized view. This

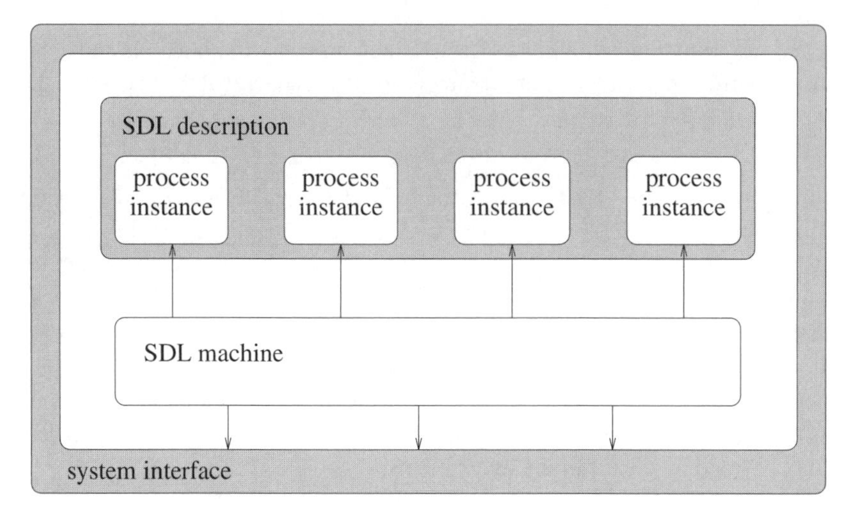

Figure 5.2 Three classes of entities important for an implementation derived from SDL

takes into account the fact that in typical implementations larger parts of an SDL description are implemented on a single host rather than distributed over various geographical locations.

5.2.1 SDL Descriptions, Semantics and Interfaces

In order to clarify the implementation of SDL descriptions, we discriminate between the following three classes of entities (see also figure 5.2):

- the *process instances* that represent instances of the SDL processes as defined in the SDL description,
- the *SDL machine*, i.e. everything that can be considered as being beyond the SDL description, but results from the underlying semantics of SDL, and
- the *system interface*, i.e. the interface to the environment which is outside of the scope of SDL.

Each SDL process represents an Extended Finite State Machine (EFSM). Thus, each instance of an SDL process executes the EFSM defined by the SDL process with its instance-specific data. Note that the EFSM, or in other words the SDL process as such, does not represent any physical data objects. Data objects are all tied to the process instances.

The EFSMs represent the visible parts of the SDL description, rather than the underlying semantics of SDL. Conversely, the SDL machine represents the underlying semantics of SDL. The SDL machine interprets the SDL description and interacts with the system environment. The system interface is responsible for supporting the interaction of the SDL system with the external

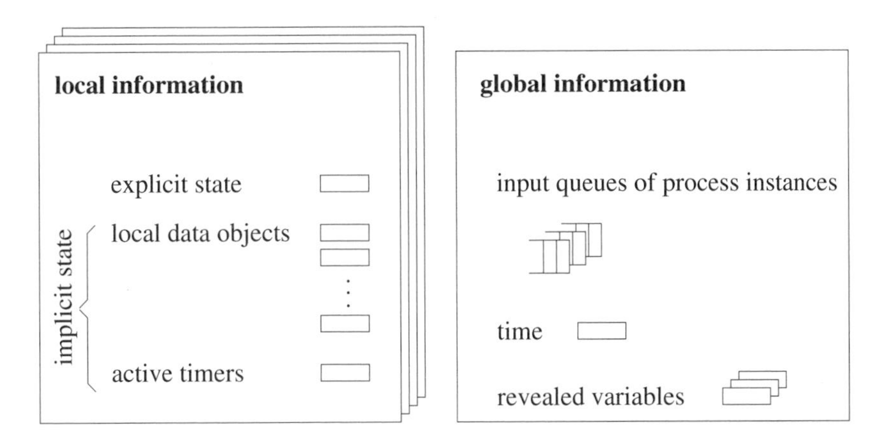

Figure 5.3 The two classes of data objects present in an implementation derived from SDL

world, i.e. the parts that do not understand the communication mechanisms of SDL, such as, for example, SDL signals.

5.2.2 Data Objects of an Implementation Derived from SDL

So far, we have not explicitly discriminated between data and control. Concerning the visibility of data in an implementation of an SDL description, there are two main classes of objects. The two classes are depicted in figure 5.3. These are

- the *local data*, i.e. all data objects that are visible only within the owning process instance and, thus, are solely accessed by this process instance,[4] and
- additional *global data* (possibly kept separate from the process instances) which are visible and accessible by more than one process instance.

Thus, we can discriminate between objects that maintain local data only, and objects requiring some global visibility, i.e. that may be accessed by more than one process instance. Note that our definition of global information represents the minimal set of information for which some global visibility is required. In other words, it defines the minimal set of data that have to be accessible to more than one process instance. As we will see later, additional data objects may be defined as being global in order to improve the efficiency.

Additionally note that our definition of global information is independent of the ownership of the data. Thus, global data objects may be associated or owned by specific process instances. Or in other words, the data may be bound to the lifetime of a specific process instance. The most

[4]Note that this view differs from the view given in section 4.1.3.6 where the data objects associated with a process instance were described. This is because we talk about the implementation here, while in chapter 4 we were concerned with the language concepts.

important example of this is revealed variables, which are owned by a single process instance but accessed by several process instances.

5.2.2.1 Global Information

As noted above, some information beyond the single process instances is present in implementations of SDL descriptions. The additional information reflects all data that have to be accessible to more than one process instance. For this reason, we define these data as global information.[5]

There are several types of data objects belonging to this class. These are

- input queues holding signals and timeouts,
- the time, and
- data objects owned by some process instances but declared as being readable by other processes (revealed variables).

The type of access to the different data objects varies. For example, the input queue supports a put and a (sophisticated) get operation. The put operation may be executed by any process instance. Conversely, the get operation is executed by the receiving process instance that is associated with the queue only. The opposite is the case for the time variable, which is set by some time manager and which may, in principle, be read by any process instance.

Note that timers themselves are not considered global here. This is because they can in principle be maintained locally. By definition, a timer is set, reset and tested (active construct) by exactly one process instance only. However, a process instance is not necessarily able to test at regular periods of time whether a timer has expired. In this case, some timer manager has to be able to access active timers to ensure an efficient implementation.

5.2.2.2 Process Instances and Local Data

Important constituents of a process instance with respect to its implementation are its local data and the transitions of the EFSM with their respective trigger conditions.

Local data Each process instance maintains the following local data (see also figure 5.3):

- the explicit state of the EFSM it executes,
- the values of the data objects defined within the process instance (or at a lower level), excluding any data that require global visibility, and
- the data concerning timers of the process instance that are currently active.

Note that the actions performed by the EFSM may access global data as well.

[5]Note that this does not mean that there is physically shared memory accessible to the different EFSMs accessing these data. The definition only refers to the fact that there is some global visibility compared to the purely local data only accessible by the owning process instance.

Transitions and transition triggers of the EFSM Besides the state of the data of a process instance, the EFSM executed by a process instance defines

- a prioritized set of expressions that specify under which conditions to trigger a transition of the EFSM, and
- a transition associated with each of the trigger conditions.

The transitions executed by an EFSM may be triggered by various trigger conditions. We will return to the details in section 5.2.2.3.

Each transition comprises a set of actions that may

- consume some input, i.e. a signal or a timeout,
- produce some output or, more generally, execute some actions that may change the state of global data, i.e. changing a revealed variable, creating a new process instance, or simply sending a signal to another process instance, and
- modify the local data of the process instance, i.e. the explicit state and the state of local data objects.

Note that the first two classes of actions described involve access to data defined as global, while the last class refers to local data.

5.2.2.3 Trigger Conditions

In SDL, the trigger conditions of an EFSM may depend on the state of a large variety of data objects. Trigger conditions may depend on local information of the process instance itself as well as on global information accessible to a set of process instances.

Definitions Before going into the details of trigger conditions, let us define what exactly is meant by the term *trigger condition*.

> A trigger condition of an EFSM represents a precondition that, when evaluating to true, based on data accessible to the specific process instance, allows the process instance to execute an explicit transition.

Thus, a true trigger condition implies that the process instance is allowed to execute (fire) the explicit transition associated with the trigger condition.

Explicit transitions are defined as follows:

> Explicit transitions are transitions that are explicitly given in an SDL description (basic SDL). Explicit transitions include any additional transitions resulting from the transformation of full SDL, i.e. using extended language constructs, to basic SDL. Explicit transitions exclude any actions manipulating the input queue, i.e. saving or discarding signals.

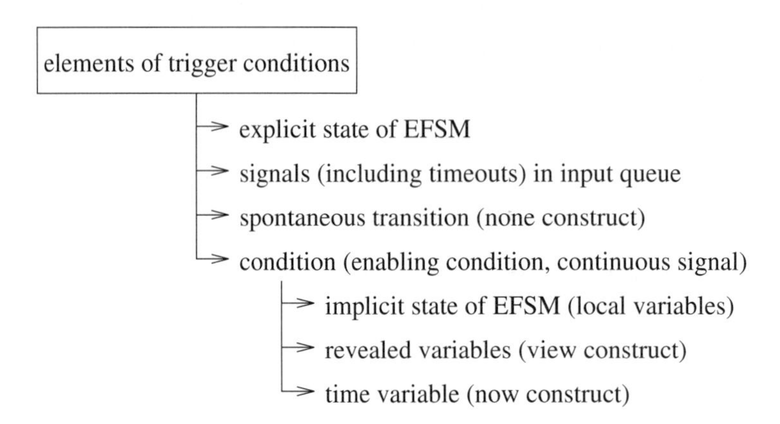

Figure 5.4 SDL elements relevant for transition triggering

The exact understanding of the trigger conditions of an EFSM executed by a process instance is especially important for an efficient implementation. For this, it is crucial to know

- which entities modify data that contribute to a trigger condition, i.e. which process or activity can set a trigger condition or contribute to this, and
- when exactly and under which circumstances this trigger condition is set.

Note that for an efficient implementation it is not sufficient to know that some data or variables which are part of an expression that constitute a trigger condition are changed. In other words, a modification of such a variable does not necessarily imply that the trigger condition is set. In principle, a variable can be changed to a value that does not result in a trigger condition to be evaluated to true.

The exact trigger conditions of an EFSM depend on the specific SDL description, i.e. the specific SDL process it implements. Thus, if this information is known and can be restricted by a specific SDL description or a specific set of SDL processes, the number of tests of a trigger condition can be reduced. As we will see later, this is a major source of improvement of the efficiency of the implementation.

As described in section 4.1.3.3, SDL supports rather complex and powerful trigger conditions. For example, the presence of a certain signal may be combined with some condition imposed on local or global (revealed) data. The possible SDL elements that may have an impact on the trigger condition of an EFSM are displayed in figure 5.4. Since spontaneous transitions are intended only for SDL descriptions used for simulation and verification purposes rather than for the generation of product code, they will not be considered any further in this context.

Possible elements of trigger conditions of an EFSM include local data of the process instance as well as some of the global information. The data elements that may constitute a (part of a) trigger condition comprise

Triggering element	Modifying action	Kind of trigger	Writing entities	Reading entities
explicit state	local transition	synchronous	single	single
implicit state	local transition	synchronous	single	single
signal (input queue)	local transition	synchronous	multiple	single
	nonlocal transition	asynchronous		
	environment	asynchronous		
timeout	time passage	asynchronous	single	single
spontaneous transition	time passage	asynchronous	single	single
revealed variable	nonlocal transition	asynchronous	single	multiple
time	time passage	asynchronous	single	multiple

Table 5.1 Triggering elements and their properties

- the explicit state (EFSM state) of the process instance (local data),
- the implicit state of the process instance as defined by the values of its local data objects,
- the state of its input queue (global data), i.e. the presence of signals or timeouts, and
- other global information such as the time or the state of variables revealed by other processes.

However, note that not all arbitrary combinations are allowed. For example, a trigger condition may not be the combined presence (and consumption) of a number of signals.

The important properties of the different *triggering elements* are summarized in table 5.1. In the table, the *modifying action* identifies the class of activity that may change the trigger condition. The activities that may have an impact on a triggering element are:

- *local transition*, i.e. the execution of a transition of the process instance to be triggered,
- *nonlocal transition*, i.e. the execution of a transition of some other process instance,
- *environment*, i.e. input received from the environment, and
- *time passage*, i.e. the advancement of the time.

The *kind of trigger* discriminates between synchronous and asynchronous as follows. *Synchronous* trigger elements are all elements which can be modified only by the process instance to be triggered. Thus, these elements are never changed while the process instance is waiting for a trigger condition to get true. An important consequence of this is that these elements have to be tested only after some advancement of the process instance waiting for the trigger has been achieved. In other words, the triggering element has to be retested only after the process instance has executed a transition.

Asynchronous trigger conditions are conditions that may be changed independent of the execution of the process instance waiting for the trigger condition to get true. Thus, these conditions have to be tested independently (or asynchronously) from progress of the process instance itself. In other words, the respective triggering element has to be retested independently of the process instance to be triggered. The most obvious example of asynchronous triggering elements are signals sent by other process instances.

Other examples of asynchronous triggering elements are conditions used by a continuous signal or in an enabling condition. This is the case when the view or the now construct is used within these conditions. Asynchronous triggering elements resulting from such conditions are called *asynchronous conditions* in the following.

In order to clarify the synchronization problem with the different triggering elements, table 5.1 also specifies the number of modifying entities that read and write the respective triggering elements. The number of readers and writers is an important aspect in implementing synchronization.

5.2.3 A First Glance at the Execution of SDL Systems

So far, we have omitted the discussion of controlling the execution of the process instances, and of the active entities in an implementation derived from SDL. Numerous alternatives exist, and we start here with a simple implementation alternative.

5.2.3.1 Process Instances

A basic approach is to consider the process instances as concurrently executing entities. Additional entities may be present to implement interfacing with the environment. All entities operate on their local data as well as on the common global data. This is graphically depicted in figure 5.5.

The structure of the code that implements a process instance is outlined in figure 5.6. Each of the process instances monitors the global data that are relevant to its trigger conditions (input queue, time, viewed variables) and periodically checks whether one of its trigger conditions holds. Note that monitoring of the local data is not needed as long as the process instance does not execute a transition. This is because these data are not changed unless the process instance executes a transition.

In case a trigger condition holds, the process instance selects and executes the respective transition. In case no trigger condition holds, the selected action depends highly on the selected implementation, i.e. whether polling or some kind of blocking is used. This relates to the approach employed to monitor the trigger conditions and will be discussed in more detail in section 5.4. In our simple example, we assume that the process instance suspends its execution to give other process instances a chance to execute.[6]

[6] Note that in order to ensure the correct insertion of timeouts in the input queues, each signal in the queue has to be associated with a time stamp. Thus, timeouts can be inserted in the queue according to the time of their expiration.

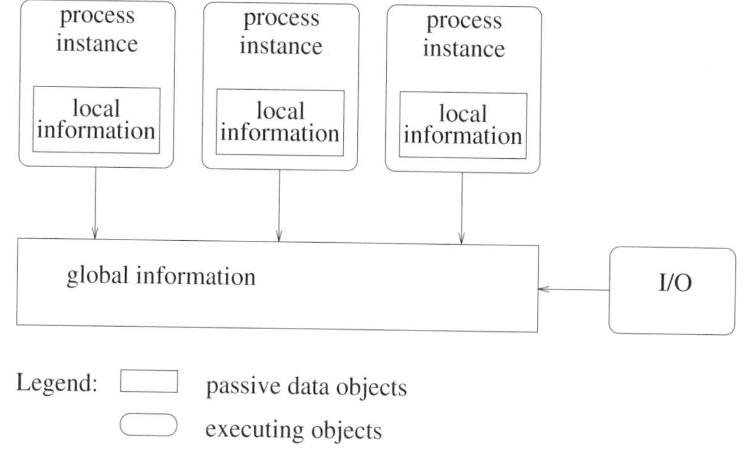

Figure 5.5 A simple approach to implementing SDL

```
(1) FOREVER
(2)   action:= test_trigger(PId)
(3)   IF action ≠ NULL
(4)     execute(PId,action)
(5)   suspend()
```

Figure 5.6 Outline of the code executed by a process instance upon activation (semi-active waiting)

5.2.3.2 Interfacing with the Environment

The implementation of the entities responsible for I/O could be similar. The respective entities periodically check their interfaces and process new input. Alternatively, the I/O entities may be triggered upon the arrival of external input, e.g. by an interrupt.

Note that there is no direct interaction between the different active entities, i.e. between the different process instances themselves and between the process instances and the entities handling I/O. For example, no synchronization is needed besides the one to ensure consistency of the global data. Any interaction between the entities is indirectly done via the global data.

5.2.3.3 Process Management

So far, we have not dealt with the question of how to exactly implement the executing entities, i.e. as active or passive instances, and how to schedule the different entities, i.e. in which order and under which circumstances to schedule the entities. In addition, we have not talked about the alternatives of interleaving of the processing of the different entities. In the following, we sketch a simple approach. Different alternatives will be discussed in section 5.6.

A simple approach to the scheduling of the entities is a nonpreemptive round robin strategy which schedules the entities (process instances and I/O) in a circular fashion. When an entity is scheduled, it checks its trigger conditions. If a trigger condition holds, the respective transition is executed. If not, the entity returns control to the scheduler. Due to the lack of true concurrency – or due to the absence from preemption – no precautions to preserve the consistency of global data are needed with this simple implementation.

5.2.3.4 Discussion

Due to the similarities of the semantics of this approach to the semantics of SDL, this simple implementation strategy can be derived from an SDL description in a straightforward way. In particular, there is no detailed analysis of the possible trigger conditions needed to derive an implementation according to this approach. In fact, the presented approach is a simple implementation of the server model introduced in section 3.3.2.1.

However, the strategy has a major drawback which is its limited efficiency. This is mainly due to the potential for repeated checking of the trigger conditions without identifying an executable transition. In order to alleviate the problem, we need to take a closer look at the potential trigger conditions present with the specific EFSMs. Thus, a goal of the optimization of the implementation is to minimize unsuccessful checks of trigger conditions. In other words, the process instance should be scheduled only when there is a clue that a trigger condition holds. The issue will be discussed in detail in section 5.4.

An additional disadvantage of this approach is that different active entities may not be assigned a preemptive priority. Thus, critical services may not be preferred over noncritical ones while these are executing a transition.

5.3 Layering, System Integration and Modularization

Before going into the details of implementing SDL, we take a look at important structuring principles that can be employed when implementing SDL descriptions.

5.3.1 Purpose of Layering

Layering as a principle to structure protocol architectures as well as its implementation have been introduced in section 3.3.1. In our context, layering is a principle that naturally follows from the usage of SDL (or a formal description technique in general). This is because SDL descriptions hide the functionality provided by the support functions described in section 3.2.3, i.e. SDL abstracts from the details of process management, timer management, process communication and buffer management.

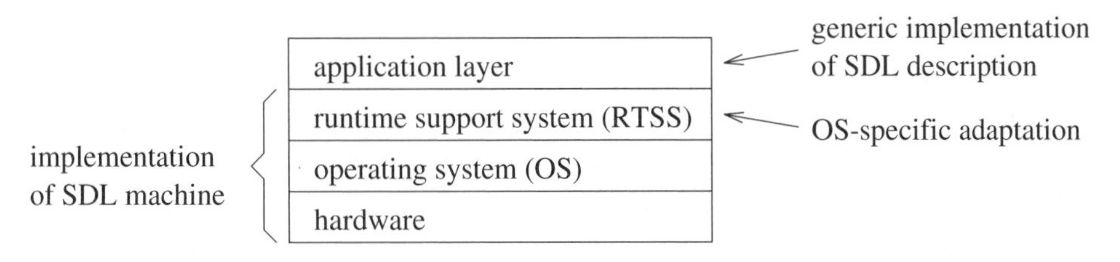

Figure 5.7 The layers of a typical implementation of an SDL description

Layering of the implementation is important in reducing the complexity of the implementation. Thus, layering supports correctness, ease of coding, readability and maintainability. The major drawback of layering is the extra overhead it adds to the implementation.

5.3.2 Layering Concept

Typically, a layered concept comprising the four layers graphically displayed in figure 5.7 is employed:

- the *application-specific part* derived from the given SDL description,
- the *RunTime Support System* (RTSS) providing an interface between the application-specific code and the target system, typically to the target operating system,
- the *target operating system* providing more generic services, and
- the *target hardware*.

Depending on the application with its specific requirements and constraints, some of the layers may be merged. An example is the merger of the RTSS with the operating system.

The application-specific part is derived by a code generator from the given SDL description. Typically, the generated code is independent of the underlying target system. Rather, the application-specific code relies on services provided by the RTSS to support process communication, as well as process and timer management.

The RTSS provides the adaptation of the SDL world to the underlying computing system. The purpose of the RTSS is to close the gap between the services required by the SDL system or its underlying semantics on one hand, and the services provided by the operating system on the other hand. Thus, in principle, the RTSS is independent of the application but specific to the underlying operating system.

5.3.3 System Integration Concepts

There is a large number of alternatives on how to map SDL descriptions on the target system. An important question relates to the specific functionality provided by the different layers. This is especially true for the services provided by the runtime support system and the operating system.

There are three main approaches to providing these services to the application-specific part. We call these approaches light integration, tight integration and bare integration.

The major differences between the integration approaches lie in the manner in which the services for the management of the process instances and the timers are implemented, and how the communication between the process instances is implemented.

5.3.3.1 Light Integration

With light integration, most of the services needed to execute an SDL description are implemented directly in the RTSS, rather than relying on services provided by the operating system. With light integration, communication between the process instances as well as the management of the process instances and timer management are implemented in the RTSS itself.

An example of a layered implementation according to the light integration principle is sketched in figure 5.8. With light integration, the complete SDL system is implemented by a single process of the operating system. This single process includes all process instances as well as the different support functions, e.g. management of process instances and timers, communication between process instances, etc.

A big advantage of light integration is that it requires only minimal support from the operating system. In addition, the presence of all process instances in a single common operating system process allows for direct access between the different SDL objects. Thus, the process instances may communicate via shared memory. Additionally, the approach grants the RTSS full control of the execution of the process instances, or the SDL system in general. This allows for the application of a set of optimizations not otherwise possible.

5.3.3.2 Tight Integration

With tight integration, the RTSS maps the service requests issued by the application-specific code as directly as possible on services provided by the target operating system. Thus, process communication, process management and timer management are almost completely handled by the operating system itself.

An example of a layered implementation according to the tight integration principle is outlined in figure 5.9. With tight integration, each process instance is implemented by a separate process of the operating system. Thus, the management of the process instances is mainly implemented by the operating system. A major drawback is the additional overhead introduced by the high number of operating system processes. This is especially true where traditional operating systems are employed.

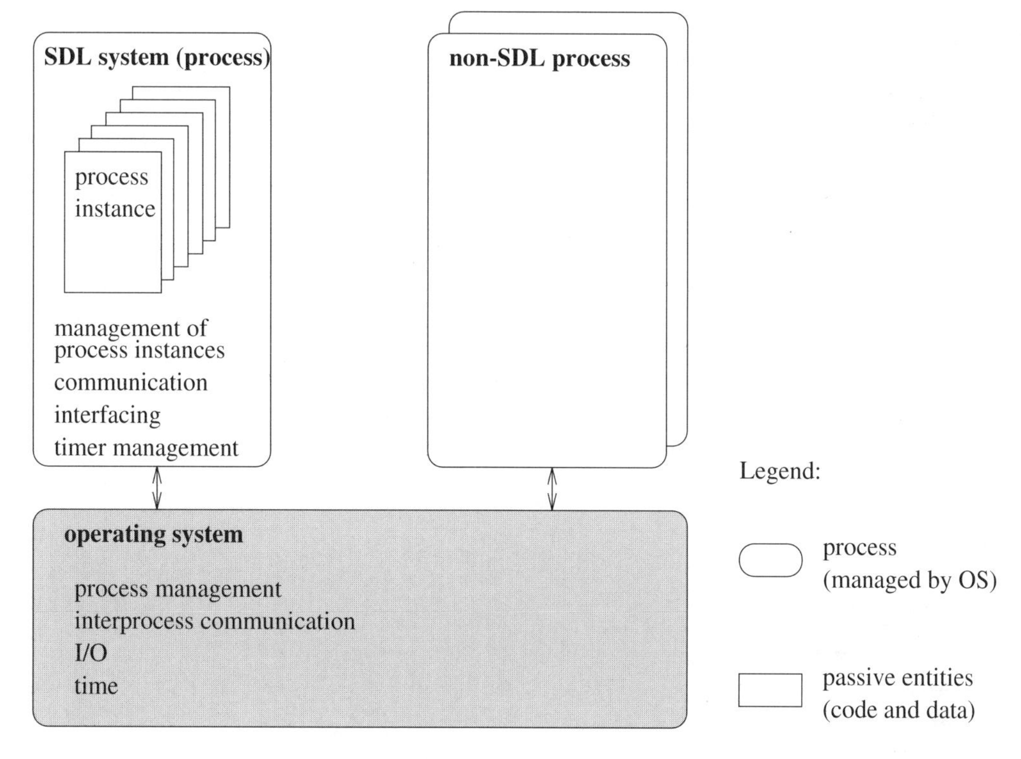

Figure 5.8 Light integration

There are important differences between tight integration approaches, depending on whether the operating system supports the use of shared memory between processes.

A major advantage of tight integration is that it is able to support preemption between processes. Thus, the fast handling of urgent activities can be supported.

5.3.3.3 Bare Integration

With the bare integration approach, no operating system at all is used. Instead, the RTSS is directly executed on the naked target hardware. This approach is often employed with microcontrollers where the SDL system is the only application run on the processor.

An example of a layered implementation according to the bare integration principle is sketched in figure 5.10. With bare integration, the SDL system is implemented by a main procedure executing in a loop until the system terminates. As in the light integration approach, the main procedure calls the procedures implementing the process instances.

As a result of this integration, the RTSS can rely on very limited functionality only, i.e. services directly provided by the hardware. In this respect, this approach resembles the light integra-

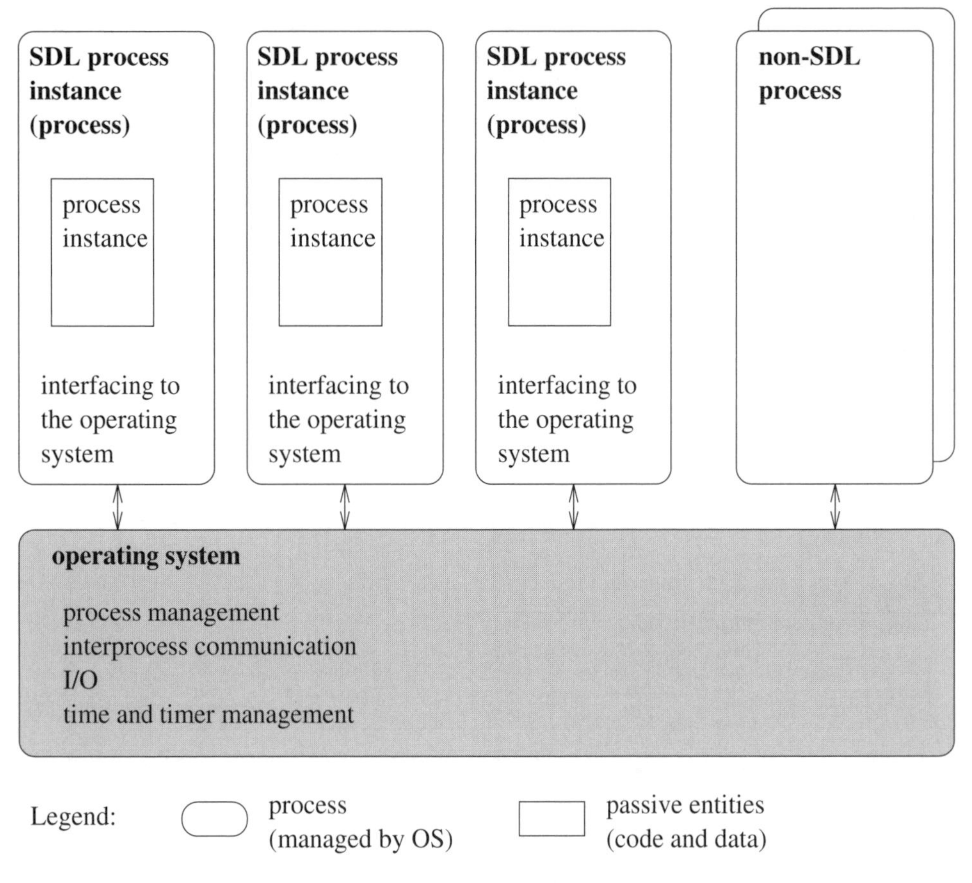

Figure 5.9 Tight integration

tion approach. A big advantage of bare integration is the high efficiency resulting from the elimination of the operating system. Another advantage is that with this approach, polling of some trigger conditions, i.e. external input or timeouts, does not pose a problem since the RTSS is the only process running on the system. Thus, there is no risk that busy wait reduces the processing resources available to other activities.

5.3.4 Functional Modularization

Hierarchical layering as described above supports structuring of the system such that the complexity of higher layers is reduced by basing its functionality on services provided by lower layers. In addition, hierarchical layering supports reuse. Unfortunately, in the context of SDL, the distribution of the functionality to the layers is an issue that depends highly on the circumstances

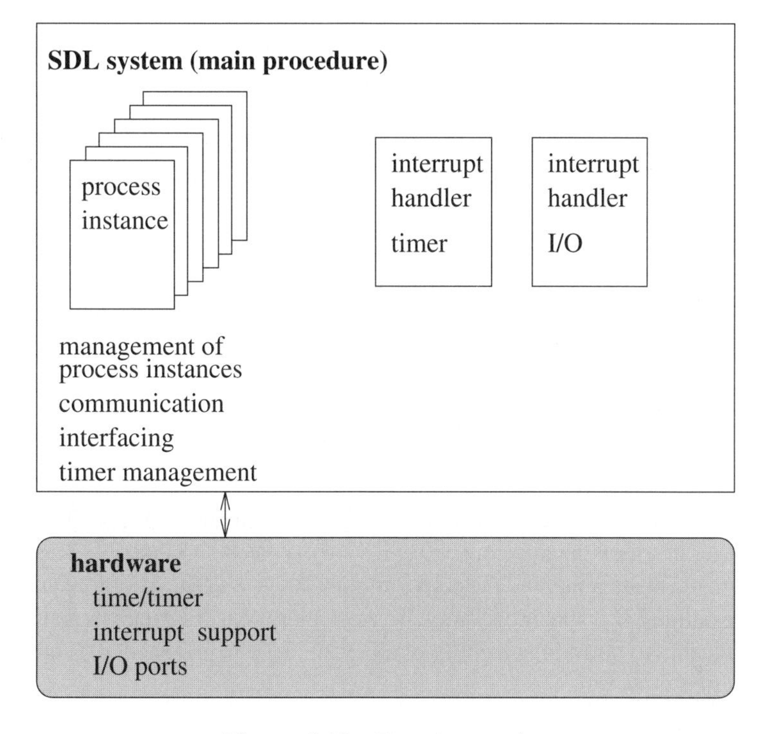

Figure 5.10 Bare integration

of the specific problem at hand. Thus, no single solution will suit all applications. In addition, many design decisions exist that are rather orthogonal to the mapping of the functionality on layers. In order to keep our discussion independent from these decisions, i.e. from the mapping of functionality on layers, we organize the discussion according to a set of functional modules.

Functional modularization aims at the structuring of the system in a set of building blocks where each building block or module implements a coherent set of functions. Modularization is more appropriate to structuring and describing the implementation of SDL in a general manner. In our discussion, the implementation is divided into the following five modules:

- The EFSM module contains the application-specific code specifically derived from the given SDL description.
- The PIM (Process Instance Management) module provides the routines needed to manage the process instances present in a running system, i.e. to create, schedule and terminate them.
- The CBM (Communication and Buffer Management) module is in charge of communication between process instances. Thus, it provides the routines needed to move signals in the system.

- The ENV (ENVironment) module provides the routines to communicate with the environment of the SDL system, i.e. provides the communication interface.

- The TM (Timer Management) module contains the routines needed for the management of SDL timers and of time in SDL in general.

A major part of the remainder of the chapter will be devoted to describing the functionality provided by the different modules, along with the respective design alternatives. In order to clearly identify the module to which a routine belongs, all routines visible outside the module are prefixed with the respective abbreviation, e.g. EFSM or PIM.

5.4 Process Models

A central issue of the implementation of SDL is the process model employed. The process model defines the manner in which the different activities are executed and interact. The process model is concerned with the management of the active entities in the system. It comprises the scheduling of the executing entities, and synchronization between them. An important topic in the context of SDL is the approaches to notifying an entity of an event, e.g. the presence of a signal. This forms the major focus of this section.

5.4.1 General Issues

Active entities in a system What the active entities in a system are is mainly a matter of the viewpoint on the system. For example, for an application program employing a user-thread package (or an SDL RTSS), the threads (or SDL process instances) are the active entities. From the viewpoint of the thread library itself (or the RTSS), each thread (process instance) is merely a procedure and a chunk of memory to implement a stack. From the view of the operating system each application program is represented by just a single process, rather than a set of threads.

States of active entities Independently of the precise definition of the active entities, each of these entities is in one of the following three states:

- *active*, i.e. currently executed,

- *ready*, i.e. ready for execution, or

- *blocked*, i.e. not ready for execution.

The three states are relevant to processes handled by the operating system itself as well as to entities managed by the RTSS.

States of SDL process instances Besides the state of the active entities in the system, each process instance in a running SDL system is in one of the following two states:

- *firable*, i.e. one of its trigger conditions holds, or
- *not firable*, i.e. none of its trigger conditions holds.

In order to distinguish these states from the explicit states of SDL processes, the states are called *trigger states* in the following. Note that the states of the active entities in the system and the trigger states of process instances are very different. The states of the active entities are an implementational issue while the two trigger states of process instances represent conceptual issues of SDL, independent of any implementation.

For example, the fact that an active entity is in the ready state does not necessarily mean that the process instance it implements is firable. In addition, there may be more than one process instance that is handled by a single active entity, e.g. with the light integration approach. Thus, it is important to keep the discussion of the trigger states of the process instances separate from the states of the active entities in the system.

Scheduling The task of the scheduler is to manage the states of the active entities. Scheduling is about the selection of one of the ready entities and its execution (active state) until some future event requires a change. Scheduling is related and highly influenced by synchronization. Synchronization constraints may cause an active entity to be blocked (i.e. moved from the active to the blocked state) or unblocked (i.e. moved from the blocked to the ready state).

Scheduling and synchronization have a major influence on the efficiency of the implementation. Thus, they will be a major focus of our discussion.

Typical schedulers employ a set of queues to manage their entities. There is at least one ready queue. There may also be a ready queue for each supported priority level. In addition, a set of queues that maintain blocked entities is typically employed. Usually, each of these queues handles a specific synchronization entity, e.g. implements a semaphore or a similar construct.

Scheduling in the presence of synchronization constraints Various concepts to implement synchronization exist. As an example consider semaphores. Typically, there is one blocked queue maintained for each semaphore. The blocked queue contains all processes that are currently blocked on the `DOWN(S)` operation. With the execution of an `UP(S)` operation, exactly one blocked process is removed from the blocked queue and inserted into the ready queue.

Note that the queuing strategy of the blocked queue is not necessarily FIFO. For example, unblocking could as well be according to the process priorities of the blocked processes, i.e. to select the process with the highest priority first.

Interleaving of executing entities Depending on the actual scheduling strategy, the execution of two entities managed by a specific scheduler may be fully interleaved at any arbitrary point or only at specific points of the code. In case preemptive scheduling is supported, full interleaving is possible. The same holds in systems where physical parallelism is supported, i.e. with multiprocessor systems.

Alternatively, interleaving may be limited to well-defined points. This is the case with user-thread libraries for single-processor systems which do not support preemption of another thread within the same process. A similar approach with limited and controlled interleaving is typically employed with runtime support systems that manage SDL process instances. With these systems, the state of the entity implementing a process instance, i.e. active, ready or blocked, is changed only after a transition has been completed and not during its execution.

The main difference between the interleaving modes relates to the extra measures necessary to ensure consistency of data where arbitrary interleaving is possible.

A main advantage of a high degree of interleaving (preemption) is that other activities which require immediate action may be scheduled immediately and independently of the state of the currently executing entity. This allows the system to quickly respond to external input, or to overload or other exceptional situation. However, the price for this is the need for extra measures to ensure consistency. Consistency may be ensured by using a lock when operating in a critical section, or, on a monoprocessor, by temporarily disabling interrupts.

Process management and layering In section 5.3, layering as a structuring principle for the implementation of SDL descriptions was introduced. As already noted in section 3.3.2, process management may be an issue in various layers of the implementation. In fact, with layered approaches, process management may be structured in a hierarchical fashion. With the implementation of SDL in a layered approach, the management of the executing entities, e.g. implementing process instances, may be distributed over more than one layer. An example of this is given in figure 5.11.

In this example, a set of SDL process instances is directly managed by the runtime support system. Thus, the RTSS implements scheduling, synchronization and communication for these process instances. In the system, the RTSS as such, together with the process instances it manages, is implemented as a process of the operating system. This process along with other processes run on the same hardware, i.e. representing other parts of the SDL description or not, are managed by the operating system. Thus, the operating system again contains a scheduler and supports mechanisms for synchronization and communication. This together builds a two-level hierarchical scheme.

5.4.2 Waiting for Events

In implementing process instances of an SDL description, different strategies to wait for events exist. The possible events in the view of SDL are the presence of a trigger condition of a process

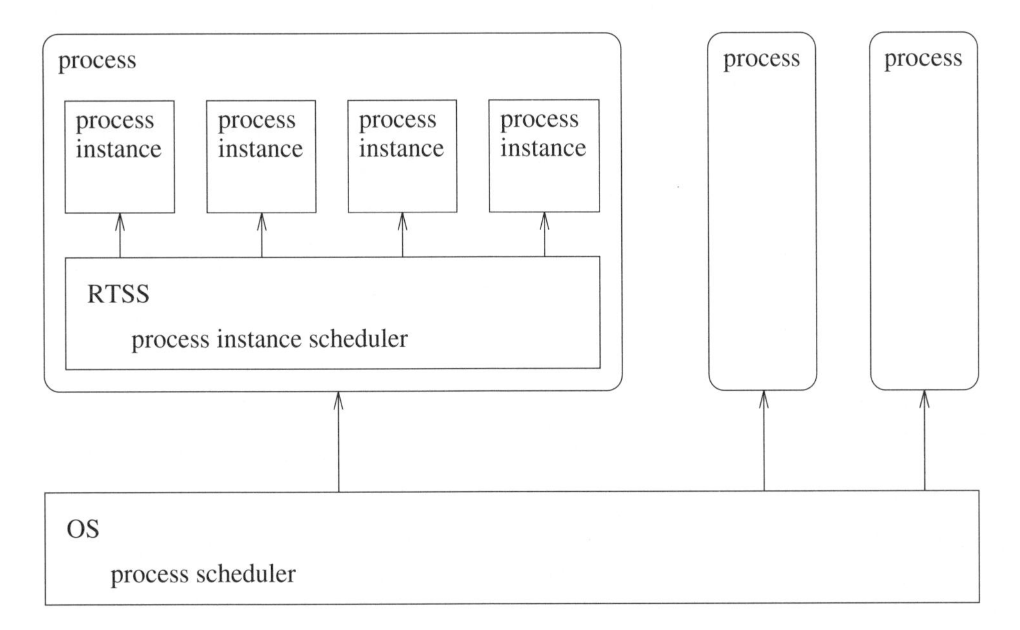

Figure 5.11 Hierarchical scheduling in a layered implementation of an SDL description (light integration)

instance, which allows the instance to execute a transition, or just the fact that there is a chance that a trigger condition holds, which has to be tested.

An active entity implementing one or more SDL process instances has to be enabled when one of the above conditions holds. Enabling gives the process instances (or any other entity that tests for triggers) a chance to perform the test and possibly to execute a transition.

There are different strategies for active entities to wait for such an event to happen. The principal alternatives are

- active waiting,
- semi-active waiting, and
- passive waiting.

We will discuss the three strategies in turn. For the discussion, we assume that each active entity, e.g. a process managed by the operating system, implements a single process instance of the SDL system. However, in principle the active entity may implement a set of process instances as well.

5.4.2.1 Active Waiting

Active waiting executes a loop subsequently testing the state of some data objects. Thus, it repeatedly polls the data objects. This is often also called busy waiting. The loop is terminated only

```
(1)  FOREVER
(2)     action:= NULL
(3)     WHILE action = NULL             /* loop until trigger condition holds */
(4)       action:=                      /* test trigger conditions */
              EFSMtest_trigger(PId)
(5)     EFSMexecute(PId,action)         /* execute transition */
```

Figure 5.12 Active waiting

when the values of the data objects satisfy some condition.

In the context of SDL, the data objects represent trigger conditions of the process instance. The application of active waiting by an entity implementing a process instance is outlined in figure 5.12. The access to the (polled) data objects is performed by the `EFSMtest_trigger` procedure which also implements the test of the trigger conditions of the given process instance. The `WHILE` loop terminates as soon as a trigger condition that evaluates to true is identified, i.e. when a firable transition is present.

The advantage of active waiting is its close accordance with the semantics of SDL. With this approach, a complete test of the trigger conditions of the respective process instance is performed independently of the kind of triggers employed, e.g. the presence of a signal or the evaluation of a continuous signal.

However, the approach has the drawback of consuming resources while waiting. Thus, other entities may be prevented from doing some useful work. As a consequence, active waiting is acceptable only if the waiting times are short or when there is no other work to be processed. Thus, the approach is not suitable for the implementation of a number of process instances each implemented by a different process. However, the approach may be acceptable if all process instances are implemented by a single process, e.g. with bare integration.

5.4.2.2 Semi-Active Waiting

Like active waiting, semi-active waiting also executes in some loop which subsequently tests the state of some data objects. However, unlike active waiting, the execution is suspended after an unsuccessful test of the data object. Thus, semi-active waiting gives other entities a chance to proceed with their processing activities.

The application of semi-active waiting by an entity that implements a process instance has been outlined in figure 5.6 already. Semi-active waiting relies on a facility that more or less periodically schedules the different entities to allow them to poll the data objects. For example, the active entity implementing the process instance should suspend itself in case none of its trigger conditions holds. This is needed to ensure that other process instances get a chance to tests their trigger conditions and to execute transitions.

```
(1) FOREVER
(2)   block(PId)
(3)   action:= EFSMtest_trigger(PId)  /* test trigger conditions */
(4)   IF action ≠ NULL
(5)     EFSMexecute(PId,action)       /* execute transition */
```

Figure 5.13 Passive waiting

5.4.2.3 Passive Waiting

The major goal of passive waiting is to eliminate unnecessary and time-consuming tests of some data objects. The basic idea is as follows: an entity is enabled, i.e. put into the ready queue only if there is work to do, otherwise it is in the blocked state. Thus, the entity relies on some other entity to enable it.

In the context of SDL, an active entity implementing a process instance is enabled if there is an executable (firable) transition or alternatively if there are changes to the data objects relevant to its trigger conditions. The structure of the respective code as executed by an entity implementing a process instance is outlined in figure 5.13. The entity blocks on some synchronization construct, e.g. a semaphore, until explicitly woken up by another entity. Upon wake-up, the entity tests its trigger conditions. As we will discuss in detail below, unblocking of the entity does not necessarily imply that a trigger condition holds.

5.4.3 Trigger Testing Strategies

In the section above, we have described possible mechanisms employed by active entities to wait for an event, e.g. for the presence of a trigger condition, or just the modification of a data object relevant for a trigger condition. Here, we deal with the question of where to actually test the trigger conditions.

There are two principal alternatives to test the trigger conditions of a process instance. These are *consumer-based testing*, which associates the test of a trigger condition with the transition it triggers, and *producer-based testing*, which associates the test of trigger conditions with the actions that modify the objects that have an impact on trigger conditions.

5.4.3.1 Consumer-Based Testing

With the consumer-based approach, testing the trigger conditions and executing the respective transition are done by the same entity. Thus, the consumer of a trigger condition, i.e. the process instance that executes a transition, also implements the preceding test of the trigger conditions to find the executable transition. In other words, the consumer tests if a transition is firable.

This is the most obvious implementation of SDL descriptions. The approach has been applied with the synchronization strategies given in the examples above, i.e. in figures 5.12 and 5.13. It supports testing and execution of a transition in the very same entity. Thus, the process instance which tests its trigger conditions does not need – with some exceptions – access to data of other process instances.

5.4.3.2 Producer-Based Testing

With producer-based testing, the entity that sets a potential trigger condition is also in charge of performing the test to exactly determine if the trigger condition really holds. Thus, producer-based testing of trigger conditions employs the test in the same entity in which the trigger condition is set, i.e. by the entity that changes the trigger condition rather than the instance that is actually triggered. In other words, the producer, i.e. the entity that changes the objects relevant for triggering, tests if this change results in one or more firable transitions in the system.

For example, with signal communication, the test whether a transmitted signal triggers the receiving process instance is implemented within the sending process instance rather than the receiver.

The testing entities (producers) may be

- a process instance, e.g. when outputting a signal or modifying a revealed variable,
- the timer manager, i.e. when a timer expires, or
- the entity handling external interfaces, i.e. when an external input arrives.

In addition, the process instance executing a transition is testing its own trigger conditions after completion of the transition. This ensures that the process instance remains enabled in the case that a trigger condition still holds.

With producer-based testing, the different trigger conditions of a process instance are tested in a distributed manner by different entities.

Depending on the kind of trigger conditions to be tested and the set of trigger conditions used by the triggered process instance, the test of the trigger may range from a simple to a rather complex matter. For example, consider a simple SDL process which is triggered solely by signals and where any signal triggers an explicit transition. In other words, any signal is consumed in any state, and no save constructs are used. In this case, any insertion of a signal in the input queue of the receiving process instance triggers the execution of the instance. Thus, the event of inserting a signal in the input queue may be used to enable the receiving process instance. As a result, no test of a trigger condition is needed at all.

Note that the implementation of the test for triggers and its relation to scheduling influences the efficiency and responsiveness of the system. As we will see below, producer-based testing allows scheduling (or enabling) of the process instances to be tightly coupled with the presence of a firable transition.

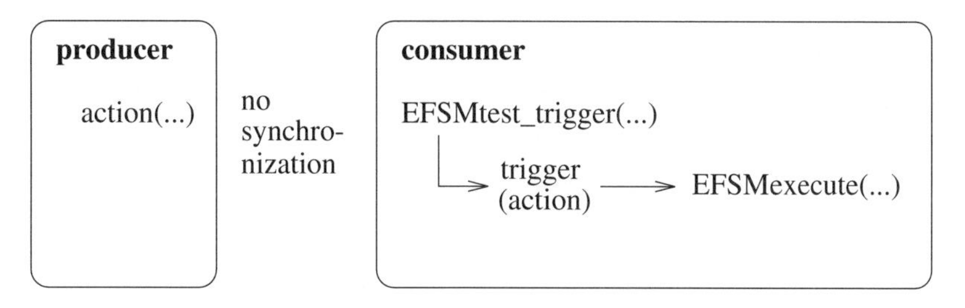

Figure 5.14 Trigger-independent enabling

5.4.4 Combining Trigger Testing with Enabling Strategies

Enabling and the test for trigger conditions are highly related issues. Enabling defines the action necessary to move an entity, e.g. representing a process instance, from the blocked to the ready state. Enabling and scheduling of an entity are needed to give the process instance it implements a chance to execute a transition. Depending on the strategy for the test of trigger conditions, enabling may also be needed to allow a process instance to test if one of its trigger conditions holds.

Concerning the interrelation between the enabling of a process instance and the presence of a trigger condition of the process instance, there are the following three alternatives:

- *trigger-independent enabling* enables the entity implementing a process instance independently of the presence of a trigger condition,
- *clue-based enabling* enables an entity when there is a clue that the process instance it implements may be firable,
- *trigger-based enabling* enables the entity only if the process instance it implements is definitely firable.

We will discuss the three enabling strategies in turn.

5.4.4.1 Trigger-Independent Enabling

Trigger-independent enabling combines active- or semi-active waiting with consumer-based testing. With trigger-independent enabling, the process instance is always enabled or is enabled within regular time periods. In the first case, the implementing entity is always in the active or ready state, independently of the presence of triggers. In the second case, the process instance is enabled by some other enabling entity independently of the presence of trigger conditions. In either case, some kind of polling is employed. The approach is outlined in figure 5.14.

The disadvantage of the approach is the potential risk of performing numerous tests of the trigger conditions before finding a trigger condition that allows a transition to be executed, or in other words, the risk of actively testing the trigger conditions without having a clue whether a

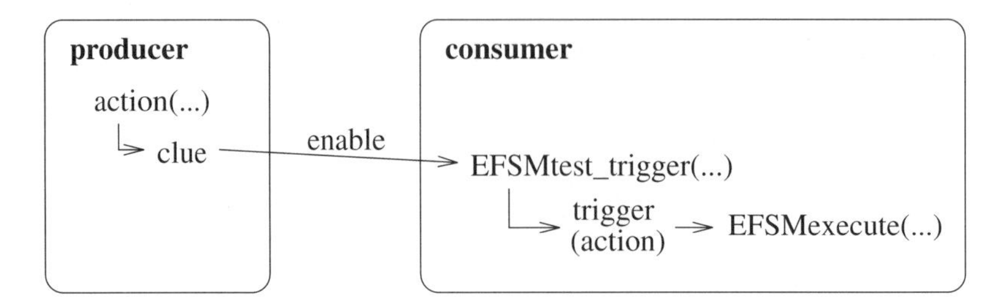

Figure 5.15 Clue-based enabling

trigger condition evaluates to true. In the worst case, $n - 1$ process instances of an SDL system consisting of n process instances have to be tested for the presence of a trigger condition before the trigger condition in the nth process instance is found. This is very inefficient and may require many tests until an executable transition is found. Especially when many process instances are present in a running system, this may resemble the search for a needle in a haystack. In addition, depending on the implementation of the process instances, a lot of overhead for context switches may be involved.

However, the appropriateness of the approach depends on the specific SDL description and the context in which it is used. Semi-active waiting may be useful where the number of unsuccessful tests of the trigger conditions is approaching zero. Another potential application area is systems where the cost of executing transitions is far above the cost of testing the triggers.

As noted before, active waiting is applicable when the system does not have any other useful work to do. Under these circumstances, active waiting does not prevent other entities from doing useful work. This may be the case with dedicated systems or with direct hardware implementations.

It is important to note is that with this approach, the process instances do not require access to data local to other process instances. Or in other words, the process instance that executes an action that results in setting a trigger of another process instance does not need to access any data of the triggered process instance. Thus, the approach requires minimal interaction and minimal sharing of data.

5.4.4.2 Clue-Based Enabling

Clue-based enabling combines passive waiting with consumer-based testing. A process instance is enabled only (i.e. moved to the ready state) when there is a clue that one of its trigger conditions holds. Or in other words, the triggered process instance relies on another entity to provide the clue. This moves the responsibility for enabling process instances to the entities that are responsible for setting the trigger, e.g. the sender of a signal, etc. The approach is displayed in figure 5.15.

Note that the overhead for testing of the trigger conditions may be minimized if the testing process instance has a knowledge of the clue responsible for enabling it.

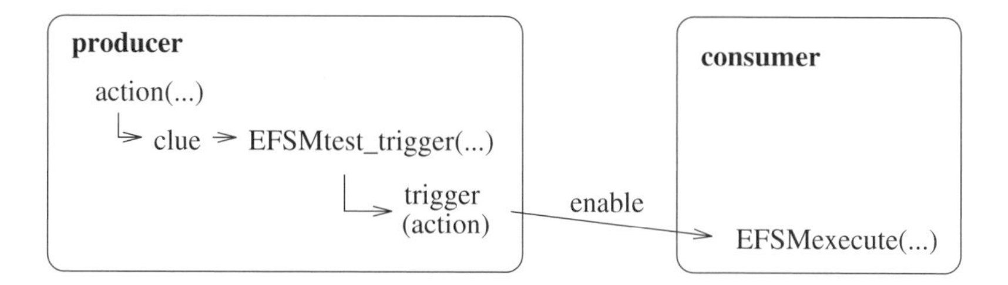

Figure 5.16 Trigger-based enabling

Like the above approach, the triggering process does not necessarily require access to the data of the process instances it triggers.

5.4.4.3 Trigger-Based Enabling

Trigger-based enabling combines passive waiting with producer-based testing. With trigger-based enabling, a process instance is enabled only if a trigger condition holds. Thus, the ready state of the entity is equivalent to the firable state of the respective process instance. This approach is graphically displayed in figure 5.16. Unlike clue-based triggering, no test of the trigger condition by the triggered process instance is needed. Instead, this is fully done within the triggering process instance.

An advantage of this is that the process instance setting the triggering condition knows exactly what to test. Thus, only a subset of the potential trigger conditions of the consumer have to be tested. However, testing the trigger conditions of another process instance requires full access to its local data, e.g. its state (explicit and implicit), and to its input queue. Thus, memory sharing between the process instances is needed, except for very simple SDL descriptions.

This is an extreme view and not easy to implement in the general case. Before we discuss some details of the approach below, let's review the potential triggering elements of a process instance described in section 5.2.2.3. The problem is the diverse trigger conditions supported by SDL.

In order to understand the problem, let's revisit the actions performed by the producer of a trigger. The important questions are when and by which action can a process instance be triggered. Thus, the focus of our attention is not the triggered process instance but the actions that may produce a trigger. The alternative triggering actions may be

- a *process instance*, i.e. by sending a signal or by changing the value of a revealed variable viewed as part of a continuous signal or enabling condition,
- the *executing process instance itself*, i.e. entering the next EFSM state for which an input signal is present in its input queue or for which an enabling condition or a continuous signal holds,

- the *timer manager* by issuing a timeout, or by simply advancing the time in case the time is part of a continuous signal or an enabling condition, and
- the *environment*, i.e. by sending a signal.

In particular, the trigger conditions that may arise due to the use of the continuous signal construct, i.e. the modification of a viewed variable or the access of the time in this construct, are problematic. This is due to the fact that exactly checking whether trigger conditions hold requires access to the data segment of the process instance to be triggered. In other words, it is much easier to provide a clue each time a trigger condition of a process instance may have arisen rather than to provide a definite statement. Obviously, an implementation of an exact test is difficult where the process instances do not have access to a shared memory area.

A related issue is that the trigger condition may no longer hold when the process instance is finally scheduled. This is because the value of the non-local variable viewed by the process instance may be changed again by the revealing process instance before the execution of the viewing process instance is commenced.

In case a given SDL description uses signals as the sole triggering mechanism, i.e. there is no revealed variable or access to the time as part of a continuous signal or an enabling condition, and if timers are handled asynchronously outside the process instances, a pure implementation of trigger-based enabling is possible. When the above restrictions to the SDL description and the implementation strategy apply, each process instance may block on a binary semaphore associated with it. The semaphore blocking the receiving process instance may be unblocked by any entity that executes some action that results in a trigger for the blocked process instance. Examples are process instances that send a signal to the blocked process instance, an interface process that sends a signal, or the timer manager that inserts a timeout signal into the queue. Instead of a binary semaphore, other synchronization constructs can be employed.

5.4.4.4 Combining Different Enabling Strategies

Trigger-based enabling may not be appropriate in all cases. The same may be true for clue-based enabling. On the other hand, trigger-based or at least clue-based enabling should be applied wherever possible to minimize overhead. In principle, the different enabling strategies can be combined as well. Thus, different strategies can be applied to handle different kinds of triggers in a system.

The trick with combining enabling strategies is to minimize the cases where

- a process instance is enabled, and
- a final analysis of the trigger conditions by the enabled process instance shows that the process instance is not firable.

In other words, the trick is to move towards the trigger-based enabling strategy as much as possible.

Routine	Function	Calling entity	Usage
EFSMinitialize_PI(PId)	initialize a PI	PIM	SDL create construct
action:= EFSMtest_trigger(PId)	test trigger conditions of PI	PIM EFSM	main loop EFSMexecute() routine
EFSMexecute(PId,action)	execute a transition of PI	PIM	main loop

Table 5.2 Routines provided by the EFSM module

Due to the restrictions imposed by the application of active and semi-active waiting, polling should be implemented in a centralized manner. In other words, the polling of all triggers for which trigger-based or clue-based enabling cannot be implemented should be done in a central place, i.e. in a single loop. Crucial for the performance is how often a polling cycle is executed. Alternatives range from a polling cycle after the execution of each transition in the system to a polling cycle only when there is no other executable transition in the system. The second case clearly minimizes overhead. Even though the second approach also conforms with the semantics of SDL, it may not necessarily be what the user requires concerning non-functional requirements. One of the drawbacks of the second approach is a delayed issuing of timeouts.

5.5 Extended Finite State Machine

The extended finite state machines implementing the SDL processes are the basic elements in an implementation derived from SDL. Each SDL process represents an EFSM. The EFSMs are implemented in the EFSM module. Here we deal with the issue of implementing the transitions of the EFSM including the test of their trigger conditions. Everything that goes beyond this, i.e. the usage of timers, input and output of signals and other communication mechanisms, is deferred to the underlying layers. These additional services are provided by the other modules of the implementation and are made available to the application-specific code by means of procedure calls to the RTSS.

As pointed out before, there are two major activities of the EFSM (or a process instance that executes the EFSM). These are the test of the trigger conditions of the process instance, and the execution of the transitions.

5.5.1 Functionality and Routines

The basic routines provided by the EFSM module are summarized in table 5.2. The abbreviation PI stands for Process Instance.

Typically, the three routines are provided for each SDL process. However, for the purpose of simplicity of our discussion, we start with the assumption that a single instance of each of the three routines exists. Thus, each routine represents a generic implementation of a specific task, e.g. testing triggers or executing a transition, for all SDL processes present in the system. We will come back to this difference as we go into the details of the implementation of the routines.

The functionality of the three procedures is as follows.

Initialization of process instances The `EFSMinitialize_PI(PId)` procedure initializes a newly created process instance. The routine is called by the `PIMcreate_PI` procedure during the creation of a new process instance. The procedure is needed to initialize the data structures maintained by the new process instance.

Testing triggers The `EFSMtest_trigger(PId)` routine provides the test of the trigger conditions of a process instance, i.e. tests if the instance is firable. If a trigger condition holds, the procedure returns a reference to the respective action to be executed. If no transition can be executed, it returns the `NULL` value.

Note that the procedure actively manipulates the input queue. For example, signals that are neither consumed nor saved in the current state of the process instance can be discarded right away. However, the procedure does not remove the signal from the input queue which is consumed by the specified `action`.

Executing a transition The `EFSMexecute(PId,action)` procedure executes a transition that has been selected by the `EFSMtest_trigger` procedure. This includes the removal of the triggering input signal from the input queue, where applicable.

Interpretation versus code generation As described in the following, the procedures are generic, i.e. are able to execute arbitrary SDL processes. As a result of this, the actual EFSMs that are executed are specified by data rather than by code. Thus, we assume a kind of interpretation of the EFSMs rather than implementing each EFSM by a separate piece of code operating on the instance-specific data.

The purpose of this kind of description is mainly for reasons of understandability. It allows the description to be generic and independent of a specific SDL example process. The interpretation-based approach covers all possible cases that may be encountered with an SDL description (basic SDL) using a single implementation. Conversely, with the code-base implementation, the implementation of each SDL process is different.

Even though the interpretation-based approach is a legal way of implementing SDL, this is not recommended for reasons of efficiency. In practice, a code generator typically derives a specific set of procedures for each SDL process. For example, for the SDL process named `p_type`, a routine `EFSMexecute_`*`p_type`*`(PId,action)` is derived. Thus, only code segments that do apply to the specific SDL process are generated.

Deriving the specific procedures for each given SDL process from the generic (interpretation-based) code discussed here is straightforward. For this, the information currently stored as data has to be transformed to code within the procedures.

5.5.2 Implementation of Transitions

The execution of transitions is implemented by the `EFSMexecute` routine(s). Typically, a routine implements all transitions of an SDL process. Alternatively, a routine may be provided for each transition of an SDL process or for parts of a transition.

Note that the usage of the term 'transition' in our context is different from its usage in automata theory. In automata theory, a transition is defined by two states, i.e. the current state of the automaton and a unique successor state. Conversely, in SDL a branch within a transition may occur, e.g. based on the implicit state of the EFSM. Thus, the successor state is decided while the transition is executed, not before. In our context, we define a transition as the set of actions associated with a trigger condition, rather than associated with two states. Thus, in our case, a single transition may result in different successor states.

As outlined in section 5.2.2.2 already, each transition comprises a set of actions, possibly including the consumption of some input, the generation of some output and the modification of the local data of the process instances, i.e. the explicit state and the state of local data objects. In the following, we focus on the implementation of actions that require interaction with other modules of the implementation.

Communication by signals is implemented by issuing calls to the CBM module (see section 5.7 for a detailed description of the procedures).

Signal input Input of a signal is implemented by a set of operations including a call of the `CBMremove_signal(PId.queue,signal)` procedure to remove the specified signal from the input queue, followed by the `copy_signal_`*`s_type`* operation to copy the signal parameters and related information into the local variables maintained by the process instance. In addition, the `CBMfree(signal)` has to be issued to free the signal buffer for further usage.

Input from the environment and input of timeouts can be implemented the same way.

Signal output Output of a signal is implemented by the following sequence of actions. First, a signal buffer for the signal is acquired issuing a call to the `CBMallocate(size)` procedure. Next, the buffer is filled with the signal type, the signal parameters and related information by the `fill_`*`s_type`* operation. If the PId of the receiver is not known, it is derived by issuing a call to the `PIMget_address` procedure. Then the signal buffer containing the signal is inserted into the input queue of the receiving process instance by calling the `CBMinsert_signal(PId.queue,signal)` procedure.

Depending on the strategy for trigger testing and enabling (see section 5.4.4), additional operations may be performed to enable the receiver of the signal. With clue-based enabling, a call of the `PIMenable(PId)` procedure provided by the PIM module is used to enable the receiving process instance to test its trigger conditions. If it is already enabled, the call has no effect.

With trigger-independent enabling, no further action is needed. This is because the approach relies on the receiver to independently test its trigger conditions.

With trigger-based enabling, the receiver may be enabled only if it is able to execute a transition. Thus, a test of the trigger condition of the receiving process instance is employed by issuing a call to a specialized version of the `EFSMtest_trigger` procedure. The procedure enables the receiver only if it is able to execute a transition. If the receiver is already enabled, testing can be further simplified. However, note that in general this requires full access to the data of the receiving process instance.

Modification of revealed variables When revealed variables are modified, this may represent the setting of a trigger condition of a process instance viewing the variable (asynchronous condition). Depending on the enabling strategy employed to notify the triggered process instance, i.e. trigger-independent, clue-based or trigger-based enabling, different actions similar to the output of a signal are needed.

A prerequisite for clue-based and trigger-based enabling is the maintenance of a list for each revealed variable used in an enabling condition or a continuous signal. The list for a specific revealed variable refers to all process instances that are viewing the revealed variables in an enabling condition or a continuous signal.

With clue-based enabling, enabling of viewing process instances is solely based on this list. In other words, each process instance contained in the list for the respective revealed variable is enabled. With trigger-based enabling, an additional test of the trigger conditions is performed for each of the process instances in the list. Only the process instances that are detected to be firable are enabled.

Which of the given strategies performs best depends on the usage of revealed variables.

Process creation and termination Creating a new process instance is implemented by calling the `PIMcreate_PI(p_type)` procedure provided by the PIM module (see section 5.6), followed by a set of actions similar to the output of a signal. This allows parameters to be passed to the newly created process instance and to enable its execution.

Process termination, i.e. the execution of the stop construct, is implemented by a call of the `PIMstop_PI(PId)` procedure also provided by the PIM module.

Timers The manipulation of timers is directly implemented by the respective routines provided by the TM module. The routines to set, reset and test a timer are described in section 5.9.

5.5.3 Implementation of Trigger Testing

As discussed in section 5.2.2.3, identifying trigger conditions is rather complex for the general case. This is due to the peculiarities of SDL which does not allow a simple mapping of the signal name and the current state of the process instance on a transition to be executed.

For the reasons given above, we start with the description of an interpretation-based implementation of the `EFSMtest_trigger` procedure. Based on this, we discuss code-based implementations of the procedure.

5.5.3.1 Interpretation-Based Implementation

An outline of a possible implementation of the EFSMtest_trigger(PId) procedure, to test if a trigger condition holds, is given in figure 5.17. The procedure is intended to be used with the consumer-based approach to trigger testing as described in section 5.4.3. Note that the procedure is optimized for readability rather than for efficiency. Optimizations will be discussed later.

```
(1)  action:= NULL
(2)  signal:= CBMfirst_signal(PId.queue)     /* get first element */

     /* scan input queue for trigger condition */
(3)  WHILE signal ≠ NULL
(4)    IF signal ∈ regular_input(PId.state)
(5)      action:= get_action(PId.state,signal)
(6)      RETURN(action)
(7)    IF signal ∈ conditional_input(PId.state)
(8)      IF cond:= enabling_condition(PId.state,signal)
                        /* an enabling condition holds */
(9)        action:= get_action(PId.state,signal,cond)
(10)       RETURN(action)
(11)     ELSE   /* implicit save */
(12)       succ_signal:= CBMnext_signal(PId.queue,signal)
(13)   IF signal ∈ discard_input(PId.state)    /* discard signal */
(14)     succ_signal:= CBMremove_signal(PId.queue,signal)
             /* remove signal from input queue and retrieve succeeding signal */
(15)     CBMfree(signal)    /* free signal buffer */
(16)   ELSE   /* explicit save */
(17)     succ_signal:= CBMnext_signal(PId.queue,signal)
(18)   signal:= succ_signal    /* proceed with successor */

     /* scan continuous signals for trigger condition */
(19) cond:= first_condition(PId.state)
(20) WHILE cond ≠ NULL    /* scan input queue */
(21)   IF continuous_signal(PId.state,cond)    /* condition is true */
(22)     action:= get_action(PId.state,cond)
(23)     RETURN(action)    /* continuous signal identified */
(24)   ELSE   /* condition is false */
(25)     cond:= next_condition(PId.state,cond)
(26) RETURN(NULL)    /* no trigger condition identified */
```

Figure 5.17 Outline of the EFSMtest_trigger procedure

The `EFSMtest_trigger` routine works as follows. Lines (3) to (18) operate on the input queue to identify a trigger condition related to an input signal. If no trigger associated with an input signal can be found, continuous signals are processed in lines (19) to (26).

The operation of the input queue works as follows. In each iteration of the first `WHILE` loop, a signal from the input queue is inspected. Note that for this, the signal remains in the input queue. When a trigger condition is identified, the `EFSMtest_trigger` procedure terminates, returning a reference to the executable action to the caller. If a signal does not result in a trigger condition, the next signal from the queue is inspected.

For the test of the trigger conditions, the set of input signals of the SDL process is partitioned in three subsets:

- `regular_input(PId.state)` (line (4)) contains all signals which unconditionally prompt a transition in the given state,
- `conditional_input(PId.state)` (line (7)) contains all signals used within an enabling condition, i.e. which require that an additional conditional expression evaluates to true to prompt a transition in the given state,
- `discard_input(PId.state)` (line (13)) contains all signals not specified in the given state, i.e. neither consumed nor saved.

Depending on the set to which the given signal belongs, different tests are performed. Note that the set to which a specific signal belongs to may depend on the current (explicit) state of the process instance.

Lines (4) to (6) handle signals that unconditionally trigger a transition. In case the signal is part of the `regular_input` set of the state, the respective transition is selected and returned to the caller. If this is not the case, enabling conditions are processed in lines (7) to (12). If the signal is combined with enabling conditions (`conditional_input(PId.state)` set), the respective conditions are tested. If an enabling condition evaluates to true, the trigger condition holds and the respective transition is returned. If not, the signal is kept in the queue, i.e. saved, and the next signal is selected for testing.

If none of the above is the case, the signal is either discarded or saved (lines (13) to (17)). If the signal is part of the `discard_input` set, the signal is removed from the input queue and the signal buffer is freed. If not, the signal is saved, i.e. remains in the input queue.

If no signal that results in a trigger condition is present in the input queue, continuous signals are tested in lines (19) to (26). This is done in the `WHILE` loop, which processes the continuous signals in the order of their priority.

Table-based approach In section 3.2.2.2, the table-based implementation of communication protocols was introduced. The table-based approach can be considered as a special case of the interpretation-based code. With the table-based approach, each tuple consisting of an input signal and a state of the EFSM is mapped on an element of a two-dimensional array. The array contains references to the actions to be performed for the different tuples, i.e. the different combinations of input signal and state.

The table-based approach has been proposed for the implementation of simplified SDL descriptions. The approach has been quoted as being more time-efficient than code-based implementations [MaRa99]. However, with highly optimizing compilers, the code-based approach may be preferable due to the fact that these compilers automatically transform code with nested `IF` or `CASE` statements to a table in assembler where appropriate. A more detailed discussion of the application of the table-based approach to a limited subset of SDL can be found in [BrHa93]. The table-based approach is applied by the SDT Cmicro code generator [Tele98].

5.5.3.2 Handling Priority Input

Priority input in Z.100 Priority input is not part of basic SDL. Instead, Z.100 defines priority inputs by providing the transformation rules for the transformation of priority inputs to basic SDL. Even though this is a principal solution, it is not very efficient. This is because the transformation is based on the introduction of an additional signal. The additional signal is output by the process instance to itself whenever the process instance enters a state where a priority input is defined. If services are used, Z.100 introduces a new signal for each service.

In addition, the definition of priority input in Z.100 may not be necessarily what the user expects. The transformation results in the following behavior. If a priority input is present in the input queue when the process instance enters the respective state, it is processed. If regular input is present only, this is processed too. However, if the queue is empty, the transformation rules force the process to wait for any triggers specified in the state. If several input signals, including priority input signals, arrive later in this state, processing of the signals is based on their order of arrival rather than their priority.[7] Under the assumption that a trigger causes an intermediate firing of the respective transition, the transformation given in Z.100 works as expected. However, for practical cases, the immediate firing cannot be assumed. Thus, a different behavior may result. For this reason, the implementation of priority input based on the transformation rules is not advisable.

Direct implementation of priority input For the reasons given above, we propose a direct implementation of priority inputs rather than applying the transformation rules. To start with, we consider a peculiarity of the priority input concept supported by SDL. In SDL, whether an input is handled with priority depends on the state of the EFSM. For example, a signal may be defined in a priority input in one state, handled as a regular signal in another state, and possibly discarded in a third state. This complicates the implementation of priority inputs.

Separate queues for priority inputs If the usage of priority input is uniform over all states of the EFSM, a much simpler implementation is possible. Or in other words, a signal defined

[7] Note that our interpretation of the Z.100 standard is based on a revised version of the standard. The transformation rules as defined by SDL 96 are even more imprecise.

by a priority input in one state is defined as priority input in all other states, too. In this case, a solution based on two input queues, one for priority signals, the other for regular signals, can be employed. This is because in this case, the decision whether a signal is consumed with priority is independent of the state of the receiver. Thus, a signal in the new priority input queue will remain in this queue until consumed. Especially, there is no need to move a signal to the regular queue, and vice versa, when the process changes its state. This implementation allows for a fast access to priority signals without the need to scan the other inputs present.

Implementation of priority input in the general case The implementation of priority inputs for the general case depends on the strategy for trigger testing (see section 5.4.3). With consumer-based testing strategies, the `EFSMtest_trigger` procedure can be extended to employ a scan of the input queue for priority input signals before processing with normal operation. This can be implemented by adding an extra `WHILE` loop between lines (2) and (3) to the algorithm outlined in figure 5.17. The extra scan of the input queue is needed here because the sender of a signal cannot indicate whether the signal it sends is to be processed by a priority input or not. This is because this may depend on the state of the receiver which is not known to the sender in the case of consumer-based trigger testing.

With producer-based trigger testing, an optimization is possible. In this case, an extra priority flag could be employed to indicate whether an input queue contains a priority input in the current state. If the priority flag is already set when the producer processes a signal defined as a priority input signal in the current state of the receiver, the producer just inserts the signal in the input queue. If the flag is not set, the producer sets the flag after insertion of the signal in the input queue. As a further optimization, the respective action of the receiver may be specified in a data structure associated with the input queue of the receiver.

In addition, solutions between the two approaches described above are also possible. With consumer-based trigger testing, an optimization is possible if the producer provides some clues. However, this requires access to the current state of the receiver. In this case, the sender can set a priority flag if the receiver is in a state in which the transmitted signal is consumed by a priority input. This tells the receiver to scan its input queue for priority inputs first before looking for regular inputs or other triggers. If the priority flag is not set, the extra scan can be skipped.

Optimized scheduling Priority inputs are often used to handle some kind of exceptional situation. Thus, the fast processing of priority inputs may be of vital importance. Fast processing of priority inputs can be supported with producer-based trigger testing. Thus, process instances receiving a priority input signal are scheduled with priority, i.e. processed before other process instances that do not hold a signal specified in a priority input of the current state.

5.5.3.3 Code-Based Implementation

As pointed out above, the interpretation-based implementation has been discussed mainly for the purpose of understanding the basic principles and issues involved with the derivation of code from SDL processes. This approach allows a discussion of important implementational issues in a generic way.

With the code-based approach, the routines discussed above are derived specifically for each SDL process. In other words, each SDL process is represented by a set of routines. The set comprises at least one routine for each of the following activities:

- initialize a process instance created for an SDL process,
- test (some) trigger conditions of a process instance, and
- execute transitions of an SDL process.

Depending on the implementation of triggering, different procedures to test different trigger conditions may be provided. Similarly, the transitions of an SDL process may all be implemented by a single procedure, or split over several procedures, e.g. a dedicated procedure per transition or parts of a transition.

The derivation of the specific procedures for each given SDL process from the more generic interpretation-based code discussed above is straightforward. In order to directly derive the code to implement an SDL process, the code given in figure 5.17 has to be expanded with the specific information provided by the given SDL process. In addition, any code segments not relevant for the specific SDL process are removed.

An example for the transformation of the code outlined in figure 5.17 is given in figure 5.18 for a specific SDL process.

For the given example, we assume an SDL process with four states (state_1 to state_4), processing two different input signals (signal_1 and signal_2) altogether. With the code segments given in the figure, we concentrate on processing signal_1. The respective excerpt from the SDL process is shown in figure 5.19. In state_1, signal_1 (lines (7) and (8)) or one of two continuous signals (lines (23) to (27)) may trigger a transition. In state_2, an enabling condition is imposed on the input of signal_1 (lines (9) to (13)). In state_3, signal_1 is not specified. Thus, it is discarded in this state (lines (14) to (16)). In state_4 (lines (17) and (18)), signal_1 is explicitly saved, implementing the save construct.

Note that some of the procedure calls given in figure 5.18 are typically replaced by inline expansion of the respective code. This should be applied to all procedures that are called once in the program only, i.e. procedures specific for the given SDL process.

Action- versus state-oriented approach The outlined code employs the action-oriented approach. In this case, the code is structured according to the signals to be handled. Thus, the first `CASE` statement selects the signal while the second `CASE` selects the state.

Alternatively, the code may be structured according to the states of the EFSM. In other words, the first `CASE` statement selects the state, and the second `CASE` the signal. We call this approach state-oriented.

Provided blocking is done outside the procedure, i.e. in the underlying runtime system, differences between the two approaches are minor. However, major differences exist when blocking is

```
(1)  action:= NULL
(2)  signal:= CBMfirst_signal(PId.queue)     /* get first element */

     /* scan input queue for trigger condition */
(3)  WHILE signal ≠ NULL
(4)    CASE signal OF
(5)      signal_1:    /* process signal_1 */
(6)        CASE PId.state OF
(7)          state_1:    /* regular input */
(8)            RETURN(action_1)
(9)          state_2:    /* enabling condition */
(10)           IF enabling_condition_1(PId)
(11)             RETURN(action_e1)
(12)           ELSE    /* implicit save */
(13)             succ_signal:= CBMnext_signal(PId.queue,signal)
(14)         state_3:    /* discard signal */
(15)           succ_signal:= CBMremove_signal(PId.queue,signal)
                   /* remove signal from input queue and retrieve succeeding signal */
(16)           CBMfree(signal)    /* free signal buffer */
(17)         state_4:    /* explicit save */
(18)           succ_signal:= CBMnext_signal(PId.queue,signal)
(19)     signal_2:          /* process signal_2*/
(20)       ...          /* some action for each state */
(21)   signal:= succ_signal    /* proceed with successor */

     /* scan continuous signals for trigger condition */
(22) CASE PId.state OF
(23)   state_1:
(24)     IF continuous_signal_1(PId)
(25)       RETURN(action_c1)
(26)     IF continuous_signal_2(PId)
(27)       RETURN(action_c2)
(28)   state_2:
(29)     ...
(30) RETURN(NULL)    /* no trigger condition identified */
```

Figure 5.18 Outline of the code-based implementation of the `EFSMtest_trigger` procedure

directly implemented within the procedure for trigger testing. With the action-oriented approach only a single blocking construct is employed in the procedure. Conversely, with the state-oriented approach, blocking may occur in various places in the procedure. Thus, the state of the process

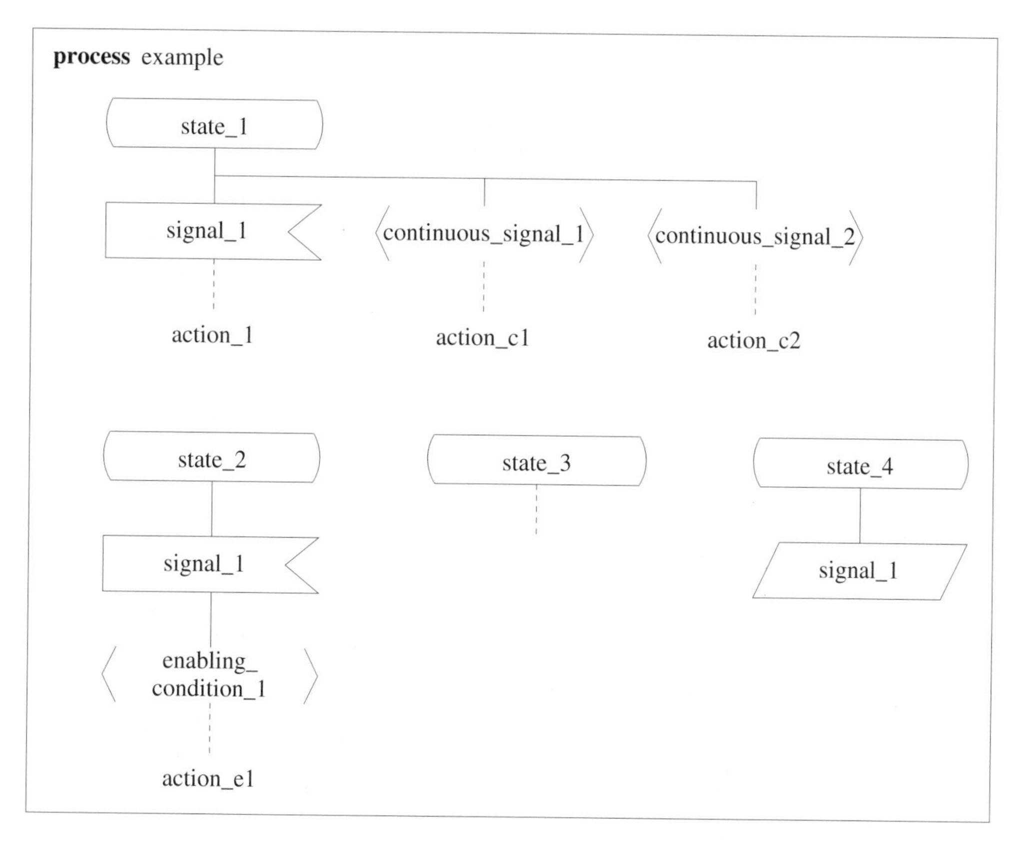

Figure 5.19 Excerpt from the SDL process

instance is represented by the position in the program, i.e. the program counter. This implies that each process instance is implemented by a separate entity of the runtime system, e.g. a thread or process. For a detailed discussion of the differences in the case of direct blocking within the procedure, we refer to [BrHa93].

5.5.4 Optimization of Trigger Testing

Testing the trigger conditions of SDL processes can be a complicated and time-consuming issue. In this section, we discuss optimizations for the general case as well as for the case where the SDL description, or parts thereof, is restricted to a subset of Z.100.

5.5.4.1 General Optimizations

Here, we discuss optimizations which can be performed without limiting the SDL language and which do not rely on the availability of shared memory.

Optimized testing sequence With most implementations a small part of the code is executed most of the time. Other parts are often needed only to deal with exceptional cases, and are not crucial to performance. Thus, it makes sense to concentrate on the critical paths.

With code-based implementations, the signal and the current state have to be selected employing a hierarchy of `CASE` or `IF` statements. When the critical paths of the system are known, the hierarchy of `IF` statements can be organized in such a way that the transitions on the critical path are selected right away without going through all the other `IF` statements testing for rare cases.

Minimization of reprocessing of conditions With the algorithms given above, enabling conditions and continuous signals are retested with each call of the `EFSMtest_trigger` procedure. However, continuous signals and enabling conditions have to be reevaluated only when there is a change to their constituents. Thus, retesting of continuous signals and enabling conditions is not needed – see below for exceptions – unless the respective process instance has executed a transition. To deal with this, a flag that indicates whether the `EFSMtest_trigger` procedure is executed the first time after a transition has been completed helps to eliminate unnecessary testing. Exceptions where more frequent testing may be needed are asynchronous conditions, i.e. continuous signals and enabling conditions that are based on variables that can be changed asynchronously, e.g. by another process instance.

Optimized handling of saved signals A source of inefficiency of the approach given above may result from the fact that the input queue is completely scanned with each call of the `EFSMtest_trigger` procedure. The problem arises when (implicitly or explicitly) saved signals are present in the input queue. Or in other words, signals are present in the input queue that can be neither discarded nor consumed at the moment.

Due to the semantics of SDL, a signal saved in one state remains in the queue until consumed or discarded in another state of the SDL process. Thus, there are two events that may force revisiting a saved signal, namely change of state, or the modification of an enabling condition. In the first case, all saved signals have to be revisited. In the second case, only implicitly saved signals have to be revisited. Provided the enabling condition can be changed only when the process instance executes a transition, no retesting of saved signals is needed unless the process instance executes a transition. This does not hold for asynchronous conditions, i.e. where enabling conditions may be asynchronously changed by other instances.

Summarizing this, revisiting saved signals is needed only when the process instance changes its state, or, if the saved signal is used with an enabling condition, when the enabling condition is changed.

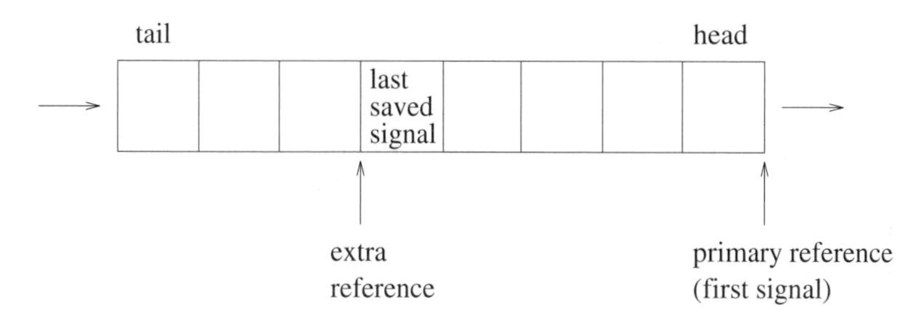

Figure 5.20 Optimized handling of saved signals

We propose to maintain an extra reference on the input queue that refers to the signal following the last saved signal. This is depicted in figure 5.20. This reference can be used to operate on the queue in the `EFSMtest_trigger` procedure. Logically, this can be considered as partitioning the queue into two parts, the front part containing saved signals, and the tail containing new input signals. The reference is reset to the head of the input queue when the process instance changes its state, or when an enabling condition is changed.

Resetting the reference due to changes to an enabling condition can be further optimized. As an alternative to the above approach, the reference may be reset only if changes to an enabling condition allow the processing of a previously saved signal. This prevents unnecessary retesting of saved signals in the input queue. However, it relies on the triggering entity or a periodic activity to perform the test of the enabling conditions. For this memory-sharing between the process instances is required.

Clue-based enabling Above, we have seen the complexity of testing triggers. This is caused by scanning through a large set of trigger conditions. The complexity and, thus, inefficiency which result from this are to a large extent caused by the fact that the producer of a trigger condition does not inform the consumer of the details of the trigger, for example when a signal is inserted in the input queue or changes are made to a revealed variable.

Improvement can be achieved if a clue is provided to the enabled entity that provides information which allows the reduction of the set of trigger conditions to be tested. This represents an extension of the clue-based enabling described in section 5.4.4. Thus, only the relevant parts of the `EFSMtest_trigger` procedure have to be executed. Even though a shared-memory-based implementation of this is preferable, in principle this information can be provided in a message-passing system as well.

5.5.4.2 Optimizations for Shared Memory Implementations

Provided the process instances have access to shared memory, some more optimizations can be implemented.

Producer-based trigger testing In section 5.4.3, producer-based testing of trigger conditions has been proposed for the case that the process instances share common memory. Thus, the producer of a potential trigger directly tests the potential consumers of the trigger, i.e. if the change of the input to the trigger conditions actually results in a trigger condition. Only in this case are the consuming process instances enabled, employing fully trigger-based enabling as described in section 5.4.4.

Where an SDL process defines a large set of complex trigger conditions, producer-based trigger testing may be faster. This is because the producer owns the clue what to search for. For example, consider a signal sent to another process instance. In this case, the sender has to execute only one iteration of the first `WHILE` loop given in figure 5.17, i.e. lines (4) to (17). In other words, only the actions for the specific signal that is transmitted have to be tested.

Another advantage of this approach results from the fact that the trigger conditions of the consumer are tested right away. Thus, the consumption of the transmitted signal can be preferred by dynamically changing the priority of the consuming process instance. For example, the priority of a process instance may be increased when a signal arrives which is defined in a priority input.

Direct deletion of unsaved signals An additional advantage of producer-based testing arises where signals are discarded rather than consumed. In this case, the approach allows the elimination of the actual transmission of the signal. This is possible when the test for trigger conditions is performed before the signal is transmitted. Thus, signals identified as being discarded at the receiver side can be discarded by the producer right away. This saves allocating a signal buffer, copying the signal parameters and related information into the buffer, and inserting the signal into the input queue of the receiver.

Dynamically optimized signal communication Again assume the producer-based testing of trigger conditions. If a signal is sent to a process instance with an empty input queue, the signal may be copied from the sending process instance into the receiving process instance right away, without employing a signal buffer. Thus, one of the two copy operations to transmit a signal between two process instances can be eliminated. This allows the cost of copying to be reduced, and eliminates the allocation and deallocation of a buffer.

However, note that the approach may change the semantics where priority inputs are used. Consider a priority input that arrives after a regular signal has been directly copied into the process instance and before the process instance is scheduled for execution. In this case, the later priority input is handled after the preceding regular input signal.

5.5.5 Optimized Implementation of EFSMs for Subsets of SDL

Numerous additional optimizations can be applied where the language features used in the SDL description or in specific SDL processes are restricted to a subset of SDL. Examples are SDL processes that do not employ save constructs, priority inputs, or asynchronous conditions (use of the view or now construct in an enabling condition or a continuous signal). In these cases, a number of additional optimizations can be applied. This is an important strategy for improving the performance of implementations derived from SDL descriptions. Unfortunately, the flexible application of optimizations to SDL processes that restrict themselves to specific subsets of SDL is not well studied yet.

Many more optimizations are possible. However, these optimizations are often restricted to very special cases. Examples are the merger of the EFSMs of a set of SDL processes to a single EFSM, or the merger of producer-based trigger testing with the direct execution of the tested process instance.

Where applicable, the merger of a set of SDL processes provides a large potential for performance improvements. However, note that the merger works only for rather static cases, i.e. there is no dynamic creation or termination of some of the process instances involved. In special cases, the merger of variables used in different process instances may be possible. In other words, communication by signals is replaced by direct access to a single memory location by a set of process instances. Thus, instead of explicitly transmitting a signal, access to the respective data of the signal is provided by directly accessing the respective variable(s) of the sending process instance. However, note that a thorough analysis is needed to ensure semantic correctness of this kind of optimization.

The merit of specific optimizations also depends on the specific context or application area in which SDL is used. We will further elaborate on the issue in section 5.12, where the applicability of optimization techniques known from the manual implementation of protocol architectures will be evaluated.

5.5.6 Mapping SDL Processes on Code and Data Segments

As pointed out in section 5.2, with the implementation of SDL processes, code and data objects are involved. We will discuss the two issues in turn. We start with the discussion of the implementation of non-object-oriented SDL and then sketch the implementation in the presence of object orientation.

Data tied with the lifetime of a process instance In SDL, the code executed by the process instances of an SDL process set is identical. However, each process instance needs a data segment to store the data specific to the process instance. In fact, all data used in SDL can at the conceptual level be considered as being local to some process instance.[8] As has been pointed out in section 5.2.2 already, some of these data are also accessed from outside the specific process instance. Obvious examples are the input queues or revealed variables (reveal/view concept).

The data that have to be allocated for each process instance include

- the current state of the process instance (state of the EFSM), and

[8]Note that even though this view is consistent with the concept of SDL, it is not very appropriate where the derivation of an efficient implementation is a major goal.

- the state (or value) of its local variables including the predefined variables to store various PIds.

Additional information (or data) which is also specific for each process instance is

- the state of its input queue, and
- the state of its timers.

However, unlike the data above, this information is often kept globally, separate from the specific process instance it conceptually belongs to. The organization of the data in a joint manner for all process instances of an SDL system or a part of an SDL system that resides on a single host allows access to the data by various entities. As we will see later, this is important in supporting an efficient implementation.

Embedding of the code of the EFSM The code representing the EFSM (or the SDL process) is typically accessed by a procedure which represents the entry point into the processing of the EFSM. The same procedure is executed by all process instances of a process set. However, note that the procedure executed on behalf of different process instances has to operate on different data segments, i.e. one data segment for the specific process instance.

Implementing data As outlined above, each process instance requires specific data throughout its lifetime, e.g. to maintain its explicit and implicit state. Additional memory may be needed during the execution of the process instance, e.g. to store local variables of a procedure that is called during the execution of the process instance. In general, the size of the additional memory varies over the execution, e.g. depending on the nesting of procedure calls. With sequential programming languages, this is implemented by a stack that stores the local variables of the called procedures. In this case, the data are stored according to the order of the procedure calls.

Conversely, process instances in SDL represent concurrently executing entities. Thus, the execution of one thread of control, i.e. a process instance, can be suspended to give another thread a chance to proceed. This is the case when a transition is completed, or in some implementations also during the execution of a transition. As a consequence of this, a separate stack is needed for each concurrent thread. In our case, this means (explicitly) maintaining a stack per process instance. This allows support of SDL procedures with explicit states, i.e. a transition having its end point within a procedure.

Under certain circumstances, the implicit stack, i.e. the stack maintained by code automatically generated by the compiler, can be used instead. Where this is applicable, explicit management of a set of stacks is no longer needed. Instead, this is implicitly supported by the underlying compiler that derives target code. However, this requires that all data that have to be stored by a process instance between two invocations, i.e. the execution of two transitions, are static in size, or at least of bounded size. For example, recursive calls of SDL procedures with states cannot be supported in this case.

Object orientation So far we have assumed that the SDL description does not employ object-oriented features, i.e. does not employ inheritance and specialization. In implementing object-oriented behavioral descriptions, there are three major alternatives.

One approach is to flatten the object-oriented structure before code is produced. Then the approach described above for flat SDL descriptions can be applied. Flattening may employ code inlining or copying to deal with inherited code segments. Alternatively, inherited code may be incorporated by procedure calls. The disadvantage of copying (or inlining) of code into different instantiations of an entity, e.g. a block or process, is the growth of the size of the code. However, the advantage should be an improved runtime behavior, especially where the inlined code segments are small.

An alternative approach is to directly implement the object-oriented structure of the SDL description employing an object-oriented programming language. Thus, inheritance and specialization are inherently supported by the underlying programming language. However, differences between the object-oriented concept employed by the programming language and object orientation in SDL have to be dealt with. The mapping of object orientation in SDL on the C++ programming language is described in [BrHa93].

A third alternative is the explicit use of data structures to store the object or inheritance structure of the SDL description. For example, consider a process type that is subsequently refined by derived process types. Or in other words, the code implementing a final process instance may be distributed over (or inherited from) several places. Due to the necessary traversing of the object structure during execution, the runtime behavior of the system may suffer.

5.6 Management of the Process Instances

Process Instance Management (PIM module) is in charge of organizing the execution of the process instances in a running implementation of an SDL system. As we have seen in section 5.3.3, there are several approaches to integrating the SDL system into the computing system. As a result of this, two main concepts to manage the process instances of the SDL system exist. The alternatives are management of the process instances by the runtime support system itself (light or bare integration) or directly by the operating system (tight integration).

Unfortunately, the issue of the management of the process instances is closely related to other activities, e.g. communication, buffer management, timer management and interfacing to the environment. In this section, we focus on the major process management issues and try to keep these as independent as possible from the related issues which will be discussed in the subsequent sections.

5.6.1 Data Structures

The most central logical data structures maintained by the manager of the process instances, i.e. independently of its implementation by the RTSS or the operating system directly, are

Routine	Function	Calling entity	Usage
PIMinitialize_system()	initialize SDL system	PIM	main procedure
PId:= PIMcreate_PI(p_type)	create PI	EFSM PIM	SDL create construct system initialization
PIMstop_PI()	terminate PI	EFSM	SDL stop construct
PIMenable(PId)	enable the execution of the given PI	EFSM TM ENV PIM/EFSM	SDL output construct expiration of timer arrival of external input asynchronous condition arises
PId:= PIMget_ready_PI()	get an enabled PI	PIM OS	light or bare integration tight integration
PId:= PIMget_address(...)	get address of PI	EFSM/ENV	SDL output via ... construct

Table 5.3 Routines provided by the process instance manager (PIM) module

- the process instance table (dynamic or static size, depending on the dynamics of the SDL description) maintaining instance-specific information needed for the management of the process instances, and
- the scheduling queues, especially the ready queue; additional queues for blocked process instances are (implicitly) employed where scheduling and synchronization are managed by the operating system.

However, note that the physical presence of the different logical queues described above depends on the selected implementation strategy.

5.6.2 Functionality and Routines

Independently of the chosen integration strategy, we discuss the principal tasks of the manager of the process instances. These are to a large extent independent of the implementation. The process management routines or activities required by an implementation of an SDL description are given in table 5.3.

System initialization The `PIMinitialize_system` routine is called during the initialization of the system. Its purpose is to initialize the data structures of the process instance manager (PIM). In addition, it is used to create the initial number of process instances as specified by the SDL description.

Creation of a process instance The `PIMcreate_PI(p_type)` procedure implements the create construct provided by SDL. It creates a new process instance (PI) of the specified SDL process `p_type`. In addition to the call of the procedure during the execution of a transition by an EFSM, the procedure is used by the `PIMinitialize_system` procedure itself to create the initial process instances. The `PIMcreate_PI` procedure allocates and initializes the data structures needed by the process instance, and maintains the data structures for process instance management in the RTSS and possibly in the operating system kernel where the tight integration approach is chosen. In the latter case, also an operating system process is created that executes the new process instance.

Depending on the dynamics of the given SDL description, different implementations can be chosen. If the dynamic creation of process instances is not used at all, the required data structures can be allocated statically and initialized during system startup. If the maximum number of process instances for each process set is known, the maximum number of created process instances can also be allocated statically and partially initialized during system startup. Thus, the remaining actions to be performed for a dynamic creation during runtime are minimized.

Termination of a process instance The `PIMstop_PI` procedure implements the stop construct provided by SDL to terminate the calling process instance. The procedure is called from the EFSM during the execution of a transition. The procedure deallocates the data structures held by the terminating process instance and updates the data structures of the process instance manager itself. In the case of tight integration, the respective operating system process terminates itself.

Enabling a process instance The `PIMenable(PId)` procedure enables the execution of the given process instance. This allows the enabled process instance to be selected by the process instance manager for execution. A typical implementation of the `PIMenable(PId)` construct is to insert the given process instance into the ready queue. Depending upon the exact strategy for trigger testing and enabling (compare sections 5.4.3 and 5.4.4), the insertion into the ready queue may occur at any time when there is some clue that the process instance might be executable. An example of clue-based enabling is the insertion of a process instance into the ready queue upon the arrival of a signal independently of whether the signal actually triggers a transition, or, alternatively, is discarded due to a missing save on the signal. Similarly, a process instance may be enabled due to the modification of a revealed variable it views.

As an alternative to the insertion of process instances into the ready queue based on clues, an exact strategy may be employed as well. With trigger-based enabling, the process instance is inserted into the ready queue only if it is sure that a transition is executable. In other words, the trigger condition has to be exactly tested before the process instance is inserted into the ready queue. The advantage of this approach is the fact that no retesting of the trigger condition is needed when the enabled process instance is scheduled for execution.

There may be various activities within an SDL system that may result in the insertion of a process instance into the ready queue. Thus, the `PIMenable(PId)` procedure is called from various modules. These are the interface module upon the receipt of an external input, the timer manager upon expiration of a timer or the advancement of the time, or any process instance that is sending a signal or is changing a revealed variable.

As we will see in detail later, the actual implementation of the `PIMenable` construct depends highly on the selected integration strategy. With light intergration, the described ready queue is implemented as a data structure of the RTSS. In the case of tight integration, the ready queue is implemented and maintained by the operating system. In order to manipulate the ready queue (and associated blocked queues) semaphores or similar constructs are typically employed.

Fetch an enabled process instance The `PIMget_ready_PI` procedure returns the PId of an enabled process instance to the caller. Thus, the procedure removes the process instance from the ready list. Depending on the enabling strategy, the given process instance may be executed right away, or a test of the trigger conditions may have to be performed first.

Retrieval of addresses of process instances The `PIMget_address` procedure retrieves the address (PId) of a process instance where not explicitly specified. The procedure is used to resolve the address used in output constructs, i.e. the address of the receiving process instance, where the PIds cannot be resolved statically during compilation time, e.g. where the via construct is used. The procedure implements a lookup of the address information maintained by the process instance manager.

5.6.3 Process Instance Management by the Runtime Support System

Having looked at the principles, we separately discuss the two fundamental approaches for the management of process instances, namely management by the runtime support system (light or bare integration), and management by the operating system (tight integration).

With the implementation of the management of the process instances directly by the runtime support system, the following principles apply.

System integration All process instances present in the system as well as the support functions needed to implement the SDL machine are implemented within a single sequential program. This is graphically displayed in figure 5.21. The sequential program comprises two parts, the RTSS and the application-specific part. The RTSS directly implements the functionality of the SDL machine. This includes the modules for the management of the process instances (PIM), communication and buffer management (CBM), timer management (TM), and the interface to the environment (ENV).

The program may be implemented as a process of the operating system (light integration), or possibly as the main loop executed on the naked hardware (bare integration), e.g. a microcontroller.

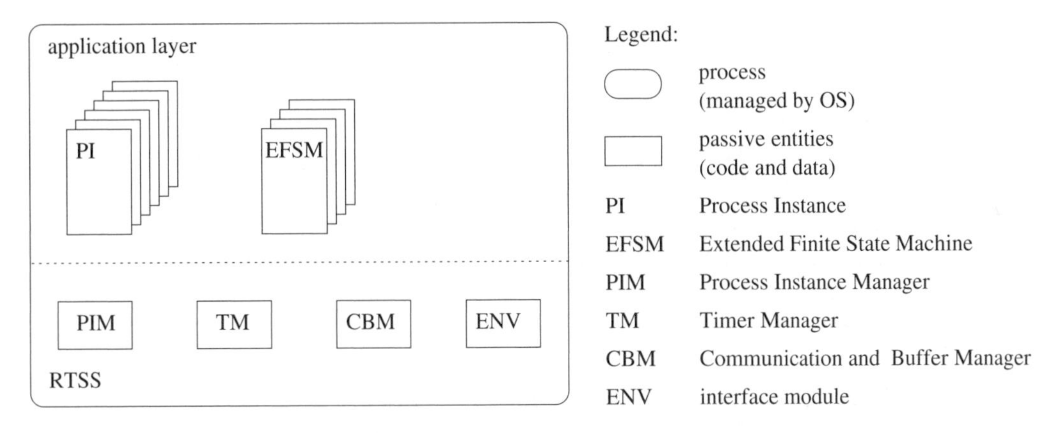

Figure 5.21 Structure of the sequential program to implement the SDL system

```
(1) PIMinitialize_system()          /* initialize the RTSS data structures and */
                                    /* create initial process instances */
(2) FOREVER
(3)    PId:= PIMget_ready_PI()      /* fetch process instance from ready queue */
(4)    IF PId ≠ NULL                /* test for availability of an enabled PI */
(5)      action:=                   /* select executable transition */
           EFSMtest_trigger(PId)
(6)    IF action ≠ NULL             /* test for presence of an executable */
                                    /* transition */
(7)      EFSMexecute(PId,action)    /* execute transition */
(8)    TMtimer_manager()            /* handle expired timeouts */
(9)    ENVenv_manager()             /* handle communication with the */
                                    /* environment */
```

Figure 5.22 Example of the main loop (scheduler) executed by the RTSS with light integration

Execution control The RTSS directly manages the execution of the process instances, including creation and termination. An example of the main loop executed by the RTSS is outlined in figure 5.22. The given example employs the clue-based enabling strategy.

After initialization, the RTSS executes an infinite loop. With each iteration of the loop, an enabled process instance is selected and, provided an executable transition exists, is executed. With the given enabling strategy, the execution may include implicit transitions executed by the `EFSMtest_trigger` routine. After this, the `TMtimer_manager` checks for expired timeouts. Additionally, the `ENVenv_manager` procedure handles external communication including the transformation of input data into SDL signals.

Note that the implementation of the scheduler by the RTSS is based on the assumption that none of the above procedures ever blocks. Due to the implementation of the whole SDL system by a single process, blocking in any procedure results in blocking of the process. Thus, the presence of blocking could completely invalidate the design. In principle, blocking can occur due to active or passive waiting on an event. Blocking must not occur within the interface routine, e.g. when the environment is not ready to provide a signal, or to consume one. Blocking (active or passive) is acceptable only where it can be ensured that the event on which the SDL system is blocking is the only event that can occur in the current state of the SDL system which can trigger an execution. For example, blocking at an external input port may be acceptable only when it is certain that no other event may occur that triggers a transition, e.g. the expiration of a timer.

Data With the exception of the procedures `TMtimer_manager` and `ENVenv_manager`, the procedures fully operate on data structures local to the executing process, i.e. data structures maintained within the RTSS or the application-specific part itself. An important example of this is the ready queue maintained by the RTSS. This increases the flexibility for the optimization of the implementation and also reduces overhead for the management of the process instances.

The organization of the ready queue is important to the performance of the implementation. If the ready queue is mainly used as a FIFO queue, a simple linked list with an additional reference to the last element may be sufficient. This works fine when all process instances share the same priority. However, the application may demand random access as well, e.g. when priority is supported. Random insertion and removal of an element from the ready queue can be implemented efficiently when a doubly linked list is employed, i.e. each element maintains a forward and a backward link.

Implementation of the PIM routines The creation of a process instance by the `PIMcreate_PI(p_type)` procedure can be implemented as follows. First an entry for the process instance table which is maintained by the PIM module is created. Next, the memory area required by the process instance is allocated from the heap and initialized with the data specific for the process instance. Then a signal with the actual parameters of the process instance is generated and transmitted to the new process instance. This is done in a way similar to the transmission of regular signals. Thus, a signal buffer is acquired, filled and finally inserted into the input queue of the newly created process instance. In addition, the new process instance is enabled. When the new process instance is finally selected for execution, it retrieves the signal and executes its initial transition similar to any other transition.

The `PIMinitialize_system` procedure allocates and initializes the data structures maintained by the PIM module, especially the process instance table and the ready queue. Then the initial number of process instances for each SDL process are created using the `PIMcreate_PI` procedure described above.

The `PIMstop_PI` procedure removes the process instance from the process instance table and frees the memory held by the process instance. Note that some care is needed to ensure that later access to the PId, which may still be referenced by other process instances, results in an error rather than referencing a new instance that reuses the same identifier.

```
(1) PIMinitialize_system()          /* initialize the RTSS data structures and */
                                    /* create initial process instances */
(2) FOREVER
(3)    PId:= PIMget_ready_PI()      /* fetch process instance from ready queue */
(4)    IF PId ≠ NULL                /* test for presence of an executable transition */
(5)      EFSMexecute(PId)           /* execute transition */
(6)    TMtimer_manager()            /* handle expired timeouts */
(7)    ENVenv_manager()             /* handle communication with the */
                                    /* environment */
```

Figure 5.23 Main loop of the RTSS with trigger-based enabling

The `PIMenable(PId)` procedure enables the specified process instance for execution. Due to the sequential nature of the implementation, the enabling of process instances is fully implemented in the RTSS employing shared memory. A process instance can be enabled by simply inserting its reference into the ready queue maintained by the PIM module.

The `PIMget_ready_PI` operation fetches (and removes) an enabled process instance from the ready queue. If no executable process instance is available, the procedure immediately returns the `NULL` value.

Enabling process instances Due to the complexity of the trigger conditions of SDL (see section 5.2.2.3), the implementation may employ a clue-based enabling strategy where there may be cases where a process instance is inserted in the ready queue which in fact is not firable. In this case, the `EFSMtest_trigger` operation just removes the process instance from the ready queue and (unsuccessfully) tests its trigger conditions. Examples where this approach is used are the receipt of a signal in a state in which the signal is not defined, i.e. does not prompt a transition, or, alternatively, the signal is defined to be saved.

An example of the main loop with trigger-based enabling is outlined in figure 5.23. In this case, a process instance is enabled only if an explicit transition is present. Thus, no call of the `EFSMtest_trigger` procedure is needed after the respective process instance has been inserted in the ready queue.

Note that with trigger-based enabling, the `EFSMexecute` procedure is called without a parameter that provides the selected action. Instead, the respective information as provided by the `EFSMtest_trigger` procedure is stored in a variable maintained within the PId data structure.

Interleaving With the implementation of the SDL system as a single sequential program, the interleaving of the execution of process instances is fully controlled by the RTSS. Thus, there is no interleaving of the execution of a process instance other than explicitly defined by the RTSS and the application-specific part. In general, interleaving may take place only when an application-specific procedure returns, or instead calls a procedure of the RTSS which itself calls an application-specific procedure that executes another (interrupting) process instance. In other words, any interrupting entity relies on the cooperation of the interrupted entity to give up the control by calling a procedure of the RTSS or by simply returning from a procedure called by the RTSS. Typical implementations (as shown in figure 5.22) support this interleaving only when the process instances are in an explicit state (EFSM state). Thus, the transitions can be considered as atomic by design. Note that this may change as soon as any activity of the SDL implementation is removed from the sequential process into another process.

Synchronization Due to the sequential nature of the implementation, no extra operations for synchronization are needed to ensure consistency. This is changing as soon as there are any asynchronous activities, i.e. another process or an interrupt that operate on the same data.

Prioritization Even though full interleaving, or preemption, is not supported, the implementation can easily support (non-preemptive) prioritization of process instances. Prioritization can be implemented by organizing the ready queue as a prioritized queue, e.g. employing a sub-queue for each priority.

Communication Communication between process instances makes extensive use of shared memory. Again, no extra synchronization is needed to ensure consistency as long as the data structures are not accessed by any other asynchronous entity (see section 5.7 for details).

Timer management With the simple implementations outlined in figures 5.22 and 5.23, timers as well as interfacing with the environment are directly handled by the same operating system process (or main program) that is executing the process instances and the RTSS. In this way synchronization problems are eliminated. However, the implementation has to employ a polling approach to handle timeouts and external input. In the implementations given above, the timer manager is called at least once upon the execution of any transition in the system.

The `TMtimer_manager` procedure tests whether there are any expired timers and possibly advances the time accessible within the SDL system. If a timeout is found, a respective signal is inserted in the input queue of the receiving process instance and the instance is enabled (`PIMenable(PId)` procedure). Enabling of process instances may also occur when the time (now construct) is used within an enabling condition or a continuous signal.

Note that several alternatives to this simple approach are possible. For example, timer management may be removed from the main loop and implemented by an asynchronous entity, e.g. an interrupt handler or a separate process managed by the operating system. We will discuss alternatives in section 5.9.

External communication The interface to the environment is implemented by the `ENVenv_manager` procedure. Like the timer manager, the `ENVenv_manager` procedure is called upon the execution of any transition. The procedure checks whether there is external input or output to be handled. With external input, a signal is generated and sent to the receiving process instance. In addition, the receiving process instance is inserted into the ready queue of the scheduler by the `PIMenable(PId)` routine.

Again many alternatives to this simple scheme exist. We will discuss the details in section 5.8.

Discussion A major advantage of the management of the process instances by the RTSS is its potential for high efficiency. Employing a central controller implemented by the RTSS within a single address space allows the use of an adapted process instance management scheme that efficiently implements the specific needs of the semantics of SDL. For example, the RTSS knows about the exact circumstances under which the trigger conditions of a process instance hold (or are changed). Thus, complex SDL concepts such as, for example, continuous signals using viewed variables can also be implemented with relatively small overhead compared to other approaches. As we will see later, the support of the RTSS for the tight control of the schedule allows for the optimization of the communication mechanism. Thus, the elimination of copying – which is a major source of overhead – is much easier.

Another advantage of the approach also relates to efficiency. The tight control of the execution by a single sequential entity eliminates the need for employing mechanisms to ensure the consistency of the data. Rather, consistency is ensured by construction. Thus, no extra operations are needed to ensure the mutual exclusive access to data structures that are used by different process instances. However, note that this only works as long as there is no asynchronous activity interrupting the RTSS and accessing its data, e.g. a UNIX-like signal handler or an interrupt handler in a real-time operating system. Interrupt handlers are often employed to handle external communication and timeouts.

However, the approach also has some drawbacks. A major drawback is the missing support for preemption which is caused by the lack of explicit consistency control mechanisms. Thus, one process instance may not start a transition while another process instance has not completed its transition and before this process instance voluntarily returns control to the scheduler. As a result, the approach as described above does not allow an immediate reaction upon asynchronous events, e.g. the receipt of an external signal or a timeout. In particular, the asynchronous handling of external input is often needed to ensure that received data do not get lost.

Following a similar argument, the given approach is also not suited for the concurrent processing of the process instances on a parallel machine. However, the handling of asynchronous events, i.e. timeouts or external input, as well as parallel processing can be supported by introducing some mechanism to ensure mutual exclusion. For this, similar mechanisms as supported by thread libraries implemented in user space can be used. Thus, the RTSS (or the process implementing the SDL system) can be interrupted to handle external input or timeouts. As a result, external input data can be retrieving instantly from the interface. However, the data cannot be processed by any process instance before the interrupted transition has been completed.

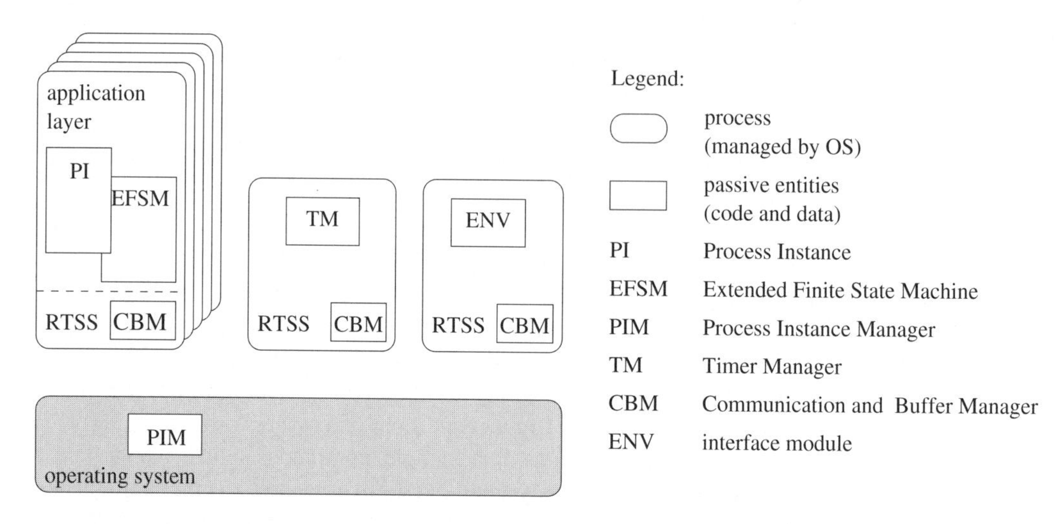

Figure 5.24 The three classes of asynchronous entities present with tight integration

5.6.4 Process Instance Management by the Operating System

System integration With tight integration, each process instance is implemented by a separate process of the operating system. Thus, process instance management is mainly implemented by the operating system rather than the RTSS. With tight integration, typically this also holds for other support functions such as timer management and communication with the environment. As a consequence of this, the RTSS itself implements only a small part of the functionality it provides to the application-specific part.

In the following, we assume that an operating system is employed which supports memory sharing between processes. Thus, signal communication can be implemented based on copying rather than employing interprocess communication mechanisms provided by the operating system.

Figure 5.24 shows the three different classes of asynchronous entities employed with this approach. These are the process instances, each implemented by a separate process, the timer process which implements the timer module and one or more entities supporting interfacing with the environment (ENV module).

Various alternatives for the implementation of the timer manager and the interface functionality exist. Note that the entities do not necessarily have to be implemented by processes run by the operating systems. Important here is that these support functions are implemented by asynchronous entities such that their execution may interleave with other activities in an arbitrary order.

Execution control As noted above, each process instance is implemented by its own process run by the operating system. Thus, the operating system directly manages and controls the execution of the process instance, including its creation and termination. In addition to the asyn-

```
(1) EFSMinitialize_PI(PId)                  /* initialize the process */
(2) FOREVER
(3)   block(PId)                            /* wait for enable */
(4)   action:= EFSMtest_trigger(PId)        /* select executable transition */
(5)   IF action ≠ NULL                      /* test for presence of an */
                                            /* executable transition */
(6)     EFSMexecute(PId,action)             /* execute transition */
```

Figure 5.25 Outline of the code executed by each process that implements a process instance of the SDL system

chronous execution of the process instances, we assume that timer management and interfacing to the environment are also handled by asynchronous activities. Thus, any of these activities may interleave each other. Scheduling and interleaving are controlled by the operating system based on the scheduling strategy it employs.

The actions performed by the operating system scheduler are similar to those performed by the scheduler of the RTSS as outlined above. Differences are the possibility of arbitrary interleaving of the executing entities, and the direct handling of the ready queue by the operating system, including its manipulation using blocking and unblocking system calls.

The structure of the code executed by each process that implements a process instance of the SDL system is outlined in figure 5.25. After initialization, the process executes in a loop until it terminates itself. In each iteration, the process blocks on some synchronization construct until enabled by another entity. Upon enabling, the process tests the trigger conditions of the process instance it implements and executes a transition if a firable transition has been identified.

The figure outlines the implementation of clue-based enabling as described in section 5.4.4. Trigger-based enabling is also possible, provided shared memory is available. In this case, the test of the trigger conditions is no longer needed in the enabled process.

The synchronization element on which the process blocks, if not enabled, is associated with (owned by) the process instance which is executed by the process. One synchronization element, e.g. a binary semaphore, is needed for each process instance in the system. This allows other entities to enable a specific process instance. This is implemented by the `PIMenable` procedure, by issuing a system call to unblock the respective process.

Note that additional synchronization constructs are needed in the system to ensure data consistency.

Implementation of the PIM routines Assuming the use of shared memory accessible by all involved processes, the following implementation strategy can be employed.

The first process generated for the SDL system calls the `PIMinitialize_system` procedure to allocate and initialize the data structures maintained by the RTSS. In addition, the procedure (repeatedly) calls the `PIMcreate_PI` procedure to create new processes to set up the initial number of process instances for each SDL process as defined in the SDL description.

In addition to the call of the `PIMcreate_PI` procedure by the `PIMinitialize_system` routine, the procedure may be called by a process instance within the `EFSMexecute(PId)` procedure to dynamically generate a new process instance.

A call of the `PIMcreate_PI(p_type)` procedure results in a set of actions. First, a new synchronization element is allocated and initialized (see below for details). This is needed to support enabling and blocking of the new process instance. Then, a system call to create a new process is issued. The new process is initialized according to the specifics of the SDL process it represents. This is implemented by the `EFSMinitialize_PI(PId)` procedure. Following this, the new process blocks on the new synchronization element.

The `PIMcreate_PI` procedure itself proceeds by modifying the data structures maintained by the RTSS to reflect the new process instance. Then, an initial signal is generated and sent to the new process transmitting the actual parameters of the newly created process instance. This is done in the usual way, i.e. by using the procedures provided by the CBM module to send signals. Following this, the new process is enabled by calling the `PIMenable(PId)` procedure.

The `PIMstop_PI` procedure terminates the issuing process by issuing a respective system call. Before this, the data structures maintained by the RTSS have to be modified to reflect the changed situation.

The `PIMenable(PId)` procedure is simple. Instead of manipulating the ready queue itself, it issues a system call to unblock the process that implements the process instance specified by the PId parameter.

The `PIMget_ready_PI` procedure to select an enabled process instance as described above is no longer needed here. Instead, this is implicitly implemented by the operating system scheduler. What is needed here is a construct to block a process if it is not enabled. This is implemented by a system call to manipulate the used synchronization element, e.g. a `DOWN(S)` operation executed on a semaphore associated with the process.

Data Unlike the approach described above, the data structures maintained by the RTSS have to be accessible by a set of processes. No ready queue is needed in the RTSS. However, data structures to store information specific for each process instance have to be maintained. Thus, a process instance table is needed as before.

Enabling process instances Provided shared memory is supported, both approaches to test trigger conditions presented in section 5.4.3 can be applied.

Enabling is implemented by employing synchronization constructs provided by the operating system. For this to work, each process employs a construct on which it can block until woken up by some other entity. Examples of such synchronization elements are binary semaphores or events supported by the operating system. Thus, the process blocks itself as soon as there is no more work to do, i.e. no transition to execute. As described above, blocking is implemented in the main loop executed by the process. Unblocking of a blocked process instance (or a process) is implemented by the `PIMenable(PId)` procedure which is called by other processes that generate some work for the currently blocked process.

Interleaving Interleaving of the execution of the process instances is controlled by the operating system. Thus, mutual preemption between process instances may be supported depending on the scheduling strategy of the operating system. As a result, any interleaving of the execution of the process instances other than at the explicit states of the EFSM (and additional states introduced by the use of remote procedure calls, export/import, etc.) is possible.

With preemption, the execution of process instances may interleave at any arbitrary point. Thus, precautions to prevent the corruption of data structures by interleaving are needed when shared memory is supported. Operating system concepts supporting this are (binary) semaphores, the monitor concept, and events. Under some circumstances the temporary disabling of interrupts may also be an alternative. When interrupts are disabled, the operating system scheduler is also usually disabled. Thus, no preemption may occur.

If preemption is not allowed, the concept is similar to the light integration approach.

Synchronization Concerning synchronization, there are two major differences compared to the implementation by the RTSS as described above:

(1) Synchronization primitives are used to block and enable process instances. Thus, a blocking system call is issued by a process when there is no more work to do. In other words, the process is moved to a blocked queue. The process is unblocked by issuing some unblocking system call wherever new work for the process is generated.

(2) Synchronization primitives are used to ensure the consistency of shared data.

Thus, synchronization is needed for two purposes, namely to ensure the consistent use of shared data and for the triggering of process instances. Due to the fact that the operating system controls scheduling, mechanisms provided by the operating system have to be employed for both synchronization tasks.

Note that special care is needed to avoid deadlocks where more than one synchronization element is used in an interleaved fashion. In particular, it has to be ensured that a process that has acquired a lock (for mutually exclusive access) does not block on a second synchronization element.

Prioritization Provided consistency control mechanisms are in place anywhere interleaving may occur, preemptive priorities may be employed to ensure fast reaction to urgent events. However, special care is needed where the preempting process has access to the same protected data as the preempted process. In case the preempting process tries to acquire a lock that is already held by the preempted process, priority inversion occurs. Thus, the preempting process is blocked until the (preempted) process holding the lock releases it. If locking is implemented in the operating system by active waiting, deadlock may occur in this case. Strategies to avoid the priority inversion problem are described in [Butt97].

Communication Wherever possible, communication between processes should employ shared memory. This is typically supported by real-time operating systems. Interprocess communication primitives provided by the operating system may be used where memory sharing is not supported. However, note that there are several disadvantages when no shared memory is present. The details of implementing communication are discussed in section 5.7.

Timer management The functionality of the timer manager to handle timeouts and possibly the time accessible within the SDL system can be implemented in various ways. Alternatives are its implementation by a separate process managed by the operating system, the implementation by an interrupt handler, or the direct handling of timers by the operating system using respective system calls. We will discuss the alternatives in section 5.9.

External communication As with timer management, numerous alternatives exist to implement the interfacing with the environment. Important alternatives are the use of a separate process (with traditional operating systems), or the use of an interrupt handler (with real-time operating systems). Again, we defer the details to section 5.8.

Discussion An advantage of the implementation of process instance management by the operating system is its responsiveness to urgent events. Fast response to events can be ensured where preemptive scheduling is supported by the operating system and where urgent activities are given higher priority than non-urgent activities.

Another advantage of the approach is that it quite naturally supports parallel processing on a multiprocessor system. Once concurrency control is installed to ensure coherence of shared data, no extra measures are needed to support the parallel processing of process instances.

A drawback of the approach is its limited efficiency. The approach requires some extra overhead for concurrency control (i.e. where arbitrary interleaving is supported). In addition, synchronization may cause priority inversion and, thus, delay the processing of urgent events when the execution in mutually exclusive areas is time consuming. Even worse is the extra overhead encountered with the use of heavyweight processes, i.e. processes managed by traditional operating systems. With the use of real-time operating systems or lightweight processes this overhead can be considerably reduced.

A problem with the approach may be caused by the necessary adaptation or mapping of the services used by the application-specific part on services provided by the operating system. Depending on the specific operating system at hand, there may be a mismatch in functionality. Thus, it may be very inefficient to provide the services needed by the application-specific code based on the services provided by the operating system. Thus, supporting the full SDL standard in this way is often problematic. Even if related constructs are available, they may not be well adapted to the specific usage patterns of SDL, or the specific application. For example, the operating system does not know the circumstances under which the trigger condition of a process instance holds or has changed. Thus, complex SDL concepts, especially the use of asynchronous conditions, e.g. continuous signals employing the view construct, etc., can often not be implemented efficiently.

In addition, the use of the operating system for the management of the process instances represents a rather loose control of the schedule. Thus, the elimination of copy operations employed with signal communication is more complicated.

Implementation without shared memory If no memory sharing is supported between the processes implementing the process instances and other entities, things get much more complicated. Everything that has been implemented above by using shared memory has to be replaced by message communication mechanisms. This represents a major burden on the performance of the implementation.

Due to the much higher cost for message communication, communication between processes has to be minimized and eliminated wherever possible. In order to allow for this, typical implementations that employ message-passing are limited to a subset of SDL. Revealed variables, continuous signals, enabling conditions, saves and priority inputs are especially expensive to implement.

The problem with revealed variables is that either viewing or modifying such a variable requires the transmission of the variable to the viewing processes.

Continuous signals and enabling conditions are problematic if they employ asynchronous conditions, i.e. can be changed while the process is blocked (see also section 5.7.2.2). Examples are the use of revealed variables and access to the time. In this case, enabling (unblocking) of the process is needed to give it a chance to reevaluate the dynamic condition. Note that enabling is needed each time a dynamic constituent of an asynchronous condition is modified. Considering access to the time variable in a continuous signal, this basically results in semi-active or even active waiting.

With the use of message passing, random access to the input queue is no longer supported. This is because the input queue is an operating system entity rather than an entity of the receiving process. Thus, workarounds to handle SDL features that require random access are needed. This especially concerns the implementation of the save construct and of priority input. We will discuss this in more detail in section 5.7.

In addition, the removal of a timeout signal from the input queue as required by the reset construct in SDL is highly complicated. In this case, a completely different implementation of timeouts is preferable, i.e. an implementation no longer based on the input queue of the process.

5.6.5 Address Resolution

Communication depends on addresses known to the sender. With SDL, communication may address a specific process instance, an SDL process or just one of a set of process instances connected to the same channel. Except for some simple cases, the address of the receiver has to be looked up somewhere. For example, this may be the case when the output via construct is used or when an SDL process is specified as receiver instead of a process instance.

As noted above, the retrieval of addresses is supported by the `PIMget_address` procedure. There are alternative ways to maintain the address information depending on the overall implementation strategy. Address information may be maintained either in a table accessible to all process instances, or by an additional process in the system, i.e. an address server.

Disallow indirect addressing The simplest solution is to disallow any indirect addressing which cannot be statically resolved at compile time. Thus, the use of communication constructs is restricted such that each address can be directly mapped on a specific process instance. In this case, the addresses can be hard-coded into the generated code or directly derived from the PId data structure at runtime.

Central table The use of a central table is probably most appropriate where dynamic addressing is needed and shared memory is available. The table contains all data needed to resolve addresses, i.e. the addresses (or PIds) of the active process instances in the system and data on how these are connected to channels, etc. Note that the table has to be maintained at runtime, i.e. new entries have to be generated when a new process instance is created, and removed upon the termination of the process instance.

Separate address server in a message-passing system If no shared memory is available, this has to be implemented by some entity that maintains the table, i.e. a separate process in charge of addressing. In this case, address resolution as well as maintaining the address table requires the transmission of additional messages in the system. This results in additional overhead. For this reason, dynamic addressing is typically not supported by message-passing implementations.

5.7 Process Communication and Buffer Management

Process Communication and Buffer Management (CBM module) is in charge of transmitting SDL signals with their parameters between process instances, and with the environment. Signals may be regular signals or timeout signals. In order to transmit a signal, some buffer or memory space is needed to provide a container for the signal.

The focus of this section is on the implementation of signal communication in general. The subsequent section deals with the exchange of signals with the environment.

5.7.1 Functionality and Routines

The tasks needed to support signal communication within an SDL system can be summarized as follows:

- identification of the receiver (addressing),
- management of the memory to store the signal, signal parameters and related information,

- moving data between different process instances,
- activation of the receiver of the signal (enabling) where appropriate,
- dealing with cases where no receiver can be identified, and
- dealing with resource limitations, e.g. full input queues, or no available memory.

While the functionality of all but the last task is clearly defined by SDL, the last task is not. Or in other words, the system design is assumed to ensure that input queues never become full.

5.7.1.1 Supported Functions

In order to support communication and buffer management, we propose a small set of basic functions given in table 5.4. Since the contents of the transmitted signals are transparent to the CBM module, the auxiliary routines `fill_`*`s_type`* and `copy_signal_`*`s_type`* are not strictly part of the CBM module. However, they are discussed here because they form an integral part of process communication.

For the moment, we assume unlimited memory. Or in other words, we make the optimistic assumption that the implementation – in conjunction with the behavior of the environment – ensures that the system never reaches a state where no more memory is available.

Allocation of a signal buffer The `CBMallocate(size)` procedure is used to allocate a signal buffer of a given `size` from the heap, i.e. from an unused chunk of memory. Note that the signal may be a regular signal or a timeout signal. If the size of the parameters of the signals is fixed, the signal type could be used instead of the size.

Filling a signal buffer The `fill_`*`s_type`*`( ...)` procedure is application-specific and has to be provided for each type of signal (*`s_type`*). Its purpose is to fill the signal buffer with the needed data, i.e. signal type, signal parameters, and related data such as the PId of the sender.

Insertion of a signal in the input queue The `CBMinsert_signal(queue,signal)` procedure appends the given signal (signal buffer) at the tail of the specified input queue.

Retrieval of the references to a signal in the input queue The CBM module provides two procedures to retrieve the reference to a signal stored in the input queue. The `CBMfirst_signal(queue)` procedure retrieves the reference of the first signal in the specified input queue. The `CBMnext_signal(queue,signal)` procedure retrieves the reference to the signal in the queue that directly follows the signal referenced by the `signal` parameter.

Routine	Function	Calling entity	Usage
signal:= CBMallocate(size)	allocate a signal buffer	EFSM TM ENV	SDL output construct timer expiration external input
CBMinsert_signal (queue,signal)	insert signal in input queue of given PI	EFSM TM ENV	SDL output construct timer expiration external input
signal:= CBMfirst_signal(queue)	get reference of first signal in input queue	EFSM TM	EFSMtest_trigger SDL reset construct
signal:= CBMnext_signal (queue,signal)	get reference of next signal in input queue	EFSM TM	EFSMtest_trigger SDL reset construct
signal:= CBMremove_signal (queue,signal)	remove specified signal from input queue and return reference to next signal in input queue	EFSM TM	EFSMtest_trigger() and EFSMexecute() SDL reset construct
CBMfree(signal)	free buffer	EFSM TM	SDL input construct SDL reset construct
fill_*s_type*(signal, ...)	fill signal buffer	EFSM TM ENV	SDL output construct timer expiration external input
copy_signal_*s_type* (queue,signal, ...)	copy signal parameters on local variables	EFSM	SDL input construct

Table 5.4 Routines for process communication and buffer management (CBM)

Removal of a signal from the input queue The consumption of an input signal from the input queue is implemented by the `CBMremove_signal(queue,signal)` procedure. The procedure removes the referenced input signal from the specified input queue. In addition, the procedure returns the reference to the following signal in the input queue.

Note that the procedure allows the retrieval of any arbitrary signal from the queue. This is important for handling saved signals, timeout signals and regular signals in the same queue. Due to this requirement, the implementation of the input queue should support fast access to arbitrary elements, i.e. allow the removal of a signal from the input queue which is not the first element in the queue. Random access is required for the removal of timeout signals from the queue, i.e. when a timer is reset, and also for the implementation of saves.

Copying of a signal buffer The `copy_signal_`*`s_type`*`(PId,signal, ...)` procedure is application-specific and copies the signal parameters and the related data from the signal buffer into the local data area of the executing process instance. A specific procedure is derived for each signal type including timeout signals and signals from the environment.

Deallocation of a signal buffer The `CBMfree(signal)` deallocates a signal buffer, i.e. frees the memory for future use.

5.7.1.2 Dealing with Limited Input Queues

SDL is based on the assumption of infinite input queues. However, in practice, buffers are always limited. This has to be taken into account when implementing SDL descriptions.

There are several alternatives to handle the problem (compare also section 5.13 for a general discussion). The selected strategy depends highly on the circumstances under which the system is used.

Prevention The system and the environment are designed in such a way that the system never hits any buffer limitations during regular operation. In other words, the implementation of the SDL description relies on the environment to provide as much load as the system can handle only. Thus, the limited queues are virtually unlimited and no additional action is needed. As a result of this, the implementation behaves exactly the way the SDL description does.

If buffer overflow happens anyway, often the system is simply restarted.

Discard signals An alternative to resetting the whole system or a subsystem is to simply discard signals that cannot be stored at the moment. This approach is often employed with communication systems, e.g. when a router is overloaded. However, note that this behavior of input queues is not in accordance with the semantics of SDL. On the other hand, SDL can be used to model this behavior, i.e. by explicitly describing limited queues. Thus, the respective behavior of the implementation can be verified against the corresponding SDL description.

Blocking When an input queue is full, the sender is blocked. This may result in a back-propagation of the overload, i.e. the queue of the blocked sender itself fills up, which in turn blocks process instances sending to its input queue, and so on. Blocking on full buffers can be implemented using general semaphores.

Note that the use of blocking is dangerous as it may easily result in deadlock. Blocking changes the semantics of the system. Thus, the SDL description is no longer in accordance with the implementation. In addition, blocking cannot be described in a straightforward manner in SDL. Thus, the modeling of blocking requires considerable changes to the SDL description.

Dynamic control In communication systems none of the above approaches may be appropriate. Often the system is required to actively react to overload situations. For example, the system may start to react to overload when the queues are filling up, e.g. when a queue is more than 50% full. In this case, the environment or some interface has to be instructed to refrain from sending more signals. The best practice is to throttle the load at the source, i.e. the entities where the load is generated.

The implementation of this kind of overload handling requires the dynamic reaction of the SDL system to the load figures on the executing machine (see also section 5.13.3). Concerning input queues, an approach is to check the number of signals in a queue each time a signal is inserted. If the number of signals exceeds a certain level, a signal could be generated by the CBM module and sent to an SDL process that is in charge of overload handling. However, note that rather tight timing constraints have to be enforced for this to work. Thus, the prioritized handling of this kind of signal has to be ensured, otherwise the signal may just get stuck in another overloaded input queue.

In order to implement this, extensions and modifications to the SDL language have been proposed. An example is the introduction of priorities for signals which allows overtaking of signals in the input queue. With this approach, scheduling is based on the priorities of the signals, i.e. the signal with the highest priority in the system is processed first, independently of its position in the input queue. Another extension that has been used for overload handling is based on the extension of the priority input concept of SDL to support more than two input priorities.

5.7.2 Data Structures for the Input Queue

The implementation of the input queues is important for the efficiency of the system.

Requirements on input queues in SDL In order to cover the full SDL language, two requirements for the queue are of special importance.

- *Random access on the elements in the queue:* random access is needed to scan the queue and to remove arbitrary elements. This is due to the complex semantics of triggers in SDL.

- *Buffering of elements of variable size:* due to the fact that a parameter of an SDL signal may be of variable size, a flexible memory management scheme is needed.

These requirements are relaxed where a given SDL description restricts itself to a subset of the SDL language. Depending on the subset of SDL supported (or used), the efficiency of the implementation may differ considerably.

5.7.2.1 Shared-Memory Implementation

We assume the availability of common memory accessible to all process instances and also to the entities implementing the support functions, i.e. timer management and interfacing. In this case, the input queues can be implemented by data structures in the shared memory area.

Solution for full SDL Due to the complex semantics of SDL concerning transition triggering and, thus, the consumption of signals from the input queue, a simple linked list is often not the best solution. This is because the semantics of SDL require the removal of arbitrary elements from the queues, e.g. to handle saves or the removal of a timer signal.

In order to efficiently handle random access to the elements of the queue, a doubly linked list is more appropriate where the typical number of elements in the list is large. With such a list, each element maintains a forward and a backward reference. This allows the direct removal of an element without the need to traverse the linked list to find the preceding element.

In order to support the handling of signals of variable size, a rather general memory management scheme is needed. Examples are memory management schemes as provided by common runtime libraries, e.g. the *malloc* and *free* procedures provided in C. With the implementation of memory management, there is a tradeoff between the fast implementation of the memory management routines and the efficient usage of the available memory. The use of the routines provided by runtime libraries is not necessarily the best choice. They are often optimized for efficient memory usage rather than for speed. In addition, these algorithms are optimized for variable-sized elements. Especially where the usage pattern of the memory management routines is more restricted, more efficient solutions exist. Also be aware of the memory fragmentation that may occur over time where memory chunks of variable size are used.

Preallocation of equally sized buffers Under the assumption that most signals used in a system are of fixed or of similar size, faster schemes are possible. For example, a simple and fast memory management scheme could use a set of buffers of equal size. The size of the buffers is selected to hold the largest signal to be transmitted in the system. Thus, any transmitted signal gets a buffer of the maximal size. This considerable eases memory management. In this simple case, a simple list of all free buffers is maintained. With a request for a buffer, the first element from the free list is selected. Deallocation of a buffer is equally simple.

The big advantage of this simple scheme is that no search for an appropriate memory area is needed. In addition, no memory fragmentation may occur. However, the clear disadvantage of the approach is the pure utilization of the memory provided by each buffer. Pure memory utilization results where the average size of the signals differs from the maximum size.

Preallocation of a set of buffer pools One solution is to preallocate a number of signal buffers for each signal type or for each signal size. Thus, there is a separate buffer pool for each signal type or signal size. In this case, allocation and deallocation are implemented by simply removing or inserting a buffer into the respective free buffer list. Large signals can be handled separately, employing a conventional memory management scheme. If buffer pools are maintained for signal types, the signal identifier needed for the communication and possibly other information have to be inserted into the buffer only once during system initialization.

Another solution is to preallocate a pool of buffers of some given size. A buffer from the pool is used for each signal that is transmitted. If a signal does not completely fit into the buffer element, additional memory is acquired, employing a regular memory management scheme.

Partitioning of the buffer space A related question is the number of elements in the buffer pools, and the partitioning of the available memory space among different buffer pools. The memory space may be partitioned statically among different buffer pools or alternatively partitioned dynamically, depending on the demands of the application. Dynamic partitioning allows for a more efficient usage of the memory space. On the other hand, static partitioning is easier to handle from the control theory standpoint.

A related question is the definition of buffer pools, or in other words which communication is handled by which buffer pool. Note that this also relates to control theory. The selected solution has a major impact on the system behavior in overload situations.

Dynamic bypassing of the input queue With producer-based trigger testing in place, the input queue can be bypassed under certain (dynamic) conditions. Thus, the signal is copied into the receiving process instance right away. This is possible if the input queue of the receiving process instance is empty and the present signal actually triggers the receiver. In other words, the signal has to be consumed rather than saved (explicitly or implicitly). If the receiver also handles priority inputs in the state and the given signal is not part of a priority input, it has to be ensured that the receiver is scheduled right away.

Elimination of the input queue If it can be ensured that the input queue is always empty when a signal is received, the respective input queue can be eliminated altogether. However, note that this requires support from the scheduler. In other words, the scheduler has to ensure that the signal is consumed before the subsequent signal arrives. This only works for special cases. In section 5.12, we will discuss how and under which circumstances this can be applied to the implementation of protocol architectures.

Common input queue for all process instances In principle, the signals in a system can be maintained in a single common queue instead of distributed over a large set of input queues. The order given in the joint queue completely defines the order of the signals in the (now imaginary) input queues of the process instances. Thus, the two solutions can be considered equivalent.

A joint queue has an advantage where only simple operations on the queue have to be supported, i.e. an enqueue and dequeue operation. A joint queue is definitely not appropriate where a search of the queue is required. Simple operations on the common input queue are sufficient where the SDL description is limited to a very small subset of SDL. This is also the application area of the common queue.

In simple cases, the input queue can be jointly used to store signals and also as the ready queue of the system. Thus, a signal in the queue intrinsically enables the receiving process instance.

Concurrency control The approaches described above rely on the fact that all accesses to the common data are sequential and no interleaving occurs. For example, there are no interrupt routines in the system that access the data. This forces the implementation to employ polling to implement timers and interfacing with the environment.

In order to support interleaving of the different activities, concurrency control has to be put in place to ensure consistency of the shared data. Note that only the data structures or input queues where interleaving is possible have to be protected. For example, if an interrupt handler is employed to process external input, and provided this is the only asynchronous activity in the system that can access common data, then only the input queues accessed by the interrupt handler have to be protected. All other input queues do not require concurrency control. However, note that concurrency control is also needed where the asynchronous activities access other common resources, e.g. a common buffer pool. On the other hand, when parallel processing of the process instances is supported, all input queues have to be protected to ensure mutually exclusive access.

The input queues in the implementation can all be protected by a single synchronization element. In this case, only one active entity can operate on any input queue at any time. This may cause contention. Alternatively, each input queue can be protected by its own synchronization construct. Thus, different input queues can be manipulated in parallel.

In order to ensure consistency, simple synchronization mechanisms provided by operating systems or directly by hardware are sufficient. Examples are the use of binary semaphores, or the use of monitors. Under some circumstances, the temporary disabling of interrupts when operating in a critical area may be an alternative. However, note that disabling interrupts does not work on parallel systems.

5.7.2.2 Implementation by Message Passing

Above, two major requirements on the data structures to implement the input queue have been identified, namely random access to the queue and buffering of variable-sized elements. While the second requirement typically does not cause a problem, the requirement for random access poses a severe restriction.

Random access on input queue Random access is usually not supported by interprocess communication mechanisms provided by the operating system. Thus, some workaround is needed.

There are two major alternatives:

- eliminate the need for random access, i.e. restrict the subset of SDL such that no random access is needed, or
- maintain all received signals separately within an additional internal queue that provides random access.

Communication and enabling Another difference when using interprocess communication mechanisms provided by the operating system relates to synchronization. With message-passing systems, waiting for an input is typically tied with enabling of the receiver. Thus, the process is blocked until an input is available. This is not a problem as long as there are no other trigger conditions which are not tied to the receipt of an input signal. In this case, the implementation has to ensure that blocking is released also if any other trigger condition arises.

Restriction of SDL Remember that random access to the input queue is needed to properly handle save constructs. Saving of a signal is needed for various reasons, including enabling conditions, handling priority inputs and others. The need for random access to the input queue can be eliminated where saves (explicit and implicit) are eliminated. However, as saves are used for various purposes, this poses a severe restriction on the usage of SDL.

If some mechanism is provided by the operating system to concurrently listen to more than one input port, a restricted version of priority inputs can be supported. The restriction necessary for the usage of priority input is that any signal used in a priority input construct of an SDL process must not be used in a regular input construct of another transition of the same process. In this way it is ensured that the priority is associated with the signal itself independently of the state of the SDL process. Thus, priority inputs (signals) can be handled by a separate (high-priority) input queue (refer also to section 5.5.3.2). Conversely, if the priority of a signal depends on the current state of the SDL process, the priority of signals may be changed when moving to another state. This prohibits the use of separate low- and high-priority input queues.

Some operating systems support mechanisms to directly listen to more than one input port. An example is the select system call provided by UNIX. If such a mechanism is not provided, (software) interrupts could be an alternative to implement priority inputs while regular inputs are handled by the standard interprocess communication mechanisms. This implementation takes into account the exceptional character of priority inputs and allows these inputs to be processed with preference.

Second internal input queue In case random access to the input queue is indispensable, random access has to be implemented within the process itself. Thus, the signals received at the input port of the process are moved to a second internal queue before being processed, or in other words moved to a memory area which is randomly accessible to the process instance. Thus, the arrival of an input results in copying the signal(s) from the input port maintained by the operating system into the internal queue maintained by the process itself. However, note that this adds extra overhead to the implementation.

When all signals have been moved from the input port to the internal queue, an appropriate action is selected employing similar strategies to test trigger conditions as described in section 5.5.3. In this way, priority inputs and save constructs present with the current state of the process instance are taken care of.

Under the provision that priority input is not used by a process instance or handled separately, an optimization of the above scheme is possible. In this case, signals are removed from the input port and possibly saved in the internal queue only until a signal that triggers a transition is found. All following signals are left in the input port, rather than moved to the input queue right away. Thus, the internal input queue contains saved signals only. When the process instance enters the next state, the internal queue of saved signals is processed first, before a signal is consumed from the input port.

Asynchronous conditions and the view/reveal concept There are some more differences to the shared-memory approach. These relate to the handling of reveal/view constructs, and the implementation of asynchronous conditions, i.e. conditions that may change while the executing process instance is waiting. Examples of this are the use of the view construct or access to the time in a continuous signal or an enabling condition. These concepts are typically not implemented with approaches that do not support shared memory. This is because the cost of implementing these concepts based on message passing is very high. Thus, the designer will typically rather revise their SDL description to transform these concepts to other communication mechanisms that can be implemented more efficiently.

5.8 Interfacing with the Environment

The interface with the ENVironment (ENV module) is in charge of implementing communication with the environment of the system. The design of the interface is crucial to the design of the system. The design and implementation of the interface module highly depends on the services provided by the underlying system, e.g. whether a traditional operating system or a real-time operating system is employed. The design of the interface has a major influence on the remaining parts of the system. Thus, interfacing should be among the first things to be considered when designing a system.

5.8.1 Functionality

SDL communication mechanisms do not distinguish between communication between process instances and communication with the environment. Both are based on signals, employing input and output constructs. Thus, communication is transparent from the view of the SDL processes.

The task of the ENV module is to provide these services for communication beyond the interface of the local processing system. Thus, the ENV module provides the implementation of the signal communication mechanisms that interface to the environment. Depending on the concrete system at hand and the constraints and requirements of the environment, different implementations are possible.

So far, we have assumed that the `ENVenv_manager` procedure is called on a more or less periodic basis. We have assumed that the `ENVenv_manager` completely implements the interfacing with the environment. The idea behind this is that the procedure checks an output queue in which signals transmitted to the environment are placed, and tests if external input data are available at the interface. The respective structure of the `ENVenv_manager` procedure is outlined in figure 5.26.

Signals to be sent to the environment are placed in a special output queue by the sending process instances. This can be done in exactly the same way as the sending of a signal to another process instance. Signals in the output queue are put on the output interface to make them available to external devices.

```
(1) IF external_output()
(2)    process_external_output()
(3) IF external_input()
(4)    process_external_input()
```

Figure 5.26 Outline of the possible structure of the `ENVenv_manager` routine

If external input is available, a signal is generated and inserted into the input queue of the receiving process instance. This again can be done in a very similar way to the transmission of signals between process instances themselves.

The outlined design of the interface module is based on the assumption that polling is used to identify external input as well as output. This is not necessarily the best implementation strategy. In the following we discuss other alternatives for interface design.

5.8.2 General Design Alternatives

In designing the interface module, there are two important questions that influence the design:

- the required response time to react to events at the external interface, i.e. the allowed time to retrieve available input data or to provide output data, and
- the degree of decoupling and interaction between the I/O part and the application.

The allowed response time to retrieve available input from the interface has a considerable influence on the architecture of the interface module. The topic also relates to the question of decoupling I/O processing from the application. Decoupling is important where the absence of progress in the processing of one entity is not allowed to influence or hamper progress in another entity. An example of the absence of progress may result from blocking of a process on a system call to wait for input.

5.8.2.1 Access to External Data

Depending on the specifics of the underlying system, there are two principal approaches to acquiring access to external data. These are

- the use of communication primitives provided by the operating system, or
- the direct access of interface hardware.

The first approach is typically employed with traditional operating systems, while the direct access to the hardware is used with real-time systems. Depending on the specific processor, direct access to hardware may be based on memory-mapped I/O, i.e. the interface routines directly access an address in the system address space which represents special hardware, or use specific I/O instructions provided by the processor to access I/O ports.

5.8.2.2 Synchronization with the Application

Concerning the handling of external interfaces, there are three approaches to notifying the application of the occurrence of an event. An event may be related to input as well as to output. It may reflect the availability of input data or, conversely, indicate that the external hardware is ready to receive data. The three alternatives for the design of the entities that handle the events are as follows:

- the entity blocks until the event occurs, i.e. until external input is available or external output can be transmitted,
- the entity polls the interface repeatedly to see when an event occurs, or
- the entity is directly triggered (called) by the hardware (hardware interrupt) or the operating system (software interrupt).

We will discuss the different approaches in turn.

Blocking Blocking may occur on a system call for communication provided by the operating system or simply on a synchronization construct. In the case of a blocking system call, unblocking occurs as soon as data are available (input) or data can be provided (output). With the use of a synchronization construct, unblocking occurs when another entity enables the blocked entity.

We have seen previously that blocking may cause problems. When a process is blocked, it typically waits for a single event to occur. Thus, other events which may be relevant to the process do not result in unblocking the process. As a consequence, blocking on a single event should be used only when no other events can occur that are relevant to the process in the current state.

Polling External input or output interfaces may be polled for the presence of data or the readiness of the external device to receive data. Polling should occur at rather regular intervals to ensure the timely handling of the data.

Polling results in semi-active or even active waiting. Thus, it consumes processing resources while waiting. This causes problems in a system where several entities compete for processing resources. In order to prevent this, polling should be employed only in a main loop executed by the system. Examples of such loops are the main loop executed by the RTSS as described with bare or light integration, or the main loop (scheduler) executed by the operating system.

Hardware interrupts Instead of having the software monitoring the interface, this can be implemented in hardware as well. This is called interrupt handling. If an interrupt is raised by the environment, the hardware immediately jumps to a given memory address to execute the instructions stored there. Thus, a special procedure is executed, i.e. the interrupt routine. There are several alternatives to the implementation of the interrupt routine.

The interrupt routine could directly handle the input data, i.e. remove them from the interface and buffer them somewhere, or instruct some other entity to take some action. The first case is rather problematic as the interrupt handler may not execute instructions that itself may block. This is because blocking the interrupt routine typically results in deadlock of the system.

However, recall that some synchronization mechanism is needed to ensure consistency of the data structures. A potential solution is the use of constructs that enable and disable interrupts. Thus, a process executing in a critical section temporarily disables interrupts to ensure that data are kept in a consistent state. However, as noted before, this only works for single processor systems.

Instead of the direct retrieval of the input data by the interrupt routine, the interrupt could just enable a process that is in charge of retrieving the external input data. An example of this is the combination with blocking on a synchronization construct, as described above, i.e. the execution of an unblock operation (`UP(S)`) on a binary semaphore.

Software interrupts Another alternative is software interrupts, also called signal handlers or second-level interrupts. In this case, the (hardware) interrupt routine instructs the operating system to send a signal (note that this is different from a signal in SDL) to a specific process. This results in the operating system interrupting regular processing of the specified process and processing the specified signal handling routine instead. The signal handler is executed within the address space of the interrupted process. In our case, the signal handler can be used to retrieve data from the external interface.

However, again special care is needed where potentially blocking constructs are issued. A solution is to check if a lock is held instead of acquiring it right away. To clarify this, consider the lock being held by the interrupted process itself. In this case, the attempt to acquire the lock in the signal handling routine results in its preemption. Depending on the implementation of signal handling, this may result in deadlock. In case the preemption of the signal handler can result in deadlock, some alternative actions are needed. For example, a retrieved signal could be temporarily stored in an intermediate buffer and later moved to the shared medium when the lock is available.

5.8.3 Output

The output of an SDL signal to the environment can be implemented in a synchronous way or asynchronously, i.e. decoupled from the sending SDL process instance.

5.8.3.1 Synchronous Output

Often, the SDL output construct is directly mapped on some function or system call that immediately outputs the SDL signal. Thus, either the data are directly written to the interface hardware, or a system call provided by the operating system is used. In the simplest case, the SDL output construct is directly mapped on some `WRITE(device_addr,signal)` command that writes the given signal to the specified device.

```
(1) FOR i:= 1 TO size_of_signal(signal)
(2)    WHILE READ(device_status) ≠ READY
          /* active waiting for interface to accept new data */
(3)    WRITE(device_addr, buffer[i])
```

Figure 5.27 Synchronous output with active waiting

Often, the sender is faster than the receiver, or a signal has to be transmitted in a set of chunks instead of a single output operation. Thus, the sender may be blocked until the receiver is ready to receive more data. The outline of a simple algorithm to handle this is given in figure 5.27.

Instead of active waiting for the receiver to get ready, blocking the sender, i.e. the sending SDL process instance, may be an alternative.

5.8.3.2 Asynchronous Output

With the above solutions, the process instance that outputs a signal is blocked until the signal has been transmitted. In some cases, e.g. when the receiver is not always ready to receive data, the decoupling of the SDL output construct from the receiving instance may be advisable. In this case, the SDL output construct just places the signal in an output queue similar to an output to another SDL process. The external output queue is than processed asynchronously by some other entity, taking into account the speed and special requirements of the interface.

In this case, the same constructs as for internal signal communication can be employed to implement the external output from an SDL process, e.g. the `CBMallocate` and `CBMinsert_signal` routines provided by the CBM module.

The output of the signal to the environment by the asynchronous entity can be implemented as described above, i.e. by polling, blocking or interrupt controlled. For example with interrupts, an interrupt is raised by the environment when it is ready to receive (more) data.

5.8.4 Input

In order to process external input, similar approaches to the processing of external output exist.

5.8.4.1 Asynchronous Input

Communication in SDL is based on the asynchronous communication paradigm. This also holds for communication with the environment. Thus, the design of an asynchronously operating interface seems to be most natural. In other words, the reception of external input signals is decoupled from the processing of the signal. The decoupled entity handling the interface may employ the techniques described above to learn about an input event, e.g. polling, blocking or interrupt-controlled.

Polling The asynchronous input handler may poll the state of the input port maintained by the operating system, or, if direct access to hardware is supported, poll a special address area in memory used for I/O. Polling may be implemented by periodic calls of the `ENVenv_manager` procedure. When an input signal is detected, the procedures provided by the CBM module are employed to insert the signal in the appropriate input queue. In addition, the receiver of the signal is typically enabled.

Blocking If the entity handling external input is completely decoupled from the remaining parts, i.e. implemented by a separate process maintained by the operating system, blocking may be employed as well. Thus, a special process blocks on some input channel (socket, etc.) by issuing a blocking system call. When an input arrives, the system call is unblocked and the data are mapped on a signal and inserted in the input queue of the receiving process instance. In this case, unblocking of the special interface process is done by the operating system upon the reception of data from the environment.

With a real-time operating system, the process may block on a synchronization construct as well. The blocked process is unblocked by an interrupt handler which in turn is triggered by the environment when new data are available.

Interrupt handler Alternatively to the above approaches, an interrupt handler may be directly employed to process input. In this case, the interrupt routine directly inserts the signal from the environment into the SDL system, i.e. into the input queue of the receiving process instance. This approach is often employed with microcontrollers and with real-time operating systems.

5.8.4.2 Synchronous Input

Instead of decoupling the input from the application, a synchronous approach may be employed in special cases. The advantage of the synchronous processing of input is the minimization of overhead due to the elimination of the redirection. In other words, the input signal can be directly processed by the receiving process instance without the need to move the signal into the input queue first.

With the synchronous implementation of input, the input directly triggers the execution of the receiving process instance. Thus, the process instance blocks on the input, or, alternatively, is directly implemented as part of the interrupt handler.

Under certain provisions, this synchronous execution principle can be pushed further by integrating several process instances into a synchronous execution. This results in the application of the activity thread model as introduced in section 3.3.2.2.

5.9 Timer Management

Timer Management (TM module) is in charge of supporting timers and providing access to the current time as supported by SDL.

Routine	Function	Calling entity	Usage
TMset_timer(PId,TId,expiration)	set timer	EFSM	SDL set construct
TMreset_timer(PId,TId)	reset timer	EFSM	SDL reset construct
status:= TMtest_timer(PId,TId)	test state of timer	EFSM	SDL active construct
time:= TMget_time()	get actual time	EFSM	SDL now construct
TMtimer_manager()	check timers (and update time)	PIM TM OS	light integration separate OS process implicit by OS

Table 5.5 Routines provided for time and timer management

5.9.1 Functionality

SDL supports the following functions for timer and time management:

- set a timer to a certain expiration time,
- reset a timer, i.e. delete the timer or remove the timeout signal from the input queue,
- test the state of a timer, i.e. whether it is active or has expired, and
- read the system time.

In SDL, the notification of a process instance of the expiration of a timer is supported by inserting a timeout signal in its input queue. The actual action taken on a timeout may depend on the state the process instance is in.

Note that there may be more than one timer active for a single process instance at any time. In addition to simple timers, timers may use parameters. Thus, arrays of timers can be employed. Also note that each timer is local to a single process instance. Thus, the timeout resulting from an expired timer is inserted into the input queue of the process instance that has set the timer.

The functions needed to implement SDL timers and time are summarized in table 5.5. The `TId` parameter used by the routines specifies the timer instance; `expiration` defines the timer expiration time.

The given routines directly reflect the functionality of the respective SDL constructs to manipulate time. The only exception is the `TMtimer_manager` procedure. Time passes more or less continuously. Thus, some entity is needed to model this, i.e. to advance the time and to check for expired timers. The `TMtimer_manager` procedure checks for the expiration of a timer and possibly also advances the time implicitly used by the timer constructs or explicitly accessed with the now construct. Alternatively to the explicit advancement of the SDL time, some time variable maintained by the operating system or directly implemented in hardware may be used as the SDL time.

If the `TMtimer_manager` encounters an expired timeout, it creates a timeout signal, inserts it into the input queue of the owning process instance and enables it. Thus, the process instance is ready and may consume the timeout signal.

Problematic is the implementation of the now construct when used as an asynchronous condition, i.e. employed by a continuous signal or enabling condition. In this case, enabling of the respective process instance has to be ensured when the condition turns true.

At first glance the functionality of the `TMset_timer` procedure is not obvious. Before a timer is set, it has to be ensured that any previous activation of the timer is canceled. Thus, the `TMreset_timer` procedure is called within the `TMset_timer` procedure.

5.9.2 Data Structures

An important question relates to the implementation of the data structures for timer management. The issue is common to almost any timer management scheme independently of its implementation in the RTSS, the operating system, or by a separate process.

Central linked list Commonly, some kind of linked list is employed to implement timers. Typically, the list is organized in increasing order of the expiration time of the timers. Thus, the timer manager has to check the first element of the list only to see if a timer has expired. However, conversely to the cheap test for timer expiration, insertion of a timer in the list (SDL set construct) and also its removal (SDL reset construct) are more expensive. For these operations, the list has to be traversed until the respective element is found. This is especially expensive when the number of active timers is large, as may be the case with communication protocols.

Linked list per process instance In order to support an efficient implementation of the set and reset constructs, a direct association of timers with their respective process instances has been proposed. It allows each process to directly access and manipulate its active timers. Thus, setting and resetting the timers by the owning entity can be implemented efficiently. This is especially important for the implementation of communication protocols where the majority of timers are reset before their expiration. However, testing the timers is more expensive. For this, the timers of all process instances have to be checked periodically. Thus, the approach is especially beneficial when the period for testing the timers is large.

Central doubly linked list with reference A compromise between the two approaches is possible as well. Thus, the timer is inserted in the global timer list. In addition, a direct reference to the timer element in the list is maintained by the owning process instance. Removing a timer element from the linked list can be implemented efficiently when a doubly linked list is employed, i.e. each element maintains a forward and a backward link. Thus, test and reset operations can be implemented efficiently. However, setting, or in other words inserting a timer in the ordered timer list, remains expensive.

Obviously, the most efficient implementation of timers depends on the circumstances in which timers are used, e.g. the average number of active timers, the testing period and the timer usage patterns.

5.9.3 Implementation Alternatives

Having discussed the data structures to implement timers, we take a closer look at the way the timer manager as a whole is implemented.

The following design alternatives exist:

- timer management by the RTSS, i.e. a timer procedure tests more or less periodically whether a timer has expired, solely based on some time variable provided by an underlying layer,
- timer management completely handled by (within) the operating system,
- timer management by a separate process,
- decentralized timer management by each process instance,
- timer management by an interrupt handler, and
- the direct implementation of timers in hardware.

However, note that the different approaches are based on different prerequisites. Thus, not all alternatives are available with all system integration strategies and with every operating system or hardware.

5.9.3.1 Timer Management by Procedures of the RTSS

A simple approach that is typically employed with light and bare integration is the implementation of the timer management within the RTSS. In other words, the procedures given in the table above are directly implemented in the RTSS requiring minimal support from the underlying layers.

The sole requirement of the approach from the underlying system is access to some time base. As outlined in section 5.6.3, the `TMtimer_manager` is repeatedly called (polling) from the RTSS (process instance manager) while executing in its main loop. With each call, the `TMtimer_manager` checks whether a timer has expired. If this is the case, it reacts by sending a timeout signal to the owning process instance, and executing some operations for maintenance of the timer data structures. Sending the timeout signal can be implemented in the same way as regular signal output is handled by the implementation. Thus, typically a signal is inserted in the input queue of the receiver and the respective process instance is enabled by issuing the `PIMenable` procedure. Note that overhead for timer management as well as the accuracy of the time depends on the frequency of the calls of the `TMtimer_manager` procedure.

Rather more complicated is the implementation of the `TMreset_timer` procedure. This is due to the fact that a timer to be reset may be in three different states. If the timer has not expired yet, the timer can be simply removed from the timer list. If the timer has expired but not been consumed yet, the respective timeout signal is present in the input queue and has to be removed. If the timer has been consumed already, i.e. is neither present in the timer list nor in the input queue, no further actions are needed.

The other routines given above simply manipulate the timer data structures, especially the timer list described above. Concerning the implementation of the timer data structures, the alternatives and discussion given above apply. Thus, timers may be implemented by a central timer list or directly associated with the owning process instances.

A problem of the polling approach is that the RTSS uses up all processing resources that are provided to it even when there is no work to do. On the other hand, repeated polling can only be avoided, i.e. replaced by blocking, when some wake-up mechanism is provided. In addition, it has to be ensured that the wake-up mechanism catches all possible events, e.g. the expiration of a timeout and any external input. In other words, the case in which the RTSS blocks on a timeout while an external input arrives, or vice versa, has to be avoided. Implementing the timer manager and the interface manager in the RTSS directly, this is not trivial in the general case. Especially it requires a tight cooperation of the interface manager with the timer manager. However, exceptions exist for specific SDL descriptions that limit themselves to a subset of SDL. Below we will see other approaches to avoiding active waiting due to repeated polling.

5.9.3.2 Timer Management by the Operating System

Under the provision that appropriate and efficient timer services are provided by the operating system, timers can be directly handled by the operating system. However, note that this approach relies on the operating system to provide timer services that resemble the timer services provided by SDL. The approach fits most naturally the tight integration principle, i.e. the case that each process instance is implemented by a separate process. In this case, a timeout can be issued directly to the process that implements the process instance owning the timer.

As a consequence of the direct implementation by the operating system timer mechanism, the routines given in the table above are – with the exception of the `TMtimer_manager` procedure – directly mapped on system calls. The `TMtimer_manager` procedure is implicitly implemented within the operating system kernel.

An important issue with this approach is the mechanism used to notify a process that a timeout has occurred. Depending on the specific operating system at hand, different mechanisms are supported. For example, a process may issue a system call that is blocking the process until the timeout occurs. Thus, the process is just waiting for the timeout. No other work may be done during this waiting period. As there are typically other events in the SDL system that may trigger a transition of the process instance, this approach is appropriate for very simple cases only.

Often, operating systems also support the association of a timer with some other blocking system calls, e.g. waiting at an input port. This may be used in special cases, i.e. provided the given SDL description restricts itself to a subset of SDL. However, it does not provide a flexible way of implementing arbitrary SDL descriptions.

Most operating systems also provide some kind of asynchronous signal handling mechanism. In this case, the expiration of a timer results in the interruption of the normal flow of control of the receiving process. As a result of the timer expiration, a specific signal handling routine – provided by the user – is called by the operating system. Unlike the above approaches, this mechanism supports the issuing of a timeout while the process is in regular operation, e.g. while executing a transition of the process instance, or being blocked.

The signal handler may be used to unblock the process by calling the `PIMenable(PId)` procedure with its own PId. Depending on the implementation of the `PIMenable` procedure, this typically moves the process into the ready queue, thus, enabling the operating system scheduler to schedule the process for regular operation.

The efficiency of the timer mechanism depends on the implementation strategy employed within the operating system kernel, e.g. whether timers are optimized for set–reset patterns or not, and on the period for the testing of timeouts used within the kernel.

Another issue is that timer mechanisms provided by operating systems usually do not support parameters. Thus, only one timer may be used per process. This is in conflict with the use of timers in SDL. Thus, additional support within the RTSS is needed to map multiple timers on a single timer handled by the operating system. In other words, large parts of timer management are handled within the RTSS by maintaining a timer list for each process instance. The timer handled by the operating system itself is set to the expiration time of the SDL timer that expires first.

Due to the possible limitations of the timer mechanisms provided by operating systems, the build-in mechanisms are often only useful where the SDL description makes only restricted use of timers. Examples are the prohibition of parameterized timers, or the use of a single timer per process instance only.

5.9.3.3 Timer Management by a Separate Process

Instead of bothering with the very details and peculiarities of the timer mechanism provided by the host operating system, timers may be implemented in the user space by a separate process. This only requires minimal functionality from the operating system, i.e. just some facility to periodically trigger the timer process.

In this case, the timer process executes the `TMtimer_manager` procedure outlined in figure 5.28. The procedure blocks until it is triggered by a timeout signal from the operating system. Upon triggering, the process tests the SDL timers for possible expiration. If an SDL timer (or several) has expired, the timer is removed from the timer list, a timeout signal is inserted in the input queue of the process instance owning the timer, and the `PIMenable(PId)` routine is called to enable the owning process instance to process the timeout.

The other routines given above simply manipulate the timer data structures, especially the timer list described above. As with the direct use of timers by the operating system, special care has to be take to implement the `TMreset_timer` procedure. Here again, the timer list as well as the input queue of the owning process have to be scanned to remove the timer.

```
(1) FOREVER
(2)   wait (block) on periodic timeout
(3)   FOR EACH expired timer
(4)     remove timer from timer list
(5)     generate timeout signal
(6)     insert signal in input queue
(7)     PIMenable(PId)
```

Figure 5.28 Timer management by a separate process

Concerning the implementation of the timer data structures, again the alternatives discussed above apply. Thus, timers may be implemented by a central timer list or directly associated with the owning process instances. However, note that concurrency control has to be in place to ensure consistency of the shared timer data structures.

The efficiency of the scheme depends on the period to trigger the timer process. Thus, there is a trade-off between the accuracy of the SDL timers and the overhead involved with testing the timers.

Instead of periodic triggering of the timer process, the process may be triggered by the operating system at the earliest expiration time of the SDL timers currently set. Thus, the respective timer managed by the operating system is set to the value of the earliest expiration time of all the SDL timers the timer process manages.

Compared to the periodic approach, the number of operating system timeouts may be reduced, On the other hand, extra cost may be involved for setting and possibly resetting the timer maintained by the operating system.

With both variants, SDL timers are handled by a central process. Thus, no timer manager has to be employed by the processes implementing the process instances. The centralization results in a minimization of the effort for searching for expired timers. However, note that the central timer process results in extra overhead for synchronization between the processes implementing process instances and the timer process. Especially, the `TMset_timer` and `TMreset_timer` procedures are expensive. This is due to their shared use and to the larger number of elements managed by a central timer manager. An advantage of the approach is that only minimal support is needed by the operating system. Thus, the timer process (`TMtimer_manager` procedure) is rather independent of the operating system used.

5.9.3.4 Decentralized Timer Management by Process Instances

With this approach, each process instance is completely responsible for its own timers. No central support is provided to manage SDL timers. Support from the operating system may range from providing a simple time variable to a complete timer management scheme.

The implementation of this approach highly depends on the selected system integration strategy. With light integration, each process instance repeatedly tests its timers by comparing the expiration time with the current time. For this to work, the process instance manager has to ensure that each process instance is enabled more or less periodically. Otherwise, the process instance does not have a chance to test its timers.

With tight integration, more sophisticated support from the operating system may be employed by each process that implements a process instance.

5.9.3.5 Timer Management by a Single Interrupt Handler

The use of an interrupt handler routine to handle timeouts is very similar to the use of a separate timer management process. This approach is especially suitable with real-time operating systems, where straightforward access to the hardware is supported.

The use of an interrupt handler to handle timeouts requires some clock mechanism that is able to periodically raise an interrupt. Upon the occurrence of a clock interrupt, the interrupt handler routine is called by the operating system or directly by the hardware and takes care of testing the timer list for expired timeouts. Thus, the interrupt handler routine implements the `TMtimer_manager` procedure.

The other timer procedures are implemented according to one of the approaches described above. The procedures maintain the timer list and operate on the input queues of the process instances. The exact implementation depends on the system integration approach the timer management scheme is combined with. In principle, timer management using interrupt handling may be used with tight as well as with light or bare integration. With light integration, concurrency control has to be put in place before interrupt handling can be used. This is needed to ensure mutually exclusive access to the timer data structures. With tight integration, concurrency control is typically in place independently of the implementation of timers.

In any case, special care is needed when concurrency control mechanisms are used within interrupt handler routines. As already discussed in section 5.8.2.2, this is because interrupt handler routines as very low-level mechanisms must not block. Thus, any system calls that may block cannot be used here, e.g. the acquisition of a semaphore or waiting for an event.

5.9.3.6 Timer Management Directly in Hardware

Typical computer systems employ a single hardware timer mechanism and provide the remaining functionality in software. However, other solutions are also possible, especially where timers are critical to performance. Instead of implementing timer management somewhere in software, the direct implementation in hardware is possible as well. For example, this is supported by some microcontrollers that provide more than one timer. Alternatively to the on-chip solution, additional timers may be added at the processor interface. A comprehensive overview of the implementation of timer mechanisms in hardware is provided in [Heat97].

Implementing a timer in hardware basically requires a register addressable by the processor and a counter. Each timer may raise a specific interrupt when the counter hits zero. For each timer, an interrupt routine is provided that takes the appropriate action. Alternatively, several timers may share a common interrupt. In this case, the expired timer has to be identified using some status register.

With this approach, the `TMtimer_manager` is implemented by the respective interrupt handler routine(s). The actions performed by the interrupt routine are similar to the ones described with the previous interrupt-based approach. The `TMset_timer` procedure simply writes the number of ticks into the register which handles the respective timer. In addition, it ensures that no timeout signal is in the input queue of the respective process instance. The `TMreset_timer` procedure sets the register to zero and also removes the timeout signal from the respective input queue where applicable. The `TMtest_timer` procedure is implemented by testing the value of the timer register and possibly searching the input queue for an expired, but not yet consumed, timeout.

With the hardware implementation, the `TMset_timer` and `TMreset_timer` operation reduce to a very small number of machine instructions, provided the timer has not expired yet. If the set or reset operation has to remove a timeout signal from the input queue, much higher costs are involved. A similar argument applies for the `TMtest_timer` procedure.

The drawback of this approach has to do with the fact that hardware is not as flexible as software. Thus, one physical timer has to be available for each logical timer used within the SDL description. Since the number of hardware timers is typically very small, this highly restricts the application of the approach.

5.10 Implementation of SDL Data

As discussed in section 4.1.4, the data concept of SDL focuses on the externally visible behavior of data types, i.e. on the question of what the functionality of abstract data types is, instead of its implementation, i.e. how a certain functionality is supported.[9] Some of the data types supported by SDL can be rather abstract. Thus, they are far from typical data types as provided by programming languages. Examples of built-in data structures of SDL that do not have a natural implementation in typical programming languages are sets (powerset) and arrays indexed by characters or character strings.

Due to the abstraction, many potential alternatives for the implementation of abstract data types exist. As known from the theory of abstract data types and algorithms (e.g. see [AhHU87]), this has an important influence on the performance. Thus, the issue is important for the derivation of efficient implementations from SDL.

In implementing SDL data there are two main issues. First, the way in which the abstract data types are implemented in the implementation language is important. This includes the selection of the appropriate data structures as supported by the data type concept of the implementation language, e.g. the implementation of a set or a queue, and the implementation of the operations

[9] According to the classification of data objects in section 5.2.2, the user-defined data as described here are typically local data. However, exceptions exist where global visibility is required, i.e. where the view/reveal concept is applied to the data.

supported by the abstract data type. In addition, the exact size of data objects, e.g. of integers or reals, has to be decided.

The second issue is how to map the code and data segments, i.e. the data objects used and the algorithmic description, on instructions provided by the target machine, and how to map the data structures on the hardware entities provided by the target processor.

The first issue clearly has to be dealt with by the code generator that translates the SDL description to code in an implementation language. This will be sketched below. The second issue, i.e. how to map the code segments and data on the given target machine, is the responsibility of the compiler for the implementation language. Optimizing compilation techniques are discussed in [BaGS94, TrSo89]. An issue that can in principle be solved by the code generator or the compiler is the efficient implementation of algorithmic operations, e.g. the replacement of an addition by 1 by an increment operation, or the representation of characters by 32-bit words instead of 8-bit words for efficient processing on a 32-bit processor.

Obviously, the performance of the implementation of the abstract data types depends on the efficiency of the operations they provide. In addition, the specific use pattern for the abstract data type may be of importance. Often there is a tradeoff between the efficiency of different operations provided by an abstract data type. For example consider an ordered set. In this case there is a tradeoff between an efficient insertion of new elements and their removal. In addition, often considerable performance improvements can be achieved if the size of the data is known in advance, e.g. the maximum size of a queue or a set.

Concerning the implementation of SDL data types in an implementation language, the following cases can be identified.

Predefined data types can be directly mapped on corresponding data types provided by the implementation language. The only problem is the limited size of data types in implementation languages, which is different from the unlimited size of data in SDL. The same holds for predefined generators which can also be mapped on typical data types provided by implementation languages. However, efficiency problems may result from the unlimited nature of the data structures in SDL, i.e. the unknown size of composite objects as arrays or sets, which requires dynamic memory allocation. More efficient implementations of these data types can be derived if the maximum size of the objects is known, which allows for the static allocation of memory. Similar arguments apply to user-defined generators. Support for these is possible where the type definition is based on predefined data types and algorithmic operators.

Concerning the implementation of axiomatically defined data types (ACT-ONE), there are principal problems with term rewriting which preclude the automatic derivation of implementations from arbitrary axiomatic definitions of data types. A discussion of the automatic derivation of code from axiomatic specifications of abstract data types can be found in [MRCM96].

Very important in the communication area is the use of ASN.1 for the specification of data types. As ASN.1 data types are rather concrete, there is no problem with their efficient implementation. Problems concerning the integration of ASN.1 with SDL are rather syntactic in nature.

In addition, SDL also supports the use of data types provided in other languages. This is supported by means of an interface concept that allows one to refer to procedures and data types that are defined outside the SDL syntax. However, the integration of other languages is problematic, especially the proper interfacing between SDL data and data defined outside.

Considering the implementation of data in SDL is definitely an important issue for performance. However, the topic is rather independent of the behavioral model on which SDL is based. Thus, the efficient implementation of abstract data types is a general issue and not particularly specific to SDL. In addition, the issue is highly application-specific, i.e. depends on the specific application that is described with SDL. For these reasons, we have focused on the implementation of the abstract data types needed to implement the underlying semantic model of SDL, rather than on application-specific abstract data types. A thorough discussion of problems and issues related to data in SDL can be found in [Schr00].

5.11 Implementation of Other Features of SDL

The definition of SDL in Z.100 is divided into basic and advanced features of the language. Basic features are directly defined in the standard. Conversely, advanced features are defined by transformation rules to basic features of the SDL language.

5.11.1 Basic Features

The important basic features we have not covered yet are procedures and services. These will be covered in this section.

5.11.1.1 Procedures

Provided that procedure calls return before a transition completes, i.e. the procedure does not contain states, an implementation similar to the implementation of procedures in other languages can be selected.

However, if the procedure contains states, additional care is needed. As described in section 4.1.3.4, the set of states defined by a procedure is disjunct from the set of states of the calling entity, e.g. the calling process instance. If the procedure uses states, it can be considered as a separate EFSM. However, the input queue on which the procedure operates is the input queue of the calling process instance.

Taking this into account, a procedure call can be implemented by switching from the EFSM implementing the calling entity to the EFSM implementing the called procedure. When the procedure returns, the implementation switches back to the original EFSM. With respect to the `EFSMtest_trigger` procedures described in section 5.5.3, this results in switching from the `EFSMtest_trigger` procedure of the caller to the `EFSMtest_trigger` procedure of the called procedure. However, note that some care is needed due to the fact that procedures can be called from various entities.

In any case, it has to be ensured that each process instance has its own stack to store the states, local variables and other information used by the called procedure. An alternative to the switching of the EFSMs may be the inline expansion of procedures where no recursion is used.

5.11.1.2 Services

As described in section 4.1.3.4, services in SDL represent EFSMs, where each service instance, i.e. each EFSM, operates on a distinct set of input signals. Service instances operate in mutually exclusive mode, i.e. only one service instance of a process instance may execute a transition at a time. The service instances of a process instance share the common data of the process instance as well as the input queue associated with the process instance.

Due to the disjunct nature of the input signal sets of the services, the implementation of services is straightforward. The use of services has some impact on the test of trigger conditions as described in section 5.5.3. Thus, the algorithms given in figures 5.17 and 5.18 need some minor refinements to support services.

5.11.2 Advanced Features

Advanced features are defined by transformation rules applied to the SDL description to transform the advanced features to basic language features. In the Z.100 standard, the following language constructs are defined as advanced features:

- priority input,
- continuous signal,
- enabling condition,
- export/import, and
- remote procedures.

In addition, some constructs that represent simple shortcuts are defined as advanced features. This includes notations to specify common transitions for all states of an SDL process, or to specify a set of input signals to be saved or consumed.

For some of the advanced features, the direct application of the transformation rules provided by the standard is not appropriate. This is mostly due to its negative impact on performance, e.g. employing active waiting. This applies to the implementation of priority input, continuous signals and enabling conditions which have been discussed already.

The export/import concept and remote procedure calls have been introduced in section 4.1.3.5. The implementation of the two concepts has not been described so far. Both concepts can be nicely transformed to the basic SDL signal communication mechanism.

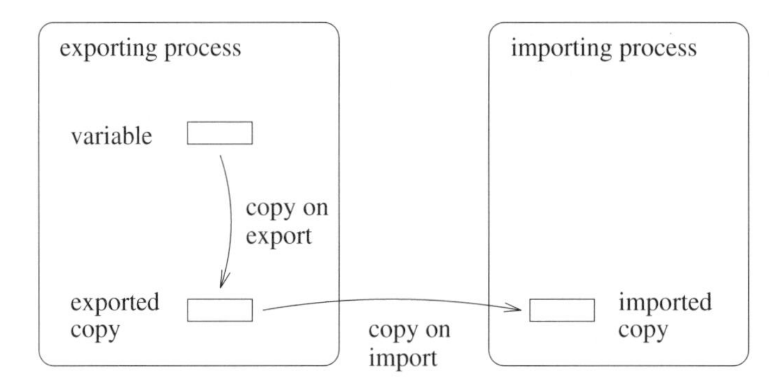

Figure 5.29 Implementation of the export/import concept by basic signal communication

5.11.2.1 Export/Import

A correct implementation of the export/import concept maintains a copy of the exported variable with the exporting process instance as well as a copy with each importing process instance. This is graphically displayed in figure 5.29. The copy associated with the exporting process is updated with the value of the original variable whenever the owning process instance executes the export construct. Conversely, the copy associated with the importing process instance is updated with the value of the exported copy of the variable whenever the import construct is used.

The export construct can be implemented by a copy operation applied to the exported data structure. The import construct can be implemented according to the transformation rules given in Z.100. The transformation rules employ a signal sent by the importing process instance to the exporter, requesting the value of its exported copy. In turn, the exporting process instance replies with a signal containing the actual value as a parameter. This can be implemented by adding the respective transitions to the exporting and importing processes. In addition, a new state is introduced with the importing process in order to wait for the reply from the exporter.

The implementation scheme ensures mutually exclusive access to the exported variable. Note that the transformation to basic signal communication bears some peculiarities which may not be expected from an export/import mechanism. Due to the fact that the import construct is implemented by sending a signal to the exporting process instance, the reply to this request may be delayed by the exporting process instance. For example, consider an exporter with a non-empty input queue. In this case, all other signals have to be processed first before the import request can be handled. This may result in a considerable delay of the import construct.

In case conformance to the semantics of the export/import concept in Z.100 is not strictly required, an alternative implementation of the import construct can be employed provided the exporting and importing process instance share a common memory area. Instead of transmitting signals, the import construct can be implemented by copying the exported value into the imported copy of the variable maintained by the importer. In order to ensure mutually exclusive access to the variable, a synchronization mechanism may be needed. However, note that this implemen-

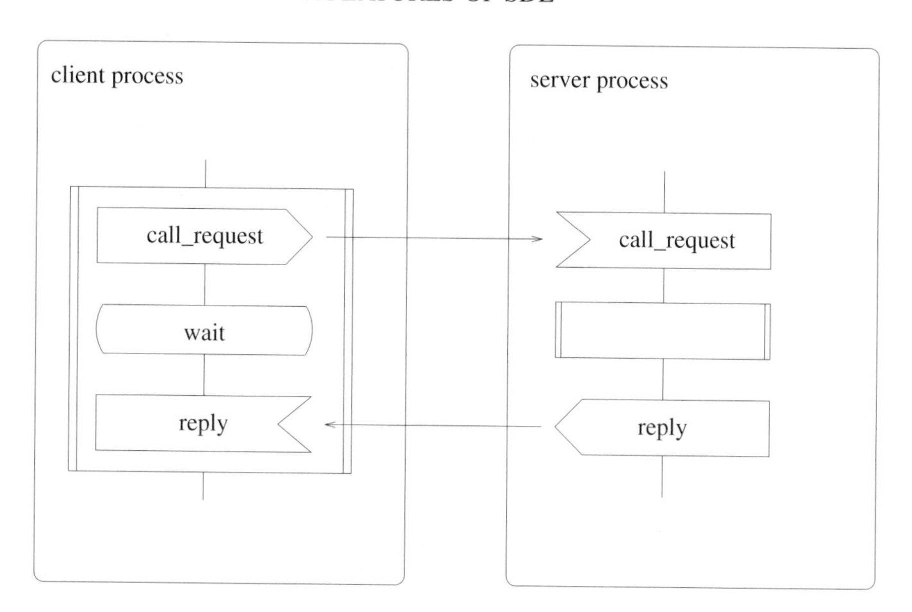

Figure 5.30 Implementation of remote procedure calls by basic signal communication

tation scheme is not in conformance with the semantics of SDL. On the other hand, copying is much more efficient than sending signals.

5.11.2.2 Remote Procedure Call

Implementing remote procedure calls, again the transformation rules given in Z.100 can be applied. This is outlined in figure 5.30. A call of a remote procedure can be implemented by sending a respective call request signal to the process instance implementing the remote procedure. The remote process instance (server) receives the call request signal, executes the respective transition and returns a reply signal with the respective return parameters of the call. In order to implement this, an additional wait state is introduced with the calling process instance (client) to allow waiting for the results of the call. In addition, some extensions to the called process instance are needed to handle the call request signal, to execute the remote procedure, and to reply to the client.

Again, mapping remote procedure calls on signal communication may not be the most efficient implementation strategy. On the other hand, the strategy conforms to the standard. More efficient implementation strategies are possible when the peculiarities of signal communication (as described above) are not important to the specific application at hand. For example, an implementation not conforming to the standard could directly implement the call. However, this requires common access by the client and the server to a shared memory area.

5.12 Optimizations for Protocol Processing

The merits of optimizing code generation techniques depend on the context in which SDL is used, i.e. the specific application area. Here, we focus on the automatic derivation of efficient implementations for a major application domain of SDL, namely for protocol processing. We evaluate the applicability of various performance optimization principles known from the manual implementation of protocol architectures. For this, the optimization techniques described in sections 3.2 and 3.3 are revisited, and their suitability for the automatic code generation from SDL is evaluated.

5.12.1 Minimizing Data Movement

Protocol architectures consist of a set of hierarchically layered protocols. Typically, each protocol layer is described by an SDL entity, e.g. a process or a block. In specifying protocol architectures in such a modular way, often rather large packets are exchanged between the active process instances of the SDL system. On the other hand, the actual amount of computation performed by the process instances that implement the protocol functionality may be rather small. Thus, the minimization of data movement provides a large potential to increase the efficiency of the derived implementation. As already discussed in section 3.3.3, the copying of packets is the most published don't of the implementation of protocol architectures. Unfortunately, code generators for SDL make abundant use of copying.

We start with an analysis of the typical copy operations employed by code automatically derived from SDL descriptions of layered protocol architectures. Then we revisit manual strategies to minimize data movement with protocol implementations (see also section 3.3.3) and discuss the applicability of these optimization strategies to the automatic code generation from SDL.

5.12.1.1 Copy Operations

Conceptually, SDL assumes asynchronous buffered communication between process instances. Employing SDL for the design of protocol architectures, the typical structure of an implementation derived from an SDL process (in SDL-like pseudo code) is depicted in figure 5.31.

The figure shows the processing of an outbound packet by a protocol instance. Typical processing steps for this case are the addition of the protocol-specific Protocol Control Information (PCI) to the packet provided by the higher layer, and the update of the protocol machine. The PCI is typically added as a header (see section 3.1.3.3 for a description of the enveloping principle).

The processing of a packet by a process instance typically involves the following four data copy operations:

(1) **Input:** The implementation of the input construct in SDL typically results in copying the packet (i.e. the signal parameters) from the input queue of the process instance into the internal address space (i.e. the local variables) of the receiving process instance.

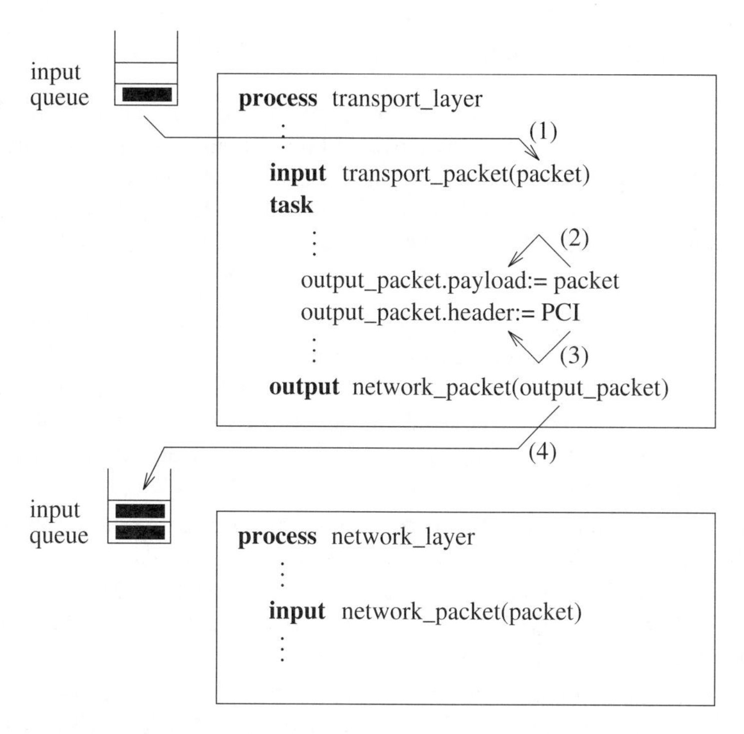

Figure 5.31 Copy operations typically involved with the processing of a process instance

(2) **Prepare output packet:** With communication protocols, typically a header is added to the packet before it is passed to the lower protocol instance (i.e. another SDL process). As a prerequisite for this, the packet is copied in a different address area to provide extra space for the PCI.

(3) **Add header:** In order to complete the output, the packet is typically prefixed with the PCI. This involves copying the PCI into the data structure holding the output packet.

(4) **Output:** Once the transition is completed, the packet (i.e. for the lower protocol entity) is transmitted to the input queue of the process instance that implements the lower protocol instance.

Summarizing our analysis of copying, it can be stated that far too many copy operations are employed to arrive at an efficient implementation. Overhead is caused not only by the mere copy operations. Often, costly buffer management operations (compare section 5.7) are employed to allocate the buffer before the actual copying takes place and to free the respective buffer after its use.

5.12.1.2 Elimination of Copy on Signal Input

Above, SDL signal communication has been identified as a major source of overhead. Signal communication in SDL typically involves copying of the data in the input queue upon execution of an output construct (operation (4)), and copying the data into the address space of the receiving process instance when the input is consumed (operation (1)). Here we focus on the elimination of the copy operation involved with the input construct.

Basic idea Typically, the buffer which holds the input signal is freed after the input is completed, i.e. after the parameters of the signal have been copied to the local variables of the receiving process instance. The basic idea pursued here is to keep the buffer storing the input signal until the consuming transition has been completed, rather than moving the data into the address space of the consuming process instance. Thus, any access to the parameters of the signal is directed to the buffer itself rather than to local variables.

In order to support this, consider the partitioning of the information transmitted between the process instances in two parts, namely a data and a control part. The data part contains the actual parameters of the signal. The control part contains the signal identification and related information as well as a reference to the data part. The data part of a signal may be of variable size. Conversely, the control part is of fixed size.

With the optimization technique, the control part of the signal is transmitted as before, employing queues. Conversely, the data part of the signal is stored in a common data area that is shared by the process instances. Thus, the data part is no longer part of the queued element and is not copied into the address space of the process instance. Instead, the data part is referenced by the process instance with the reference given in the control part.

Prerequisites The prerequisites for the application of this optimization are that

- the process instances involved have access to a common memory area, and
- the parameters of the input signal are not reused after the completion of the transition that consumes the signal.

If reuse of a signal parameter in another transition is needed, the respective parameter has to be copied and maintained within the process instance.

Application The elimination of the copy operation is depicted in figure 5.32. The figure shows the data copy operations remaining after the transformation. The application of the optimization has the following effect on the four copy operations depicted in figure 5.31:

(1) **Input:** Provided the prerequisites stated above hold, the copy of the data part (parameters) of the signal is superfluous. The control part is copied into the address space of the process instance as before.

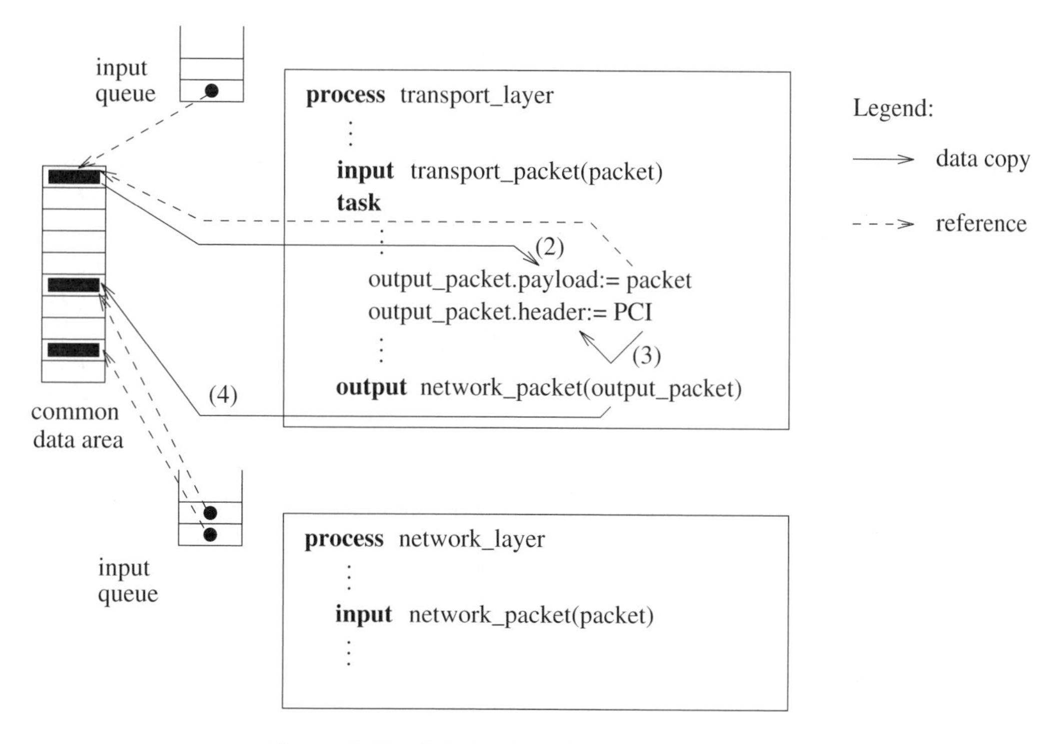

Figure 5.32 Elimination of copy on signal input

(2) **Prepare output packet:** The copy operation is modified such that instead of the internal copy, the data part of the signal is copied from the common address space to the address space of the process instance.

(3) **Add header:** Unchanged.

(4) **Output:** Analogously to operation (2), the data part of the signal is copied into the common address space. The control part is put in the input queue of the receiving process instance.

Merits Applying the optimization technique to protocol processing, one of the four operations to copy the data part can be eliminated. The application of the approach to the implementation of a TCP/IP protocol stack has shown a speedup of about 25% [LaKö99].

5.12.1.3 Common Buffer Management

Above, a first step to a common buffer management has been discussed, i.e. keeping the signal in the buffer instead of copying the signal into the address space of the receiving process instance.

However, there the common buffer was applied to a single process communication only, i.e. applied to the two process instances involved in the communication.

Basic idea As introduced in section 3.3.3.1, common buffer management goes one step further. It supports the use of a common buffer by a set of communicating SDL processes. The idea is to leave a packet to be operated on by a set of SDL processes in a common data area while manipulating and transforming it. In the ideal case, a buffer is acquired when a packet is entering the SDL system and released upon completion of the processing by the system. Thus, copy operations are only applied to the parts of the packets that are actually changed within an SDL process and not to the other parts which remain unchanged. With typical protocol processing, the unchanged parts often represent the majority of the data in a packet. In addition, this denotes the fact that the (multiple) allocation and deallocation of buffers involves considerable overhead.

Prerequisites The application of a common buffer management scheme as described above requires a set of prerequisites. As already noted above, access to a common memory area is required. A major restriction of the application of a common buffer management scheme in the context of SDL results from the notion of ownership of packets passed in the SDL system which is introduced by the scheme. In other words, for each SDL process that participates in a common buffer management scheme, it has to be ensured that at any time there is only one process instance that operates on the buffer. The respective process instance is called the current owner of the buffer. Signal communication can then be considered as passing the ownership of the buffer.

The semantics of passing the ownership of a buffer, i.e. its reference, are very different from copying the signal itself. Thus, there is often a semantic conflict that prohibits or at least restricts the use of a common buffer management scheme. For example, in SDL a signal (or a parameter of a signal) may be sent to more than one process using several output constructs. Thus, ownership would be shared between several receivers. Another problem is input parameters of signals that are reused in later transitions (as described above).

As a result of the semantic differences, a thorough analysis of the given SDL description is needed to identify the cases where the optimization can be safely applied. Concerning the automatic code generation from SDL descriptions, abstract interpretation techniques [FiHa88] can be employed to identify the cases where the application of the optimization technique does not interfere with the semantics of SDL.

Application In practice, there are three main alternatives to the implementation of common buffer management in the context of SDL:

- **Automatic analysis and code generation:** A detailed analysis of the SDL description is employed to verify that the single ownership principle is adhered to by the processes on which the optimization is applied, adjunct with the automatic derivation of the respective code. Thus, no direct user involvement is required to ensure that the semantics of SDL are adhered to.

- **Language extensions:** The SDL language could be extended to explicitly allow for the specification and use of owned references [WeEL97]. Thus, the user is in charge of ensuring the correct usage of the buffer. This is currently supported by the SDT code generator [Tele98].

- **Target language constructs in SDL:** Code generators often support the direct use of target language constructs (e.g. C code) in the SDL description. This may be employed to implement common buffer management, e.g. by using C-style pointers. However, note that in this case the user is fully responsible for consistency.

Merits The merits of introducing a common buffer management scheme are twofold. First, it allows the reduction of the number of copy operations involved with signal communication. Important is also the elimination of operations to allocate and deallocate buffers. Note that in the optimal case, buffer allocation and deallocation are needed only at the end points of processing, i.e. when a signal enters the system (allocate buffer) and when it leaves the system (deallocate buffer).

However, note that the automatic application of the common buffer management scheme to a number of SDL processes typically requires an SDL description that has been designed with the prerequisites for a common buffer management in mind.

5.12.1.4 Offset Technique

Common buffer management represents a rather general optimization technique to eliminate copying of packets that are passed between SDL processes when major parts of the packets remain unchanged.

Above, we have noted that additional measures are needed in order to apply a common buffer management scheme. One of these measures is the offset technique introduced in section 3.3.3.2.

Basic idea Unlike the techniques described above, the offset technique aims at the elimination of copy operations within the process instances. The motivation for the offset technique is the fact that often only a small part of the packet is changed, while most of the data contained in the packet are passed through unchanged.

In general, four cases for protocol processing can be identified:

(1) The input signal is passed to the output without changes. Examples are multiplexers and to some extent routers.

(2) Some additional data are added to the front and/or the tail of the packet. This is typically the case with outbound traffic.

(3) Some data are removed from the front and/or the tail of the packet. This is typically the case with inbound traffic.

(4) The output signal is completely different from the respective input signal. This is the case with protocol instances employing encoding, decoding, encryption, segmentation and fragmentation.

The last case is the only case where the data are completely changed by the process instance. All other cases pass at least a part of their input data directly to the output. These cases carry a potential for optimization by eliminating copy operations. In the first and third cases, the copy operations are completely superfluous.

In the second case, the copying of the incoming packet can be eliminated if the packet is mapped in the address space such that enough memory space is available in front (and possibly at the tail) of the packet. In this case, the PCI can be added to the payload without actually copying the payload. In order to support this for a set of protocol instances, the offset technique has been introduced.

The principle of the offset technique was described in section 3.3.3.2 (see figure 3.9). The main idea of the offset technique is to allocate at a higher protocol layer (typically the application layer) a buffer that leaves enough memory space such that all headers (or trailers) subsequently prefixed (or appended) to the packet can be directly added without actually copying the part that is passed unchanged.

Application The application of the offset technique to a process instance is shown in figure 5.33. It has the following effect on the copy operations shown in figure 5.32. Copying the packet into a new address area is no longer needed. Thus, copy operation (2) (see figure 5.32) can be eliminated. Copy operation (3) is modified such that the header is copied directly into the buffer, i.e. the common data area. If no header is added to the packet, copy operation (3) is not present. This is often the case for inbound traffic and with multiplexers and routers. Provided that there is no access to the data (buffer) after the output, and the transition contains a single output for the buffer only, copy operation (4) (compare figure 5.32) can be eliminated as well. This is often the case for protocol processing. If the buffer is output more than once in a transition, the elimination of copy operation (4) can be applied to one of the outputs only. For the other outputs, the buffer is copied to avoid conflicts.

Prerequisites In general, the automatic transformation of an arbitrary SDL description to an equivalent SDL description that applies the offset technique is rather complex. In addition, the effective application of the offset technique may require knowledge not present in the SDL description itself, for example of the probabilities for the execution of different execution branches of the SDL description. A related issue is the selection of the appropriate size of the buffer to hold the headers subsequently added. This may vary depending on the state of the process instances as well as the exact contents of the packet.

If the SDL description has been prepared to support the offset principle, the derivation of code that implements the offset technique is straightforward. The prerequisites for the SDL description to support the automatic derivation of code that implements the offset technique are as follows:

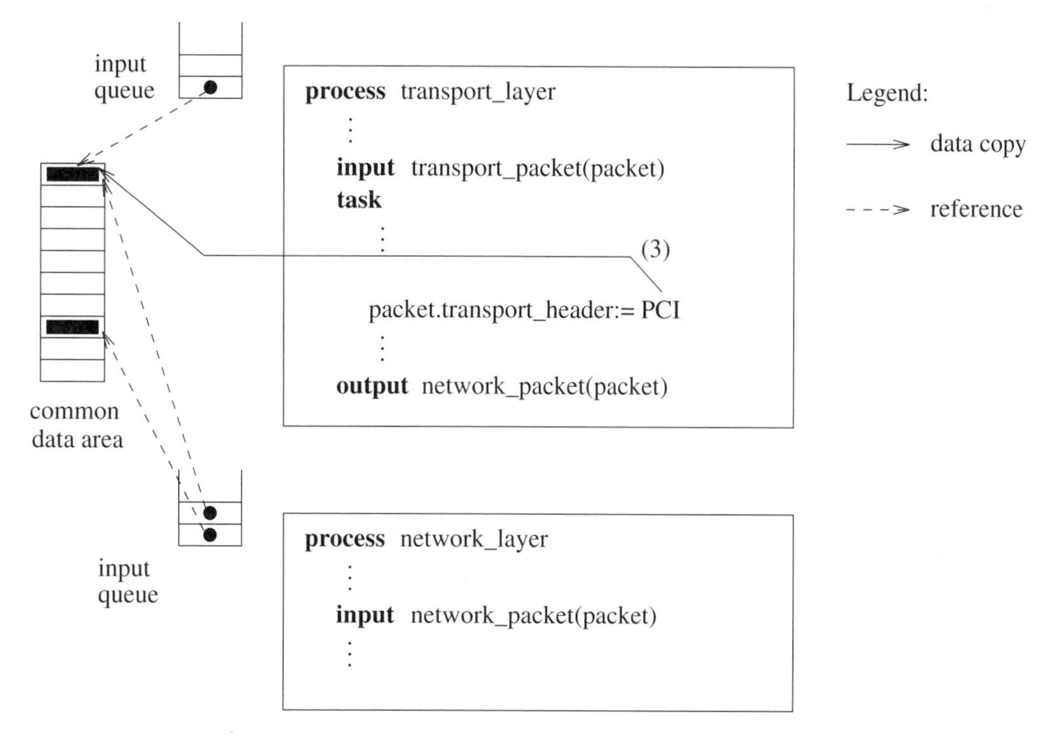

Figure 5.33 The application of the offset technique

(1) The respective buffer size has to be defined such that enough free memory space is available to allow the process instances manipulating the buffer to add the respective data (headers). If no data are added, no extra memory needs to be allocated.

(2) The manipulation of a copy of the input signal (i.e. a copy of its parameters) has to be replaced by the direct manipulation of the input signal. Thus, the input signals can be manipulated directly in the shared address space.

(3) The parameter name of the input signal has to be identical to the parameter name of the output signal on which the offset technique is applied. This allows the code generator to identify the cases where the offset technique can be applied.

(4) If several outputs of a transition specify the same parameter name on which the offset technique is applied, the offset technique may be applied only to one of the outputs. For the other outputs, the parameter has to be copied to avoid several process instances concurrently manipulating the same copy. However, since in SDL it is not possible to specify to which output reference passing is applied, the automatic application of the offset technique in the case of several outputs with identical parameter names is problematic. This is because an arbitrary selection of one of the outputs to apply reference passing may violate

the semantics. This is due to the fact that the required size of the buffer may depend on the path the packet in the buffer takes through the SDL system. Thus, a buffer may not be large enough to hold the data added to it on a different path.

(5) The signal input parameter which is passed with the output statement may not be manipulated later in the same or any subsequent transition.

Merits Depending on the application, the use of the offset technique may be a prerequisite to the application of a common buffer management scheme. Thus, the merits described for common buffer management apply.

For the joint application of the offset technique and the common buffer management scheme to an XTP implementation, an increase in speed by a factor of about 10.9 compared to an XTP implementation solely employing the technique to eliminate copying on signal input has been measured (see [HeKM97] for details).

5.12.1.5 Gather/Scatter

As described in section 3.3.3.3, the gather/scatter technique employs an approach that is radically different from the offset technique. With gather/scatter, the PCIs of the different layers are maintained and processed separately. Several practical problems raised with the gather/scatter approach have been discussed in section 3.3.3.3.

The application of the approach in the context of SDL is mainly an issue of the design of the SDL description rather than an issue of the automatic code generation. Thus, the gather/scatter approach may be applied as a technique to describe protocol processing in SDL.

5.12.1.6 Integrated Layer Processing

As described in section 3.3.3.4, Integrated Layer Processing (ILP) aims at the integration or merger of several data processing steps. Thus, ILP is targeted towards the merger of several layers of the protocol architecture. In the context of SDL, this means merging a set of SDL processes. This is not simple at all for a code generator. The complexity of this optimization goes far beyond the complexity of the optimizations described above.

An approach for the (semi-)automatic application of integrated layer processing in the context of SDL has been outlined in [LeOe96]. However, the approach described there is limited to a very small subset of SDL. Thus, major issues such as dealing with saves, etc. are not covered. Instead, the work reported in [LeOe96] focuses on the analysis of SDL descriptions for data and control dependences and the integration of data manipulation operations on a common execution path rather than the automatic derivation of code. For the automatic derivation of code that supports the integrated processing of data manipulation operations of various layers, also the protocol executions that differ from the common path have to be handled.

5.12.2 Other Sequential Optimizations

Besides the minimization of data copy operations, other optimizations have been proposed for the manual implementation of protocol architectures. Especially important is the optimization of process management and the minimization of overhead involved with it. One approach to optimize process management in the context of SDL is the activity thread model. Other techniques aim at the reorganization of protocol processing in order to minimize the protocol processing cost.

5.12.2.1 Combining Scheduling and Communication – The Activity Thread Model

In section 3.3.2, the two principal process models to implement protocol architectures have been discussed. These are the server model and the activity thread model. Concerning the derivation of code from SDL descriptions, only the server model has been discussed so far (see section 5.6).

Differences between the server model and the activity thread model Due to the semantic similarities between SDL and the server model (asynchronous computation and communication), the automatic derivation of code according to the server model is straightforward. In contrast, with the activity thread model as introduced in section 3.3.2.2, process instances are executed in a synchronous fashion. Thus, communication is directly implemented by procedure calls rather than asynchronous mechanisms.

The server model exhibits extra overhead which is not present with the activity thread model, e.g. overhead for asynchronous communication including queuing operations and overhead for process management. The problem with the activity thread model is that it is based on a semantic model (synchronous computation and communication) that differs considerably from the SDL model. Thus, special care has to be taken to correctly map the semantics of SDL on activity threads. In addition, we will see SDL concepts that cannot be implemented by the pure activity thread model.

In order to apply the activity thread model rather than the server model to derive code from SDL, the following differences between the two models have to be taken into account:

- The active elements in the server model are the protocol entities. In the context of SDL, each process instance can be considered a server, managed by the operating system or the SDL runtime support system. Thus, the execution sequence of the events is determined by the scheduling strategy of the operating system. Communication is asynchronous via buffers.

- The activity thread model is event-driven. It supports an optimized execution sequence of the events. With the activity thread model, one event is handled at a time. Processing of subsequent events is not started before the previous event has been completely processed.

The complete processing of an external input event by the system is denoted as an activity thread. Communication is synchronous, i.e. no buffers are employed. Thus, the model is well suited for the layer-integrating implementation of protocols.

Principles Several potential mappings of (a subset of) SDL on the activity thread model are possible. In the following, we describe a mapping which represents a rather strict implementation of the activity thread model. Thus, our approach is rather restricted in the subset of SDL it supports. Several other mapping strategies which represent some kind of combination of the activity thread model and the server model have been proposed, too (see [HeKM97, LaKö97]). These combined approaches mainly aim at the extension of the subset of SDL supported by the code generation strategy.

It is also important to note that the activity thread approach does not have to be applied to the SDL system as a whole. Instead it can be applied to a part of the SDL system only. Thus, an optimized implementation for critical parts of the system can be derived.

A strict implementation of the activity thread model transforms the output constructs in the SDL processes to procedure calls. Thus, instead of transmitting a signal to the receiving process instance, the respective procedure implementing the process instance is called directly. This represents a merger of scheduling and communication. In other words, the order of the output constructs encountered with the execution of the SDL description determines the schedule of the process instances.

Concerning the code generation, this means that output constructs sending signals to process instances integrated in the same activity thread are replaced by the call of the respective procedure which implements the SDL process that receives the signal. Conversely, output to process instances beyond the scope of the activity thread is implemented by regular asynchronous communication mechanisms.

Concerning communication internal to the activity thread, the next executable transition is defined by the signal itself, the state and the input constructs of the process instance receiving the signal. The sequence of the encountered output and input constructs (output – input ... output – input ...) results in a sequence of procedure calls, i.e. an activity thread. The activity thread is completed when the event has been completely processed by the respective SDL processes constituting the activity thread. Each procedure call corresponds to a transition executed on a signal input.

If the receiving process instance is known statically, the output can be directly translated to a procedure call. If addressing is dynamic, an intermediate step to resolve the address of the receiver is needed.

SDL descriptions often employ the save construct. Saves can be implemented by the activity thread model if a facility to store signals is supported. If a procedure implementing an SDL process is called and the respective process instance is in a state in which the given input signal is saved, the signal has to be stored. The call of the procedure itself returns after saving has been completed. When the process instance changes its state later on, i.e. after the execution of a transition, the saved signal is revisited as part of the call of the respective procedure and is processed as specified. In this case, a call of the procedure implementing the process instance results in the execution of two (or more) transitions, i.e. the transition resulting from the current signal and the transitions resulting from the saved signals.

Problems If the transition has a simple structure, i.e. input – action(s) – output, the simple transformation strategy described above can be applied. However, besides this simple structure of transitions, SDL also allows more complex cases:

- transitions not triggered by an input signal,
- transitions performing an action after an output,
- transitions with multiple outputs,
- transitions without output, and
- transitions with output beyond the scope of the activity thread only.

A transition without an output or with an output beyond the scope of the activity thread simply results in the termination of the activity thread. For the other cases, special care has to be taken to ensure conformance with the semantics of SDL. In particular, the simple transformation of an output statement to a procedure call is no longer possible. In the following, we discuss the semantic problems with these transition types and sketch solutions to preserve the semantics.

- **Transitions not triggered by an input signal**

  In SDL, transitions may be triggered by events other than input signals. An important class of this are continuous signals. These transitions are triggered only if the following conditions hold: first, there is no signal available that activates a transition. Second, the boolean condition constituting the continuous signal evaluates to true. Since there is no signal to be mapped on a procedure call, transitions with no input signal cannot be implemented by an activity thread. Thus, transitions with no input signal have to be removed during the refinement of the SDL description or an alternative implementation strategy has to be applied.

  Similar problems exist with transitions that are triggered based on the combined presence of a signal and other conditions, e.g. with enabling conditions. Problems with enabling conditions arise where the signal is present but the enabling condition does not hold. In this case, the activity thread has to postpone its execution which contradicts the synchronous execution paradigm. In some cases, the solution described above to deal with save constructs can be applied.

- **Transitions performing an action after an output**

  The simple replacement of an output construct by the call of the respective procedure (implementing the SDL process receiving the signal) may violate the atomicity of transitions. An example of such a system (relevant parts only) is given in figure 5.34. The example comprises two communicating SDL processes, P1 and P2. For example, P1 may implement a routing function. A signal `sig_extern` received by process P1 is routed to process P2 (i.e. `sig_a`) for processing. Process P2 replies by sending signal `sig_b` to process P1. Within P1, the variable `status` is employed to maintain the current status of the processing of the received external signal (`sig_extern`).

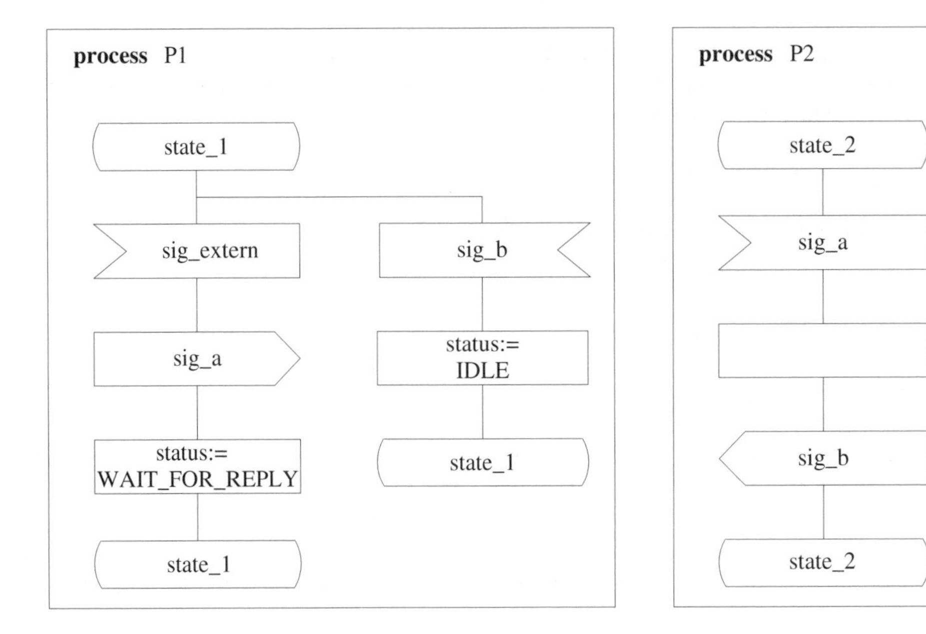

Figure 5.34 Example SDL description (incomplete)

An example execution sequence, where the simple replacement of output constructs by procedure calls does violate the atomicity of a transition, is shown in figure 5.35. In the example, the assignments to the variable `status` in process P1 are made in the wrong order.

In general, the problem may arise if two conditions hold. First, an output construct is encountered that is not the last action executed in a transition of a process instance. In addition, the same process instance is called in a recursive fashion during further processing of the activity thread.

An approach to solving the problem for most cases is to defer the procedure call until the transition has been completed. This way, the interleaving of the processing of transitions within a single process instance is prevented.

- **Transitions with multiple outputs**

When multiple output constructs are encountered during the execution of a transition, the activity thread branches. Thus, a second thread of activity emerges. Since the scheduling strategy for process instances is not defined by SDL, the order in which the processing of the two activity threads is interleaved is (often) a matter of the implementation. Thus, the completion of the first activity thread before the second one is processed is in many cases in accordance with the semantics of SDL. However, note that even if the semantics of SDL are preserved, this may not always be what the user expects.

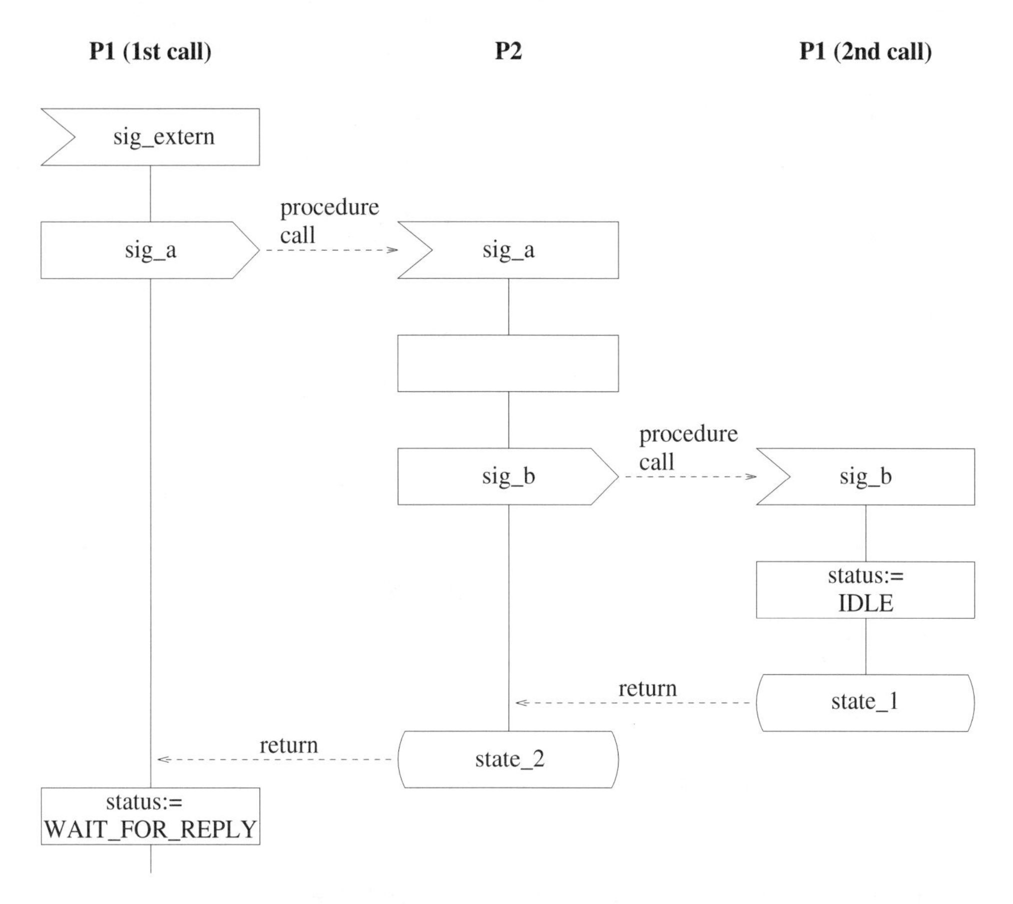

Figure 5.35 Violation of the atomicity of a transition

In addition, exceptions exist where the semantics of SDL are violated. For example, consider the (indirect) recursive call of a process instance as depicted in figure 5.36. In this case, the order in which the output of signal `sig_b` is processed by process P1 is not in accordance with the semantics of SDL. In other words, the output of the signal `sig_b` in the first call of the transition is executed after the same output construct has been executed in the second call of the procedure.

The problem can be solved by postponing the call of the procedures that implement process instances until the transition has been completed. However, this requires some kind of queuing of the output signals which in turn results in additional overhead (see [HeKM97] for a discussion).

An approach to concurrently processing several activity threads in a fair way is described in [HeKM97]. This paper also proposes solutions to some of the problems raised above. An activity

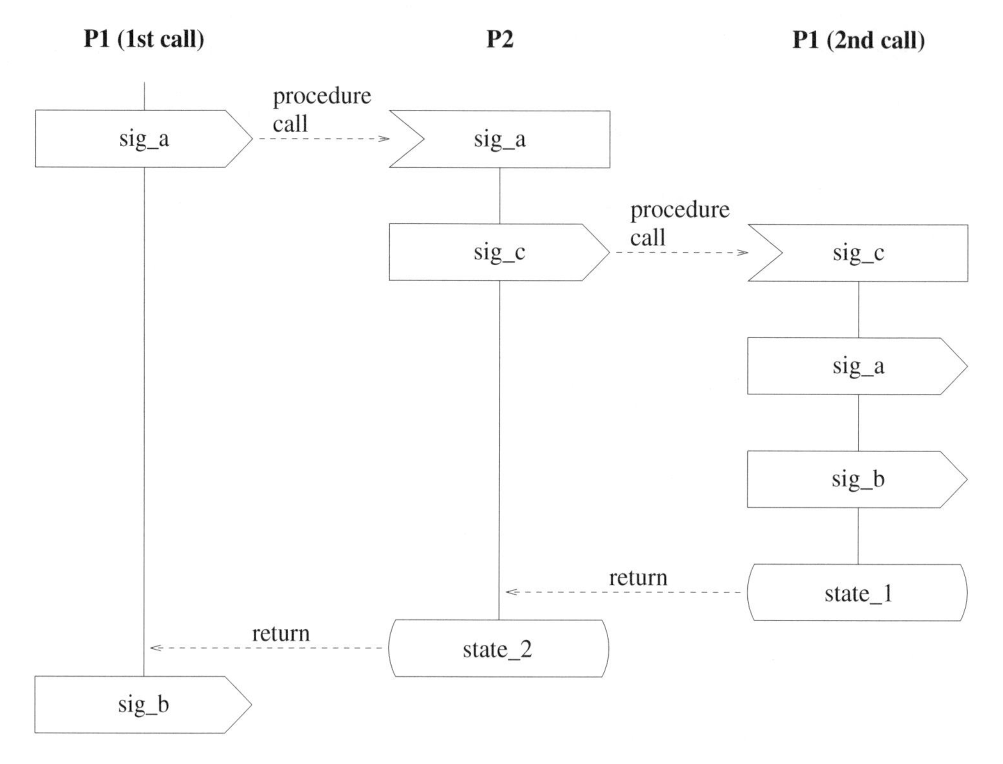

Figure 5.36 Violation of the signal order with multiple outputs

thread approach that exploits parallelism can be found in [HMKL96].

Merits The merits of the application of the activity thread model to the code generation from SDL can be summarized as follows:

- the optimization of the execution order of external events,
- the minimization of overhead for the management of the process instances, and
- the reduction of communication cost due to the merger of scheduling and communication.

The optimization of the execution order is due to the fact that an external event is completely processed before processing of any subsequent event is commenced. Thus, the response time for events can be minimized.

The performance improvements achievable with the activity thread model depend highly on the specific application at hand. The smaller the processing cost of the process instances, the larger are the improvements achievable with the activity thread model. In [HeKM97] an increase in speed by a factor of between 6.4 and 7.1 was reported for an activity thread implementation of the XTP protocol compared to a server model implementation. The application of the activity thread approach to a TCP/IP protocol stack (which is much more complex than XTP) still exhibits a doubling of speed [LaKö99] compared to the code generated by the SDT Cadvanced code generator [Tele98]. With the joint application of the activity thread model and techniques for the minimization of data movement (as described in sections 5.12.1.2 to 5.12.1.4), an increase in speed by a factor as high as 431 has been measured for the implementation of the XTP protocol [HeKM97].

5.12.2.2 Application Level Framing

Application Level Framing (ALF) has been described in section 3.3.4.1. The aim of the technique is the optimization of the design of the protocol architecture for maximum performance rather than the optimization of the implementation of a given protocol specification.

The goal of ALF is to organize the functionality of the protocol architecture in such a way that packet segmentation and reassembly are eliminated or at least minimized. Thus, the technique is targeting the protocol architecture which is reflected by the SDL specification of the protocol layers. Even though ALF has a considerable influence on the performance of the implementation, it is not an issue that can be supported by the code generator.

5.12.2.3 Common Path Optimization

As described in section 3.3.4.2, some execution paths through the protocol architecture are typically more important with respect to performance than others. Concerning SDL, this information is not provided in standard SDL. Thus, additional information has to be provided to identify and optimize the common path. An important example where this information can be used is the testing of the trigger conditions of SDL processes. As outlined in section 5.5.4, trigger testing can be organized in such a way that the conditions representing the common path are tested first. These optimizations can be done automatically when the respective information is provided.

More complicated are optimizations of the common path at the design level. For example, protocol processing may assume that the majority of the arriving packets requires no error handling. The design of the system may be based on this assumption, which often allows for the application of a set of optimizations to this common execution path. If the actual processing deviates from the common path, e.g. a transmission error is detected, a recovery mechanism is employed to backtrack and to resume regular processing. Obviously, these kinds of optimization are hard to automate as they have a considerable influence on the protocol algorithm. Thus, these optimizations are typically described at the level of the SDL description, rather than automatically performed by the code generator.

5.12.3 Exploitation of Parallelism

The different kinds of parallelism which may be exploited by protocol architectures have been described in section 3.3.5. In the following, we discriminate between two categories, namely *explicit* and *implicit* parallelism. Explicit parallelism denotes parallelism explicitly present in the given SDL description of the system. Conversely, implicit parallelism denotes parallelism not explicitly present in the SDL description.

5.12.3.1 Explicit Parallelism

Explicit parallelism in a given SDL description may be present at the level of process instances, processes or blocks. The kinds of parallelism which are typically exploited explicitly at these levels are connectional, layer, directional and functional parallelism. In addition, packet parallelism may be supported explicitly, too.

Connectional and layer parallelism are exploited quite naturally when using SDL. Connectional parallelism is supported by employing a separate process instance of an SDL process for each connection. Layer parallelism is supported by implementing different layers by different SDL processes. Additionally, directional or functional parallelism can be supported within a layer or beyond layers by employing different SDL processes for specific functional features (or traffic directions).

5.12.3.2 Implicit Parallelism

Implicit parallelism is not explicit in the SDL description. A kind of implicit parallelism that can be exploited under certain conditions is packet parallelism, i.e. the case in which several instances of a protocol instance operate in parallel on different packets. However, this only works for the rare case that no dependences between transitions exist, i.e. where the EFSMs do not maintain an (explicit or implicit) state. In addition, the exploitation of packet parallelism requires special care to ensure that the order of the transmitted SDL signals is preserved. An activity thread approach that exploits implicit packet parallelism has been described in [HMKL96].

5.13 Dealing with Implementational Limitations

As pointed out in section 5.1.2, SDL abstracts from implementational aspects of real systems. SDL abstracts from the passage of time caused by computations and communications and from the consumption of memory. On the other hand, all physical implementations consume time and memory. Thus, the implementation has to deal with these limitations, even though this is not completely specified at the abstract level of SDL.

If physical and organizational limitations are ignored, unexpected behavior may result. Potential results – if not handled with care – may be deadlocks, loss of signals or even a crash of the system.

If physical limitations are hit, this is typically a result of too much load in the system and/or of inappropriate design or implementation.

The different kinds of physical and organizational limitations that have to be dealt with when implementing SDL descriptions have already been described in section 5.1.2.2. Here, we focus on approaches to dealing with these limitations.

Different approaches to dealing with an important example of resource limitations have been discussed in section 5.7.1.2, namely handling limited input queues. Here, we generalize our discussion to arbitrary resource limitations.

5.13.1 Principal Approaches to Dealing with Limitations

There are two principal approaches to dealing with implementational limitations of real systems:

- **Prevention:** The implementation of the system and/or its environment is such that physical or organizational boundaries of the implementation are never hit. Thus, no dynamic reaction to the (impossible) case that limits are hit is needed.
- **Dynamic control:** Mechanisms are put in place that dynamically react where limits are approached. These mechanisms effectively control the system and possibly the environment such that physical or organizational boundaries are not hit or do not cause unexpected behavior.

A third approach is to ignore the problem. This is typically done by validation, functional simulation and prototyping tools. However, even though this may be tolerable for a system description, it is certainly not an alternative for the final system.

5.13.2 Prevention

5.13.2.1 Principle

Systems process the load that is provided to them by the environment. From this simple statement, an important design principle to deal with limitations can be derived. Provided the environment ensures that some maximum load delivered to the system is never exceeded, the system can be designed for this maximum load. If the system is designed for these maximum load figures, it will be able to handle all other cases as well.

In other words, the system implementation is designed in such a way that the arriving load can always be handled, i.e. overload does not enter the system (SDL part) at all.

In order to ensure that the load that can be handled by the system is not exceeded, two alternatives exist. The first is that this may be ensured by the environment of the system. In other words, it is ensured that the environment always behaves in a manner such that the maximum load that can be handled by the system is not exceeded. A second alternative is to design the system itself in such a way that only as much load as can be handled by the system is allowed to enter. For example, this is the case where polling of the interface (environment) is employed. Thus, load that exceeds the specified limit (e.g. the arrival rate) is simply ignored. This approach may be implemented by putting a shell around the implementation derived from the SDL description. The shell ensures that only load that can be handled by the implementation of the SDL description passes the shell. In other words, overload control is implemented by the shell.

5.13.2.2 Design for Prevention

In both cases, the system description as specified with SDL does not have to be adapted to deal with limited resources. Where overload can never occur, no reaction needs to be installed. Thus, provided the environment (or shell) always behaves well, the design description is equivalent to the implementation in this respect.

In order for this approach to work, the resources that handle the SDL parts have to be dimensioned accordingly, e.g. processors and queues.

5.13.2.3 Problems

Even though the preventive approach seems intuitive, there are serious problems with this. First, there may be unexpected or unforeseen load scenarios which may result in a crash of the system. In addition, the definition of the maximum load is not as simple as it looks at first glance. For example, the maximum load may result from an unexpected combination of load scenarios.

In addition, simply shielding the SDL part from overload is often not possible. For example in telecommunication systems, often complex mechanisms are employed to deal with overload. Typically, overload control consists of a set of integrated measures and actions between the environment of the system and the system itself.

5.13.3 Dynamic Control

As pointed out above, the static approach is only applicable to simple cases. In telecommunication systems, overload typically has to be handled within the system rather than at the outskirts.

5.13.3.1 Principle

With dynamic control, the limitations of the system implementation are handled within the system during runtime. There are two issues with dynamic control (see also [Butt97]):

- recognize the problems on time, i.e. an overload situation, and
- react appropriately.

In addition, an important practical question is how to describe and implement the needed actions for overload control, i.e. whether to directly integrate it into the implementation or formally specify the measures in the SDL description. Currently, SDL does not support the description of mechanisms for overload control very well. This is due to the abstraction of SDL from implementational details such as resources, which are an important part of any overload control mechanisms.

5.13.3.2 Feedback from Resources

In order to recognize an overload situation, feedback from the resources of the system is needed. From the viewpoint of the application, i.e. the SDL description, there are three principal ways to get information from the underlying resources. These are

- *synchronous feedback* when a resource request cannot be handled,
- *(cyclic) polling* of the state of the resources initiated by the application, and
- *signaling* of the state of the resources to the application.

Synchronous feedback With the synchronous approach, feedback is provided to the requester of a resource at the time the resource is requested and the request cannot be satisfied. Examples of SDL constructs where (implicit) resource requests are involved are the create construct and the output construct. In both cases, memory is required to handle the construct.

There are several problems with the synchronous approach. First, the time at which the application is informed of the problem may be too late for the application to react. For example, if an input queue is full, this may cause deadlock of the system, i.e. by entering some system state not covered by verification. A similar situation may arise with processors. An overloaded processor may not be able to react to the situation, e.g. to instruct its environment to limit the load it provides. A related problem with this approach is that the instance that requires a resource is informed of the problem. This is not necessarily the entity that is able to react or remedy the overload problem. In addition, the approach is not supported by SDL. This is because SDL does not provide the means of informing the application, i.e. a process instance, that a limit is hit.

Conversely to the synchronous approach, the other two approaches provide mechanisms to get earlier feedback, i.e. before the limit is actually hit. This allows for time to react to the problem. In addition, these two approaches allow one to decouple the entity that requires a resource from the entity that reacts or deals with the resource limitation.

Polling Polling is initiated and controlled by the application. With polling, the application periodically tests the state of the resources to make sure it is not missing a (possibly fast) increase of the load. Thus, special care is needed for the polling approach to ensure that the time between two pollings is not beyond an upper limit. This requires special care with the design of the respective polling process. Even worse, the remaining parts of the system also have to be designed carefully since the polling process may rely on other processes to give up resources.

Polling of the state of resources from the application specified in SDL is rather problematic. This is because polling has to be done in a cyclic manner to ensure rising load figures are detected on time. Due to the poor time semantics of SDL this is hard to support at the level of SDL. Thus, polling should be employed by some underlying level which can ensure that probes are taken at regular periods of time, i.e. the runtime support system or the operating system. This leads us to the signaling approach.

Signaling Unlike polling by the application, signaling is initiated by the underlying system, i.e. the operating system or the runtime support system. Signaling of the application by the resources can be done in various ways, i.e. on a regular (periodic) base, or only when specific situations arise. Note that the term signaling denotes the principle rather than the implementation which may or may not be based on SDL signals.

An example where signaling can be employed is when some threshold for the load of a resource is exceeded. Thus, some kind of signal is issued to the application if the load of a resource rises above a certain threshold or falls below a threshold. Examples where this can be applied are real physical resources such as processors, or resources with rather organizational limitations such as input queues or buffer pools.

The advantage of signaling is that the underlying system may employ a higher priority to signal an overload situation to the application quickly. In addition, the number of times the application is involved in the dynamic control is limited to the cases where overload situations arise. On the other hand, this requires the underlying layer that initiates signaling to know the exact conditions under which this takes place. Thus, the underlying layer implements a part of the overload control mechanism. This may conflict with the modularization principle.

A related problem is aggregated load figures, i.e. load figures that are derived from a set of input variables or from the values of a single variable that has been polled at different times. The issue is where to implement the aggregation of a set of measurements. As this is typically an application-specific issue, it should be dealt with in the SDL description. On the other hand, an implementation of the aggregation in an underlying layer might be preferable in order to reduce overhead.

5.13.3.3 Measures to React to Overloaded Resources

Above, we have discussed the approaches to notifying the application of a problem caused by resource limitations. Here, we discuss the approaches to dealing with the problem once it is detected.

Principal approaches In order to react to overload, the following measures may be applied:

- *reorganization of processing*, e.g. changes to the priority of processes or events to discriminate between important and unimportant events,
- *disabling of load generators*, i.e. of the sources that generate the load,
- *discarding or ignoring events*, e.g. discard signals, which represent some workload, or
- *restarting the system*, i.e. implicitly discarding internal load.

Reorganizing processing Processing can be reorganized by changing the parameters of the runtime support system or the operating system. Examples of this are changes to the priorities of the processes or to the size of buffer pools.

SDL itself does not provide support to describe the dynamic reorganization of resources. This is because the objects of the manipulation, i.e. the resources, are not defined in the context of SDL. A possible mechanism conforming with SDL is the sending of signals to the environment which are directed to the underlying resources. The signals are interpreted by the underlying layers and can be used to prompt a change of resource parameters.

In addition, SDL tools often support the integration of target code (e.g. C code) into the SDL description. Thus, system calls can be used to change resource parameters or to perform other actions related to resource limitations.

Disabling load generators Better than dealing with the limitations where they cause problems is to deal with the problem at its roots. In our context, this means disabling the entities that generate the load. This allows the system to avoid unnecessary processing, i.e. the cost resulting from this, before the intermediate results are discarded later on.

A problem with the disabling of load generators is that this typically requires interaction with the environment of the system. Thus, the disabling of a load generator is often not as fast as other approaches. However, as disabling of the load generator is often the only way to control the load in the long run, the delay involved with inhibiting load generators has to be taken into consideration when designing the mechanisms to get feedback from the resources. In other words, the feedback mechanism has to notify the application of an emerging problem in advance such that enough time remains to put the appropriate countermeasures in place.

Disabling load generators can be described at the level of SDL by sending the respective signals to the environment, i.e. to the entities that generate the load.

Discarding or ignoring events Reorganization of processing just distinguishes processing of unimportant events from the processing of important events. Discarding of events goes one step further. It discards events or, in other words, discards work that is in process. Discarding allows the system to immediately free the resources associated with the event, e.g. the memory held by a signal. On the other hand, the disadvantage of discarding is that an earlier investment in processing is discarded. Discarding of earlier work may be necessary in order to avoid a collapse of the system. However, in telecommunications, discarding events (packets) is often a rather short-term measure only. In communicating systems, discarding a packet often result in even more load in the medium term, e.g. when all the communication partners retransmit their messages after timers have expired.

Discarding of signals is supported by SDL. In SDL, input queues may be purged by simply changing to a state where certain signals are neither saved nor consumed.

Restarting the system A last resort when no other action can solve the problem is a restart and reinitialization of the system or a part thereof. However, as with the discarding of signals, this may result in even more load in the medium term. Restarting of parts of the system has the advantage that it solves otherwise irreversible problems, e.g. a deadlock caused by the lack of free buffers.

Restarting itself is not directly supported by SDL. SDL only allows process instances to terminate themselves. Thus, no process instance may terminate any other process instance or other SDL entity. However, as noted above, this can be implemented by the integration of target code into the SDL description as supported by tools, or by sending a special signal to the underlying system.

5.14 Issues Related to Code Generation

Following the discussion of the aspects and various techniques for the derivation of efficient code from SDL descriptions, we focus on issues related to the automatic code generation, namely the influence of the application of the optimizations on other quality aspects of the system, the role of the formal definition of the semantics of SDL with respect to the optimization, and methodological aspects involved with the code generation.

5.14.1 Influence on Other Aspects

A general discussion of quality issues has been given in section 2.1.1.5. Quality issues especially relevant in the context of SDL have been listed in section 5.1.3.1. In this section, we discuss important interrelations between the generation of performance-optimized code and other aspects of the product quality.

Correctness Users of SDL tools may incorrectly interpret the semantics of SDL as the semantics implemented by their favorite SDL tool. As this is only one possible interpretation, errors may result when the code generator implements a different (but still correct) interpretation. In addition, verifying the correctness of a code generator is difficult where complex optimizations are applied. This may result in a lower quality of the code generator which in turn results in lower quality of the generated code.

Robustness Robustness, i.e. the ability of the system to deal with unforeseen cases, relates to the SDL description as such and to the code generation tools. The robustness of the SDL description should not be endangered by restructuring it to provide a better base for the derivation of an optimized implementation.

As performance optimization of the generated code comprises complex transformations, there is a risk that the robustness may be reduced. This may be caused by an unexpected implementation of the semantics of SDL, selected by the optimizing code generator.

Reliability In principle, the optimized code generation may reduce reliability, i.e. increase the frequency and severity of faults. This is comparable to the application of optimizing compilers for which correctness is harder to verify. In the context of optimized code generation from SDL, it is important to formally prove the correctness of the applied optimizations in order to ensure the derivation of correct and reliable code (see below).

Maintainability and debugging Maintainability at the level of the SDL description is only influenced where the need for performance optimizations prompts changes to the SDL description itself. However, this does not necessarily have a negative impact on maintainability.

The maintainability at the level of the code derived from the SDL description is complicated, as the structure of the SDL description is not necessarily completely preserved in the target language. In addition, the generated code may vary considerably, i.e. may look very different depending on the application of specific optimizations. However, as the availability of performance-optimizing code generators reduces the need to manually modify the resulting code, this should not be a problem.

An issue closely related to maintenance is debugging. Debugging can be performed at source level, i.e. the level of SDL, or at code level. Concerning debugging at the level of SDL, the problem of relating constructs in the target language to constructs in the SDL description is complicated where restructuring optimizations are applied, e.g. where the sequence of execution of parts of a transition is different in the SDL description and the derived target code. This is a general problem of optimizing compilers and code generators. With respect to SDL, the problem may be circumvented by debugging the SDL description in simulation mode, i.e. based on an underlying code generator that employs a direct mapping of the SDL description on code.

Where the optimizations change the structure of the system, code-level debugging may be more complicated as well. On the other hand, where the optimization concentrates on special cases based on the use of a subset of SDL, debugging may be easier due to reduced complexity of the generated code or the runtime support system. For example, where no (explicit and implicit) saves are used, the respective code implementing input queues is less complex.

Reusability Reuse at the level of the SDL description is rather independent of the application of optimized code generation. Where performance requirements prompt changes to the given SDL description, this, in principle, may conflict with other goals, e.g. with reuse of the SDL description. An example are two SDL processes which are merged to a single one to support some performance requirements. Reuse at the level of the generated code (C, etc.) is restricted where interfaces to the RTSS, or the RTSS itself, is changed, in order support the derivation of efficient code.

Dynamic replacement of components Especially in telecommunications systems, where the minimization of downtimes is an important issue, the dynamic replacement of components (software and hardware) during regular operation is important. To support the dynamic replacement of software, the components, e.g. SDL processes or blocks, must have well-defined interfaces that are able to temporarily store signals which cannot be consumed while the receiving component is replaced. In addition, it has to be ensured that the underlying RTSS, which may be optimized for the specific (now replaced) SDL description, also supports the services needed by the replacing component, i.e. the SDL language features used by the new component. Thus, the granularity of the components that can be dynamically replaced depends on the optimizations that are applied and the features supported by the specific RTSS.

5.14.2 The Role of Formal Semantics

The different optimizations proposed in this chapter are – in part – highly complex and depend on specific requirements and constraints imposed on the subset of SDL to which they are applied. Thus, verifying the correctness of specific transformations is a complicated matter. Formal semantics can help with this. Unfortunately, the formal semantics of SDL 96 are not very well suited for this. However, a new semantic definition is developed for SDL 2000 [GlGP99]. The new formal semantic definition is based on Abstract State Machines (ASM) [Börg99]. ASM represents an operational semantics. So far, no experience of the application of the ASM formalism to prove the correctness of performance-optimizing transformations is available. However, experience with similar problems indicates that this should be possible. An example where a formal framework based on transition systems has been applied to formally prove the correctness of optimizing code transformations is given in [Mits94] (see [Mits99] for an overview of the approach). There, the transition system has been employed to eliminate asynchronous copy operations in conjunction with restrictions imposed on the execution order of tasks.

5.15 Implications of SDL 2000

The changes to the language in SDL 2000 have been discussed in section 4.1.5. The changes will have an impact on

- the manner communication systems are described with SDL, and
- the code generation itself.

We will discuss the two issues in turn.

5.15.1 SDL Design

New SDL descriptions of communication systems will gradually make use of the new features provided by SDL 2000. Most important from the performance perspective are shared data supported by processes, which allow protocol data units to be kept in a common place while manipulated by different entities, e.g. different protocol layers. This allows excessive data copying to be avoided, which was a major performance problem encountered with the use of SDL 96. Thus, several of the techniques important for the derivation of efficient communication systems (as described in section 5.12) can be easily applied when describing the application with SDL 2000, most notably the offset technique, the gather/scatter technique and common buffer management.

5.15.2 Code Generation

So far only limited experience is available concerning code generation from SDL 2000. Tool providers have just started to work on code generators for SDL 2000. In addition, there are still discussions on details of SDL 2000 within the ITU.

The main concept of SDL, based on the paradigm of extended finite state machines communicating via signal queues, is maintained. However, several new concepts have been added which have to be implemented by code generators, e.g. exception handling, composite states, shared data, etc. In addition, some subtle differences with SDL 96 exist that require special care.

The structuring concept described in this chapter to implement SDL 96, i.e. the division in different modules, will still suit SDL 2000. However, additions and modifications are needed. In the following, we describe the most important changes.

Concurrency and shared data SDL 2000 assumes alternating semantics for processes. This change of the execution semantics is important where processes have side effects, e.g. accessing nonlocal data. Alternating semantics of processes can be ensured in two ways, implicitly by a scheduler that schedules the state machines in a way that atomicity of transitions is ensured, or explicitly employing explicit synchronization constructs.

With light and bare integration, the RTSS implicitly controls the concurrency by executing one transition at a time. This solution is straightforward and easy to implement. With tight integration things are more complicated. When one (asynchronous) operating system process is generated for each state machine, extra synchronization constructs are needed to enforce alternating semantics, i.e. the atomicity of transitions.

As outlined above, sharing of data declared within processes can be implemented efficiently. This will not always be the case for data declared in blocks, i.e. where access to these data is (implicitly) implemented by remote procedure calls as proposed by the SDL 2000 standard. For special cases, e.g. where the atomicity of the transitions is ensured by the underlying runtime support system, more efficient schemes may be possible. However, this needs further study.

Trigger conditions and trigger testing Even with the elimination of the reveal/view concept, the performance problems caused by asynchronous trigger conditions, i.e. the repeated testing of continuous signals and enabling conditions, remain. With SDL 2000, the problem arises, among others, when nonlocal data are accesses. The solutions are the same as discussed for SDL 96. An additional problem that may arise with SDL 2000 are side effects caused by (repeated) testing of these trigger conditions.

Exception handling Exception handling requires extensions of the EFSM implementation and communication scheme to pass and handle exceptions that have been raised. In addition, an extension of the RTSS is needed to pass exceptions issued by the underlying system, i.e. the hardware or the operating system, to the SDL description.

Dynamic blocks SDL 2000 supports the dynamic creation and termination of blocks. This requires support from the RTSS or operating system similar to the approach known from process creation and termination in SDL 96.

Some subtleties arise where shared data are used in agents that are dynamically terminated. In this case, the termination of the agent also has to be tied with the lifetime of the shared data, i.e. the agent must not be discarded by the RTSS before all contained instances have been terminated.

5.16 Summary

In this chapter, a large number of potential techniques to improve the performance of code derived from SDL descriptions have been discussed. Given this vast variety, the question of when to apply which implementation or optimization technique emerges. In other words, hints are needed that indicate which techniques to apply to a specific problem, i.e. a specific SDL description at hand. Also, the potential merits, i.e. the performance gains, achievable with each optimization technique are of interest.

Providing exact rules is a complex issue. First of all, some of the decision problems discussed in the chapter have been proven to be NP-hard, e.g. scheduling and allocation issues. Thus, finding the best solution is impossible in practice. In addition, the optimizations are not necessarily independent of each other. Thus, the application of one technique may completely preclude the application of another optimization technique, or reduce its merits. In addition, the applicability and merits of the application of the optimization techniques depend on the specific SDL description at hand and its usage, i.e. the specific workload imposed on the system. Thus, an optimization technique well suited for one SDL description or even a specific workload may not perform well in another context.

As has been pointed out in section 5.1.1, not all performance problems are caused by implementation decisions. Performance problems may be due to an inappropriate SDL description, e.g. due to temporal errors in the SDL description, or inappropriate distribution of functionality in the distributed system. Others may be due to inappropriate hardware or operating systems. Concepts and tools to support the performance evaluation of systems described with SDL (and MSC) will be discussed in section 6.3. As there are many potential causes of poor performance, the identification of the cause of poor performance is an important task prior to taking some action.

5.16.1 Summary of Concepts for Optimized Code Generation

In order to enlighten the interdependences between the different optimization techniques applicable when implementing SDL, a summary of the implementation and optimization concepts with their specific prerequisites, scope and merits is given in table 5.6.

The table is organized as follows:

- The first column specifies the *implementation and optimization concepts*. The concepts are structured along 12 problem domains (shown in bold font). The numbers given in parentheses identify the sections in which the concepts have been discussed in detail.
- The second main column with its subcolumns specifies the *prerequisites* for the application of the optimization concepts. The prerequisites are specified with respect to
  - the subset of the SDL language to which the concept may be applied,
  - the target system, i.e. concerning aspects of the operating system or the hardware, and
  - dependences with other implementation concepts.
- The third main column identifies the *scope* of the optimization concepts. The scope is given with respect to
  - the SDL description to which the optimization technique can be applied, e.g. the whole SDL system or a single SDL process, and
  - the modules of the implementation influenced by the application of the optimization technique.
- The last column sketches the *merits and potential problems* of the application of the optimization techniques.

5.16.2 Identifying Performance Problems Related to Code Generation

As noted above, first the specific (or potential) performance problem of the implementation of a given SDL description has to be identified, prior to taking any action. Performance problems related to the implementation of SDL descriptions may be caused by

- high processing cost for SDL transitions, or
- high processing cost for the underlying system that implements the SDL machine and the interfacing functionality.

In the first case, potential causes may be inefficient algorithms, inefficient implementation of data types, or high cost for trigger testing.

Potential causes for the second case may be numerous, e.g. high cost for

- communication between SDL processes,
- communication with the environment,
- timer management,

- process management, or
- trigger testing.

Often, poor performance is due to support for high flexibility or high dynamics of the implementation which is not needed for the specific SDL description at hand. Examples are numerous and have been discussed in the chapter, e.g. dynamic process creation and termination, dynamic addressing, and complex buffer management due to random access to input queues and variable-sized signals. In addition to high processing cost, poor performance may also be due to inappropriate organization, e.g. inappropriate scheduling strategies.

5.16.3 Developing SDL Descriptions Suited for Efficient Implementation

As stated in section 5.1.1, many decisions beyond the mere code generation have an influence on the performance. Especially important is the SDL description provided to the code generator, i.e. the specific manner in which the problem is described with SDL.

Concerning the SDL description, there are two issues. First, the SDL description may contain functional or temporal errors or weaknesses which preclude its efficient implementation. Examples of this are inappropriate temporal behavior, e.g. caused by inappropriate timer settings, inappropriate distribution of functionality in the distributed system, or poor algorithms, e.g. poor coordination in the distribution system. The performance engineering activities described in chapter 6 can help to identify these problems.

Second, due to the fact that many optimizations discussed in the chapter are applicable to a subset of the SDL language only, the subset of SDL employed to describe the system (or parts thereof) is important. The derivation of efficient code may be precluded due to inappropriate or unreflected use of the means that are provided by SDL to describe the system.

In order to support the development of SDL descriptions with a high potential for an efficient implementation, we provide some guidelines:

- Minimize the dynamics of the SDL description, i.e. avoid dynamic process creation, the use of addresses not known at the time of code generation, and the use of signals and data types of dynamic size. Minimizing dynamics allows the SDL description to be implemented in a static way, e.g. the elimination of the need for dynamic management of memory space, and dynamic resolution of addresses.

- Avoid SDL constructs that depend on complex manipulations of the input queues, i.e. random access to the queue. These SDL constructs are saves, enabling conditions (implicit save), and priority inputs where used in a nonuniform way in different states (see section 5.5.3.2).

- Avoid SDL constructs that result in implementations where trigger conditions are repeatedly (periodically) tested. SDL constructs that typically cause repeated testing of trigger conditions are continuous signals and enabling conditions with access to non-local variables.

- Structure the SDL description in such a way that the physical transfer of data between process instances, e.g. caused by signal communication, is minimized, or can be eliminated during code generation (see section 5.12 for details). Copying data is the major bottleneck of the implementation of protocol architectures.

However, note that it is important that the code generator employed is able to exploit the simplifications and restrictions made in the SDL description.

Implementation/optimization concept	Prerequisites			Scope		Merits, problems and comments
	SDL	target system	impl. concept	SDL	modules	
system integration (5.3.3)					all but EFSM	
• light integration (5.3.3.1)	none	minimal OS support (timer and external communication)		SDL system or subset		+ small overhead for OS + tight control of execution of SDL system + direct access between different SDL objects + advantage for application of a large set of optimizations + useful where services for PIM, CBM, TM as provided by OS are expensive or not appropriate – no preemption of transitions of SDL processes by another SDL process
• bare integration (5.3.3.3)	none	minimal HW support (timer and external communication)		SDL system		+ no overhead for OS + tight control of execution of SDL system + direct access between different SDL objects + advantage for application of a large set of optimizations – no preemption of transitions of SDL processes by another SDL process
• tight integration (5.3.3.2)	limited (soft) to SDL concepts that can be directly mapped on OS concepts	OS that supports SDL concepts		SDL system or subset		+ preemption of transitions of SDL processes + parallel processing of SDL processes – overhead due to (general purpose) heavyweight OS primitives for PIM, TM and CBM – limited possibilities to exploit specifics of SDL
state automaton (5.5.3)			none		EFSM	runtime and memory efficiency depends on the details of the implementation
• code-based (5.5.3.3)						
– action-oriented	none	none		SDL process		
– state-oriented	none	none		SDL process		
• table-based (5.5.3.1)	none	none		SDL process		

Table 5.6 Summary of implementation and optimization concepts

Implementation/optimization concept	Prerequisites			Scope		Merits, problems and comments
	SDL	target system	impl. concept	SDL	modules	
trigger testing (5.4.3 and 5.5.4)						
• consumer-based (basic approach) (5.4.3.1)	none	none	none	SDL system or subset	EFSM	– potentially high number of unsuccessful tests of trigger conditions
• producer-based (5.4.3.2)	none	shared memory	light or bare integration	SDL system or subset	all	+ minimization of trigger testing
– direct deletion of unsaved signals (5.5.4.2)	none	shared memory		set of SDL processes	all	+ elimination of signal communication for signals deleted (not saved) by the receiver (rare case)
– dynamically optimized signal communication (5.5.4.2)	no priority inputs	shared memory		set of SDL processes	all	+ minimization of overhead for signal communication
• optimized test sequence (5.5.4.1, 5.12.2.3)	none	none	none	SDL process	EFSM	+ speedup for critical path or common case
• optimized handling of priority inputs (5.5.3.2)						
– separate priority input queue (5.5.3.2)	uniform usage of priority inputs in SDL process	none	none	SDL process	CBM	+ efficient implementation of input queues where a high number of elements is typically present
– priorization of scheduling (5.5.3.2)	none	control on scheduler	producer-based trigger testing	SDL system or subset	all	+ prioritized processing of PIs with priority inputs
• minimization of reprocessing of conditions (5.5.4.1)	none	none	consumer-based trigger testing	SDL process	EFSM	+ speedup in the presence of conditions (enabling conditions, continuous signals)
• optimized handling of saved signals (5.5.4.1)	none	none	consumer-based trigger testing	SDL process	EFSM, CBM	+ speedup where input queue contains many signals
• clue-based trigger testing (5.5.4.1)	none	none	consumer-based trigger testing	SDL process		+ speedup for trigger testing
object orientation (5.5.6)			none		all	
• flattened	none	none		SDL system		+ possibly improved runtime efficiency
• mapping on object-oriented programming language	none	object-oriented language		SDL system		+ possibly improved memory efficiency
• explicit use of data objects	none	none		SDL system		+ possibly improved memory efficiency
enabling strategy (5.4.4, 5.6.3 and 5.6.4)						
• trigger-independent enabling (5.4.4.1)	none	none	none	SDL process	EFSM, PIM	– potential for numerous unsuccessful tests of trigger conditons
• clue-based enabling (5.4.4.2)	none	none	none	SDL process	EFSM, PIM	+ efficient strategy where no shared memory available
• trigger-based enabling (5.4.4.3)	none	shared memory	producer-based trigger testing	SDL process	EFSM, PIM	+ efficient strategy for a subset of SDL (no asynchronous conditions) – complex to implement for full SDL

Table 5.6 (continued) Summary of implementation and optimization concepts

Implementation/optimization concept	Prerequisites			Scope		Merits, problems and comments
	SDL	target system	impl. concept	SDL	modules	
process instance management						
• by the RTSS - light/bare integration (5.6.3)	see system integration					
• by the OS - tight integration (5.6.4)	see system integration					
• preallocation of dynamically created PIs (5.6.2)	known max. number of PIs	none		SDL process	PIM	+ speedup for dynamic process creation – increased system set-up time
address resolution (5.6.5)						
• address server	none	none	none	SDL system	PIM	– costly solution due to overhead for process management and communication
• central table	none	shared memory	none	SDL system	PIM	+ more efficient solution where shared memory is available
• disallow indirect addressing (static resolution of addresses)	process addresses must be resolvable at compile time	none	none	SDL system or subset	PIM	+ high runtime efficiency
input queues						
• single-linked list (5.7.2.1)	none	shared memory	none	SDL process	CBM	+ sufficient for input queues that typically hold a small number of elements (fast insertion, slow removal)
• doubly-linked list (5.7.2.1)	none	shared memory	none	SDL process	CBM	+ appropriate for input queues that typically hold a large number of elements and where random access is important (fast removal of random elements)
• common input queue for all PIs (5.7.2.1)	small subset of SDL (no saves, priorities, conditions, etc.)	shared memory	none	SDL system	CBM, PIM	+ common input queue represents ready queue for PIM (fast solution for simple cases)
• strictly sequential queue (5.7.2.2)	subset of SDL to eliminate need for random access to input queue (no saves, signal communication only)	none	none	SDL process	CBM	+ allows use of native asynchronous communication mechanism provided by OS

Table 5.6 (continued) Summary of implementation and optimization concepts

Implementation/optimization concept	Prerequisites			Scope		Merits, problems and comments
	SDL	target system	impl. concept	SDL	modules	
buffer management (5.7.2.1)						
• preallocation of equally-sized buffer elements	none	shared memory	none	SDL system or subset	CBM	+ elimination of the need to dynamically allocate variable-sized memory from the heap for each signal to be transmitted – inefficient use of memory where signal sizes vary
• preallocation of a set of buffer pools (each buffer pool handles a specific signal type or signal size)	none	shared memory	none	SDL system or subset	CBM	+ elimination of the need to dynamically allocate variable-sized memory from the heap for each signal to be transmitted + efficient use of memory
• partitioning of buffer space - dealing with memory limitations						
– static partitioning of buffer space among queues	none	shared memory	none	SDL system or subset	CBM	– inefficient use of available memory + simplified control of overload situations
– static partitioning of buffer space among signal types	none	shared memory	none	SDL system or subset	CBM	– inefficient use of available memory + high runtime efficiency – complex control of overload situations
– dynamic partitioning of buffer space among queues or signal types	none	shared memory	none	SDL system or subset	CBM	+ efficient use of available memory – overload control is highly complicated
• dynamic bypassing of input queue	subset of SDL	shared memory	producer-based trigger testing	subset of SDL system	all	+ dynamic bypassing of buffer management when the input queue is empty (efficient solution for special cases)
• elimination of input queue	subset of SDL	shared memory	full control of schedule (light or bare integration)	subset of SDL system	all	static bypassing of buffer management + efficient solution for very special cases
communication with the environment (5.8) selected approach highly depends on the interface mechanisms provided by the target system	depends	depends	depends	SDL system	ENV	
timer management (5.9) selected approach highly depends on the functionality and efficiency of timer mechanisms provided by the target system	depends	depends	depends	SDL system	TM	

Table 5.6 (continued) Summary of implementation and optimization concepts

Implementation/optimization concept	Prerequisites			Scope		Merits, problems and comments
	SDL	target system	impl. concept	SDL	modules	
special optimizations for protocol processing						
• elimination of copy of input parameters (5.12.1.2)	no reuse of input parameters beyond current transition	shared memory	none	SDL process	EFSM	+ elimination of data copy operations for signal communication
• common buffer management (5.12.1.3)	single owner of signal buffer	shared memory	none	set of SDL processes	EFSM	+ elimination of data copy operations and buffer management operations for signal communication
• offset technique (5.12.1.4)	restricted use of data (signal parameters)	shared memory	none	set of SDL processes	EFSM	+ elimination of data copy operations within SDL processes
• activity thread model - combining scheduling and communication (5.12.2.1)	small subset of SDL	shared memory	light or bare integration producer-based trigger testing	set of SDL processes	EFSM, PIM	+ optimized execution order + minimization of overhead for process management + minimization of cost for data copying – no preemption of transitions of SDL processes by another SDL process

Table 5.6 (continued) Summary of implementation and optimization concepts

Chapter 6

Performance Engineering in the Context of SDL

The motivation for performance engineering in general has been outlined in section 1.1 and detailed in section 2.2.2 already. The major advantages pointed out are an increase of productivity, shorter time-to-market, reduced risk of project failure, and improved quality of the product.

In contrast to performance engineering, the major motivation for the use of formal description techniques results from the functional domain (compare section 1.1.3). However, the overall goals and merits of formal description techniques are very similar to the motivation of performance engineering, i.e. the increase of productivity due to the early detection of problems, decrease of the time-to-market, reduced risk of project failure, and improved product quality.

The major difference is that performance engineering is interested in non-functional aspects of the system while the focus of formal description techniques is on functional aspects. However, as performance engineering also relies on functional information of the system, an integration of performance engineering with formal specification and design languages is appealing. It inherently supports consistency between the two worlds. In addition, the joint consideration of both aspects has additional advantages not present with the separate consideration of either of the two aspects. The goal of this chapter is to show how this integration can be implemented in the case of SDL and its companions.

The chapter is organized as follows: section 6.1 discusses the application of performance engineering activities in the context of SDL and MSC. The needed extra information required to support SDL-based performance engineering activities is identified in section 6.2. In addition, the section discusses how the needed extra information can be specified and integrated with the SDL standard. Section 6.3 outlines the underlying concepts of current SDL- and MSC-based performance evaluation tools and provides a survey of the tools. Parts of the chapter are extensions of earlier work conducted with Bruno Müller-Clostermann; the sections 6.2.1, 6.3.2 and 6.3.3 include text from [MiMü99].[1]

[1] Reprinted from [MiMü99], Copyright 1999, Pages 1801 to 1815, with permission from Elsevier Science.

6.1 Basics of Performance Engineering in the Context of SDL

As described in chapter 2, performance engineering is concerned with all aspects related to the performance of the system under development. A major focus of performance engineering is on early phases of the system development. This is because of the fact that any phase in which a performance problem (or any other problem) remains undetected adds considerable cost to the product development. Thus, performance engineering activities span a large part of the system engineering process, i.e. from the analysis phase to system implementation, testing and maintenance.

As described in section 2.2.4, performance engineering is not only concerned with the pure evaluation of the performance of the system in various stages. It also deals with the optimization of the development process to derive a system with the required performance at minimal cost and with minimal risk of failure.

The decisions and development activities relevant to the performance of the final implementation in the context of system development with SDL have been discussed in section 5.1.1. Important decisions comprise the protocol specification, e.g. decisions that are made with the specification of the protocol (in SDL and MSC) by the standardization organization already, along with decisions made with the transformation of the SDL specification to an SDL description and its implementation on the target hardware with its specific operating system.

Especially with complex distributed systems, where the performance implications of the requirements and the design decisions are less obvious, the system analysis and architectural design are of great importance for the performance of the system. For these systems, a poor design may have a very negative impact on the performance of the final system. Thus, performance engineering is especially valuable and cost effective when employed during early development phases, i.e. during the definition of the requirements and the architectural design. This is in conflict with current practice, where performance issues are often deferred to the integration phase or even to a later release of the system.

6.1.1 Activities of the Performance Engineering Subcycle

A general discussion of performance engineering activities and the performance engineering subprocess has been given in section 2.2.4. Most of the principles described there also apply to the development of systems in the context of SDL and its complementary languages. However, the tight integration of performance engineering activities with the SDL and MSC specification and design techniques has implications on the performance engineering activities. Here we go through the general performance engineering activities as introduced in section 2.2.4.1 and describe the implications the use of SDL and complementary languages has on these activities.

Identify the goals of the performance engineering subcycle As described in section 2.2.4.1, the identification of the goals of the performance engineering subcycle comprises the identification of the purpose of the subcycle, the performance metrics to be estimated, the needed accuracy of the evaluation, and the kind of performance evaluation to be performed. These activities are mainly independent of the use of a formal description technique.

As we deal with distributed systems, there may be two major purposes of a performance evaluation: to evaluate the performance of the local installation, or to evaluate the performance of the system in the overall setting, i.e. to concentrate on the proper timely interaction between the distributed entities.

Study the details of the object under investigation SDL and MSC inherently support the understanding of the application, i.e. of its structure and behavior. However, a performance evaluation also requires an understanding of the workload, the available resources, and the mapping of the application entities on the resources. In addition, the comprehension of performance problems and their cause helps in concentrating on the relevant parts of the performance evaluation.

Decide on the modeling approach Concerning the application model, there are two major alternatives, SDL or MSC. The selected language depends on the goals of the performance engineering subcycle. The implications the selected language has on the evaluation are discussed in section 6.1.2. Once this has been decided, the details of integrating the performance-relevant information into the specification language have to be decided on. Currently, this is to a large extent determined by the selection of the performance evaluation tool. The goal of standardization is to support a single, flexible approach to describing performance aspects, which is supported by any SDL- or MSC-based performance tool.

Build the performance model The structure and behavior of the application are modeled by the respective SDL and/or MSC specification. Thus, no major work is needed in this respect besides selecting the parts of the specification to be evaluated and attributing them with the needed information. More work is needed to specify the workload (stimuli), the resources and the mapping of the application entities on the resources. However, this work can be reused for later performance evaluations.

Derive quantitative data for the performance model In order to complete the models, quantitative data, especially concerning the resource demands of the application entities and of the available resources, are needed. Deriving quantitative data is a hard task, especially where no previous experience with a similar product is available. The different approaches to deriving quantitative data have been discussed in section 2.3.7.2. In particular, the derivation of performance data by means of measurements can be supported by tools, i.e. to automatically instrument the SDL description, to analyze the results of the measurements, and to integrate the results of the measurements into the description [Lemm97, DaDL95].

Transform the performance model to an executable or assessable model For a performance evaluation based on SDL or MSC, the respective performance-extended SDL or MSC description is transformed to an executable or otherwise assessable model (see section 2.3). The transformation is done automatically by the tools. Examples of application models employed by tools are task graphs, process graphs, and simulation models.

Evaluate the performance model As described in section 2.4 this is supported by tools. We will survey SDL- and MSC-based performance evaluation tools in section 6.3. A validation of the results increases the confidence in the derived performance figures.

Verify the performance results against the performance requirements Provided the performance requirements are specified formally, an automated verification can – to some extent – be supported by tools. Otherwise, the results of the performance evaluation have to be checked manually with the requirements.

Take appropriate actions Once the performance evaluation is completed and results are present, the appropriate action to react to the results has to be decided on. The issue has been discussed in section 2.2.4.2.

6.1.2 Performance Engineering in the Development Process

The purpose of performance engineering activities in the different phases of the system development process and guidelines for the different phases have been discussed in section 2.2.4.4. The guidelines given there for the different development phases also apply in the context of SDL (and MSC).

As depicted in figure 2.6, the performance engineering activities executed in a certain development phase can be considered as a subcycle that is imposed on the regular development process model.

Two questions specific to the use of SDL and companions are

- how to specify the respective information needed for the performance engineering activities, and
- which language, i.e. SDL, MSC or others, to use for a specific performance engineering purpose or activity.

The first issue will be discussed in detail in section 6.2. The second issue is the focus of the remainder of this section.

6.1.2.1 Protocol Standardization and Implementation

There is a difference between the design of protocol mechanisms as such (standardization) and the development of a protocol implementation. Performance is not only an issue of the product development. It is also an issue for the development of protocol mechanisms, e.g. resource reservation protocols, flow control mechanisms, etc.

Concerning standardization, the protocol mechanisms as such are reflected by an SDL specification, possibly accompanied by a set of MSCs. For the development of protocol mechanisms, the interaction between the communicating partners and the implications this has on the performance are especially important. The availability of a performance model helps to evaluate the performance implication of the selected protocol mechanisms.

6.1.2.2 SDL-Based Performance Engineering

In our context, SDL provides the most complete model of the system. Many details that are not present in MSC are specified in the SDL specification or description. Thus, SDL provides the most accurate basis for a performance model. On the other hand, too many details may be present, which may not be needed for a specific performance evaluation.

Concerning performance engineering, SDL is especially important for the following purposes:

- the performance evaluation of protocol mechanisms during analysis (protocol specification),
- the (detailed) evaluation of designs, especially where behavioral aspects have a major impact on the performance, i.e. where the goals are an optimization of implementational parameters or the change of design parameters, or even requirements,
- the joint verification of functional and non-functional aspects, and
- performance measurements and performance debugging of the implementation.

Note that different additional information is needed for the different activities.

6.1.2.3 MSC-Based Performance Engineering

MSCs can be employed for performance engineering activities in all phases of the system development process where an appropriate set of MSCs is available. MSCs may be used for the following purposes:

- the specification of performance requirements and of the workload,
- the evaluation of the load imposed on the system (all development phases), especially of load mixtures,

- the early evaluation of the performance during the analysis phase of the system, e.g. where a respective SDL specification is not available,
- the derivation of synthetic loads for performance measurements on the target system,
- the derivation of TTCN descriptions including time requirements,
- the specification of the points of interest for performance measurements, i.e. as a base for the instrumentation of the SDL description, and
- the visualization of performance figures, e.g. derived by a model-based performance evaluation or by measurements.

Typically, MSC-based approaches focus on rather simple performance evaluations. They are not particularly suited to evaluating systems where complex synchronization and coordination problems represent a major issue or where complex workloads have to be considered. This is mainly due to the lack of respective information in MSC.

Often the joint use of SDL and MSC for performance engineering activities is appropriate, e.g. for performance debugging or a model-based performance evaluation.

6.1.2.4 Other Companions of SDL

Most SDL methodologies propose the use of an object-oriented analysis technique such as OMT or UML. Newer research (e.g. see [Pool98]) considers UML diagrams as a base for performance evaluation and works on the support of UML for performance engineering activities. Thus, in the future this approach may be used for an early performance evaluation. However, as the integration of UML with SDL is in its beginnings, the implications this has on the development process with respect to performance engineering have not been studied yet.

The use of TTCN for testing of functional aspects has been outlined in section 4.2.2. The extension of TTCN to support performance testing has been proposed in [WaGr97]. As MSCs can be mapped on TTCN, performance-extended MSCs can be mapped on performance-extended TTCN.

6.2 Integration of Performance-Relevant Information with SDL

With the integration of performance aspects with SDL, a number of questions and issues are raised. In order to support performance-related issues in the context of SDL, a study of the integration of performance and time aspects into SDL and MSC was started within the ITU-T study group 10 in 1997. The integration of performance aspects into the standards is important in extending the scope of these description techniques and in promoting the widespread use of the respective performance tools.

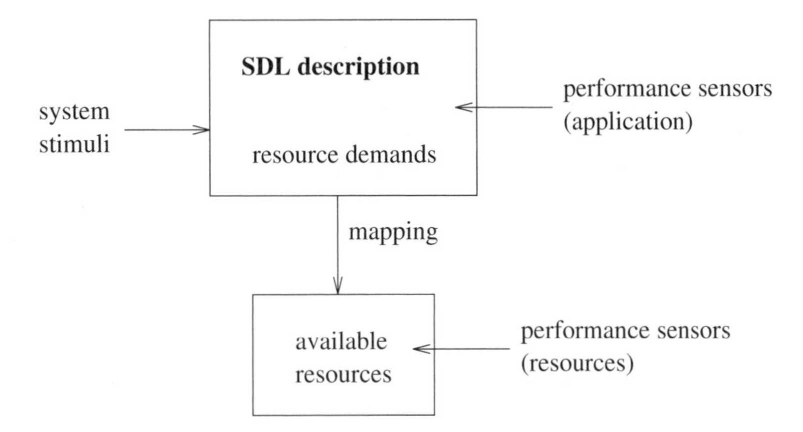

Figure 6.1 The additional information needed to support a performance evaluation

This section reports on the results of the study reached so far for an important subtopic of performance engineering, namely the performance evaluation of SDL descriptions (see also [Mits98]). We summarize the major issues involved with the association and integration of additional information needed to support the (model-based) evaluation of the performance of SDL descriptions. In addition, a proposal for standardization is outlined. The issue was also the focus of a workshop that was held at the University of Erlangen in 1998 [MiMR98].

6.2.1 Missing Information

An important prerequisite for the integrated support of performance-related activities in the system engineering process is the identification and formal specification (implicit or explicit) of the aspects relevant to performance. Thus, implementation-related quantitative properties of the system have to be derived and associated with the structural and behavioral model, most notably information describing time and resource aspects. These include performance characteristics of hardware devices, scheduling strategies, processing speeds, bandwidth of channels, and last but not least a workload or traffic characterization. An overview of the additional information and its association with SDL is depicted in figure 6.1.

Our discussion follows the terminology introduced in section 2.2.3. We also apply the classification often found in the performance world, i.e. distinguishing between the workload model, which describes the load (stimuli) imposed on the system, and the system model, which specifies the system to be evaluated. Concerning the system model, we follow the three-layer reference model introduced in section 2.3.1.3. The system model consists of three parts: the application, the resources, and the mapping of the application on the resources. In this context, an SDL (or MSC) specification represents only (a part of) the application model. The information constituting the mapping and the resource model is described neither with standard SDL nor with MSC.

6.2.1.1 System Stimuli

For a performance evaluation, some specification of the workload imposed on the system, i.e. to be handled by the application, is needed. Typically, this is described by means of stimuli. In the context of SDL, stimuli are input signals provided to the system from the environment. Adding stimuli to SDL descriptions transforms the open description into a closed description.

As described in section 2.2.3, the workload of a system may be rather complex and may comprise a set of different use cases. For the different use cases, the number of requested services as well as their type and frequency may vary. Important parameters of the system stimuli are the requested services, typically identified by SDL signals, and the interarrival times between service requests. The system stimuli induce service requests at the different SDL entities that implement the application. This in turn results in resource demands for the resources of the system (see below).

Important to note is that the workload imposed on the system under evaluation may be a dynamic parameter of the system to be evaluated. Or in other words, the load to be generated is influenced and dynamically changed by the system to be evaluated. In this case, the load generated from the environment typically depends on previous output provided to the environment.

6.2.1.2 Available Resources

Available (limited) resources denote the entities which are available to the SDL description, i.e. to handle the load resulting from serving the requests that are caused by the system stimuli. Thus, the resources can be considered as the SDL machine. The most relevant resources are the processors and the communication channels of the system. Further important resources concern the available memory to hold code and data, most notably the size of input queues. The most important parameters of the resources are their capacity and time characteristics. For processing resources, also their service strategies are of importance.

6.2.1.3 Resource Demands

Resource demands denote the quantitative requirements resulting from the execution of the SDL description, or more precisely of its implementation. Typically, resource demands are caused by stimuli which induce service requests for the resources of the system. Resource demands specify the cost caused by the implementation (or execution) of the SDL description on the specified resources. Examples of resource demands are the required processing times on processors and the required memory space, e.g. for signal queuing.

Note that unlike the underlying assumptions made by most SDL-based performance evaluation tools, a considerable part of the resource demands may result from the underlying runtime system. Depending on the application, the sum of the resource demands resulting from the implementation of the runtime system may exceed the resource demands resulting from the execution of the transitions of the SDL description [HeKM97].

6.2.1.4 Mapping

Concerning the evaluation and optimization of the system design and implementation, the mapping of the resource demands on the available resources is of importance. Employing SDL, the mapping of the SDL entities, e.g. blocks, processes, process instances, simple actions, and channels, on the available resources has to be specified. The mapping may be static, i.e. fixed during the execution of the system, or dynamic, i.e. changed between the processing of two stimuli or even during processing of a single stimulus.

6.2.1.5 Performance Metrics and Sensors

Performance metrics denote the performance figures of the system under development that are the object of the performance evaluation either as part of the goal function or as constraints. Examples of metrics are the response time to process a specific input signal, the throughput for a certain type of signal or the utilization of some resources of the system, most notably the process input queues and the load of the processors and communication devices.

In order to retrieve the values of the performance metrics, sensors are needed. Sensors may range from simple probes that simply trace the execution, to complex sensors that already provide aggregated performance figures to the outside world. If simple probes are used, most of the evaluation of the information is done outside the system. In the case of aggregated sensors, the evaluation is done within the system. This also allows the derived information to be directly reused, i.e. to dynamically influence the behavior of the SDL system based on this information. Examples where this applies are overload control mechanisms specified with SDL or systems with QoS requirements.

Sensors may be divided into application sensors and resource sensors. Application sensors focus on the behavioral specification of the system, i.e. the parts visible within the SDL description, while resource sensors monitor the underlying runtime system and the hardware.

6.2.1.6 Performance Requirements

As described in section 2.2.3, performance requirements specify the required values for specific performance metrics. Performance requirements represent constraints on the implementation of the SDL description. Performance requirements do not necessarily have to be defined formally for a performance evaluation. Typical types of performance requirements are response time and throughput figures.

6.2.2 General Issues of the Integration with Z.100

Concerning the integration of the needed extra information with formal description techniques, we focus on SDL. As MSC provides less information than SDL, additional information is needed when basing a performance evaluation on MSCs, or alternatively, compromises concerning the accuracy of the evaluation have to be made. A major problem with MSCs is the incompleteness of functional information, i.e. information solely present in the corresponding SDL specification. In principle, the needed additional performance-relevant information – as identified above – can be associated with MSC as well (e.g. see [FLMS97, Scha96, Doug98]).

6.2.2.1 Organization of the Information

During system development, similar information is needed for different purposes, e.g. some behavioral aspects of the system under study are needed for a functional analysis as well as for a performance analysis. Thus, an important goal to ease the system engineering process is to maintain a single copy of each piece of information only. This is the main motivation for the description or integration of functional and performance aspects in a single language, in our case in SDL. In order to limit the overloading of SDL with different kinds of information, tools should support different views of SDL descriptions, e.g. allow performance information to be hidden when not needed.

In addition, not all information needed for a performance evaluation should be integrated and described in SDL. In particular, information that is only loosely associated with SDL features does not have to be specified in SDL. However, this does not mean that no standardized formal language is needed to describe this information. Instead, a separate standard should be devised to describe these aspects.

Proposal for standardization We propose the following separation concerning the specification of performance-related information. The system stimuli, the resource requests and application sensors are described within the SDL description. For this, a set of extensions to the language is needed. The arguments for the integration into SDL as well as some details of the integration are given below.

The specification of the available resources, the mapping of resource requests on the available resources and resource sensors are described in a separate notation outside SDL. As described below, this supports the reuse of this information in case an MSC-based performance evaluation of the system is employed as well. While these items are closely related to each other, performance requirements are very different. A very reasonable approach to specifying performance requirements is MSCs.

6.2.2.2 Consistency with MSC

A performance evaluation may be based on an MSC specification instead of an SDL description. In this case, a sensible separation of the information needed for a performance evaluation into parts that are specific to a formal description technique and parts that are not, improves the reuse of the information.

Proposal for standardization As described above, we favor the specification of the available resources, the mapping of resource requests on the available resources and of the resource sensors by a separate notation outside SDL and MSC. The idea is to reuse this information when an MSC-based performance evaluation of the system is employed.

A problem concerning consistency is resource demands, which may be defined in MSC or SDL. The consistency of this kind of information between the SDL description and the corresponding MSC specification is desirable. However, due to the incompleteness of the MSC specification compared to the corresponding SDL description, this is a complex matter.

6.2.2.3 Level of Detail of the Performance Model

Performance evaluations are useful in different phases of the system engineering process, i.e. at any time where decisions are made that influence the performance of the resulting implementation. Since SDL supports top-level design as well as detailed design, a flexible approach that supports the specification of performance information at various levels of granularity is important. Thus, an early evaluation of the performance may be based on rather coarse performance information while a later evaluation may use very detailed and fine-grained information.

Different levels of granularity are also useful where different kinds of performance evaluations are supported, i.e. the evaluation and optimization of the local computing system or of a distributed system.

Proposal for standardization We propose a flexible approach that allows the association of cost or other performance-related information with SDL at a fine level of granularity. This is more flexible than limiting the association of performance information to SDL processes or transitions.

6.2.2.4 Kind of Performance Evaluation

Different kinds of performance evaluation techniques can be applied to a given SDL description (compare sections 2.4 and 6.3.1). So far, all tools that support a performance evaluation define their specific set of extensions to SDL to describe the information needed for the specific performance evaluation technique that is supported.

Most of the tools only support one specific performance evaluation technique. Thus, the question to what extent the needed extra information depends on the specific evaluation technique that is used is important. Or in other words, is there a generic representation of the needed extra information from which the extra information needed for other performance evaluation techniques can be derived?

Proposal for standardization The topic has not been thoroughly studied so far. However, we assume that a detailed specification of the needed performance-related information to support a performance simulation is most appropriate to support other simpler evaluation techniques as well. In addition, performance simulation seems to be the most widely accepted performance evaluation technique in the SDL community. Other simpler techniques often applied by performance engineers can probably be based on the information that is needed for a performance simulation. Thus, performance simulation should be the major focus of the integration of performance-related information in SDL.

6.2.2.5 Dealing with Behavioral Differences Between an SDL Description and an Implementation

In order to accurately estimate performance figures, the underlying performance model has to reflect the implementation rather than the SDL description. If there is a match between the behavior of the two, basing the performance model on the SDL description is not a problem. However, if there are differences that have an impact on the performance, these have to be modeled.

For example, physical systems are finite. In particular, limited buffer sizes or memory space are restrictions that may influence a system's behavior considerably. Conversely, SDL assumes input queues of infinite size. Thus, the results of a verification of a system under the assumption of unbounded queues may not be particularly useful. On the other hand, unlimited queues ease the formal verification. This is due to the fact that the limitation of queues results in additional actions which are not needed with unlimited queues.

The implications of introducing limited queues in SDL are far-reaching. The most important seems to be the question of how to deal with buffer overflow and the implication the selected solution has on the formal semantics. Alternatives for dealing with full queues are discarding the signal, throwing an exception, or restarting the system. Typically, the selected action depends on the application at hand. Thus, a flexible, user-defined approach seems to be most appropriate from the user standpoint.

Proposal for standardization Our proposal is to extend SDL to support the formal description of implementational limitations. This is because we feel that a verification of a system that greatly differs from the implementation is not particularly useful in establishing the conformance of an implementation with its specification.

We feel that the integration of these aspects in SDL is important. It allows one to actively deal with these limitations at an abstract level, i.e. allows one to formally verify the mechanisms to deal with these problems at the (abstract) level of SDL.

6.2.2.6 Semantics of Time

The time semantics of SDL are very vague. With the current definition, the time may or may not be advanced by an action or a transition. From the specification viewpoint, this is appropriate. It allows one to reason about the system independently of a specific implementation.

Conversely, a description with precisely defined time semantics has a much smaller state space. Thus, it is much easier to verify in practice. On the other hand, this kind of verification is less general. Thus, it holds for the system only under the assumption that the precisely defined temporal propositions hold.

There are two principal approaches to defining the temporal aspects more precisely. The timing may be defined either directly, stating precise time requirements, e.g. for the execution of a transition, or indirectly by specifying the temporal behavior of the underlying resources and the mapping of the SDL description on them. The indirect approach is used when employing a performance evaluation. Conversely, the specification of performance requirements represents a direct specification of temporal requirements.

Proposal for standardization As our focus is on supporting a performance evaluation, our work is mainly based on the indirect specification of the temporal behavior by defining the temporal details of the underlying resources, etc. However, as we will see below, the direct specification of temporal aspects is needed as well to precisely define the workload, i.e. the exact temporal behavior of system stimuli.

In addition, a direct specification of temporal aspects is needed to extend the scope of SDL to real-time systems, especially where hard time requirements are present.

6.2.2.7 Relation of Performance Evaluation to Object Orientation in SDL

The known approaches to performance evaluation do not deal with object orientation in SDL so far. However, since object orientation is a basic concept in SDL, the concept should also be applied for the specification of performance-related information. In principle, there seems to be no problem with the use of inheritance in the context of performance-related information. Important in this respect is the inheritance of defaults for performance-related information, which eases the task of the user for associating performance data with time-consuming actions in SDL.

6.2.3 Specification and Association of the Extra Information

In section 6.2.1, we have concentrated on the requirements for a model-based performance evaluation in the context of SDL, i.e. on the kind of information that is needed to support a performance evaluation. In the following, we focus on design issues concerning the integration or association of the extra information with SDL. We additionally refer to MSC where appropriate.

6.2.3.1 System Stimuli (Workload Model)

Several concepts have been proposed to describe the workload model by defining the system stimuli. QSDL [DHHM95] uses SDL processes (extended to model time more precisely) to describe arrival processes. PMSC [FLMS97] employs a special notation to describe system stimuli. Another approach is to have a separate workload generator linked to the SDL specification, implemented in an arbitrary (non-standardized) language or notation (e.g. [Gerl97, Roux98]).

Specification of the workload model with SDL In principle, there is no difference between a load generator that is used to generate the stimuli needed for a performance evaluation and a part of an SDL description that sends signals to another entity in the SDL description. In fact, as the load generator in distributed system is often just another entity in the system, e.g. a higher protocol layer or the peer entity, the load generator may just be another (simplified) SDL description.

Dynamic workload models In real systems, the generated load may depend on the dynamics of the executing system. An advantage of the use of SDL to describe the load generator is that it allows one to dynamically control the load generation from the application. In other words, the state of the SDL description under evaluation may control the load that is issued to it, i.e. by interacting with the load generator by standard SDL communication constructs.

Refinement of the time semantics of SDL The stimuli generated by a load generator are typically reflected by SDL signals. The generation of a signal by the load generator can be implemented based on the SDL timer mechanism. A problem with the direct use of an SDL process as load generator is the vague time semantics of SDL. Thus, the generation of a signal sent to the application at an exactly defined time is not possible. In order to remedy this, the timer mechanism of SDL has to be refined such that a transition fires at a precisely defined time and generates immediate output. For example, this has been done by QSDL, where timed states have been introduced that fire at a defined time.

Proposal for standardization For the reasons given above, we propose the use of SDL processes to specify the system stimuli. Thus, the idea is to specify the load generator as part of the SDL system and in SDL notation. This has the advantages that well-known syntax and semantics can be used. In addition, the generated load may be a function of the dynamic state of the system under evaluation.

In order to allow the generation of stimuli at precisely defined times, we propose the extension of the timer mechanism to support timed states as outlined above.

6.2.3.2 Specification of the System Model

As described in section 6.2.1, the system model for a performance evaluation comprises the description of the available resources, the resource demands resulting from the application and the mapping of the application, or more specifically its resource demands, on the available resources.

Layered performance model Concerning the description of these aspects, we propose a layered approach as introduced in section 2.3.1.3. It comprises three layers, the application layer, the mapping layer and the resource layer. The application of the reference model in the context of SDL is depicted in figure 6.2. The application layer defines the behavior of the application and, in our case, is mainly equivalent to the extended SDL description. The only exception is the description of the system stimuli in SDL, which is not part of the application layer. In order to

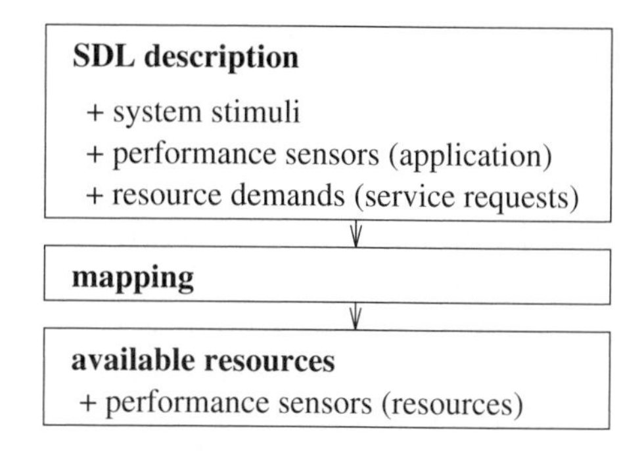

Figure 6.2 The layering concept to describe resource demands, the mapping and the available resources

support performance modeling, the application layer issues service requests (resource demands) to the mapping layer. The mapping layer defines the mapping of the service request to an appropriate resource specified in the resource layer. The resource layer itself describes the actual resources available to the application.

The layered approach supports a separation of the description of resource demands from the resources. Thus, changes to the underlying resources or the mapping can be made locally in the resource or the mapping layer, instead of changing the service requests issued by the application layer. This modular approach allows one to quickly change parameters of the available resources or of the mapping on the resources without touching the SDL description itself. In addition, it allows a change of the platform on which the SDL description is executed without making any changes in the SDL description itself.

Another advantage of this approach is that SDL and MSC can be used interchangeably to specify the application layer. Thus, a performance evaluation may be based on MSCs by simply replacing the SDL description by a respective MSC specification.

Proposal for standardization Concerning the extension of SDL for the system model, only the service requests are integrated with SDL. The other information can be hidden in a separate description. However, in order to support interoperability between tools, this information should be specified in a standardized way as well.

6.2.3.3 Available Resources

As described above, resources are different from other SDL concepts. Thus, a direct association with the SDL description is not necessary. Only a link from the SDL entities to the available resources is needed. This link is implemented by the mapping layer.

As outlined above, the available resources offer services to the application. A service request issued to the resources results in a delay in time depending on the amount of the service requested, on the service capacity of the resource, and possibly on the contention with competing service requests. Where contention is present, the service requests are processed according to a specified service strategy.

Level of detail of the resource description The level of detail of the resource description is an important issue. For an early evaluation or an evaluation where the focus is the temporal behavior in a distributed system, a coarse model may be sufficient. The known approaches range from the direct specification of absolute time delays to a precise model of the underlying hardware. In [BMSK96, Roux98], delays are specified from which the response times are derived considering concurrency within a set of processes. In [DHHM95, Gerl97], queuing models with various service strategies and priorities are employed. A detailed (exact) emulation of the underlying processor hardware is used in [StLe97].

Kind of resources In principle, there are different kinds of resources. Resources may be physical, i.e. processors, communication links or memory space, or rather more logical, such as semaphores or other synchronization constructs.

For a performance evaluation of SDL systems, typically the processors and the communication links are most important. They cause a delay of the execution of the SDL description depending on the capacity of the resource and the contention at the resource. Conversely, limited memory or buffer space typically causes a substantial delay only where the limit is hit.

The information needed for the specification of the available resources is the services they offer, their capacity and service strategy.

Proposal for standardization As outlined above, the resource description is not part of the SDL description. However, a standardized approach is needed to ensure interoperability of tools. We propose a layered approach for the description of the available resources. Thus, the resource description offers services to the upper layers.

Each resource offers a set of services. A request for a service blocks the requester for a non-negative time. The time may depend on the actual parameters given with the resource request, the specifics of the offered service, and the circumstances under which the service request has been issued.

A description of a resource comprises

- the number of servers of the resource,
- a set of services offered by the resource,
- a function for each offered service, defining the processing delay for the service as a function of its input parameters which are provided by the service request, and
- the service strategy of the resource, e.g. first-come-first-served, priority based, or others.

In order to formally specify these aspects, we propose a notation similar to the one proposed by QSDL [QUES98] (see also 6.3.2.6).

The approach allows to accurately model (or simulate) the behavior of processing and communication resources. It also supports the specification of overhead that is involved with the execution of an action or transition.

In order to handle more complex cases, e.g. resources that change their behavior dynamically, the above notation can be extended. Another advanced issue is dynamic feedback provided from the available resources to the SDL application, e.g. to indicate that the load exceeds a certain threshold. This requires some flow of information from the resource back to the SDL description.

6.2.3.4 Resource Demands

In principle, the resource demands can be given either absolutely (e.g. the delay time of a data transmission) or relatively (e.g. the volume of the transmitted data) from which the absolute time can be derived. With our layered approach, the resource demands should be specified in a generalized form, i.e. as parameters from which the specific resource demands on the selected resource are derived. Thus, the actual resource demands may depend on the decisions concerning the mapping and the specific resources that are available as well as on dynamic aspects such as the current load of the resource. With our layered approach these details are hidden from the SDL description.

Level of detail of the resource demands In principle, the granularity of the parts for which the resource demands are specified may differ. Alternatives known from literature are the association of resource demands to actions or complete transitions in SDL [Roux98], or to add special service requests to arbitrary points in the SDL specification [DHHM95, Gerl98].

Association of resource demands with the behavioral specification There are three principal approaches to associating the resource demands with the behavioral description:

- A new SDL construct (to request a service from a resource) is introduced that can be integrated at almost any point in the control flow.
- Service requests are associated or tied with regular actions in SDL. In this case, the semantics of the service request may depend on the action the performance information is associated with, e.g. different for a procedure call and an output. For example, a resource request associated with a procedure call models the cost of executing the procedure. Thus, the delay of the control flow by the respective resource demand is modeled. Conversely, the service request associated with an output construct may represent the delay of the signal transmission. Thus, there is no delay of the control flow within the executing transition.
- Alternatively, service requests can be associated with structural components, e.g. with channels, processes, services or procedures.

It has been argued that, in order to clearly discriminate between behavioral parts of the SDL description and performance-related information, service requests should be associated with SDL constructs, rather than introducing a new construct. On the other hand, service requests may use parameters (variables) as input which are clearly part of the behavioral specification. Thus, the direct integration of service requests into the (sequential) flow of control clearly defines the temporal order of the use of the variables passed to the service request. In addition, tools may hide performance-related information where the information is not needed or complicates the description.

Resource demands for implicit actions Resource demands may also result from implicit actions not explicitly modeled by an action in the SDL description. Examples are the transmission of signals or overhead resulting from timer management, buffer management or the testing of continuous signals. Focusing on the association of resource demands with the SDL specification itself, a major cause of resource demands is ignored, namely the cost involved with the runtime system, typically comprising the SDL runtime support system and the operating system. Measurements have shown that the overhead in executing an SDL specification (i.e. the SDL transitions) may be much larger than the cost of the SDL transitions itself [HeKM97]. Thus, the overhead of the underlying runtime system is an important issue and should be modeled appropriately.

In performance evaluation, there are two basic approaches to including overhead. Overhead is directly added to the resource demands of the actions themselves, or alternatively, overhead is modeled by an abstract overhead factor that reduces the capacity of the resource which is available to the application. A problem may be the fact that the overhead typically depends on the load of the runtime system itself, e.g. the number of processes or timers it has to handle.

Another requirement is that resource demands for communication (delay) should be described in a straightforward manner. This is currently not the case with most of the available performance tools which focus on the resource demands of the transitions rather than on communications. Thus, they require a refinement of a delayed channel by an SDL process which issues service requests to the model of the communication resource.

Hierarchy and defaults The association of information concerning the resource demands to the respective time-consuming actions of the SDL system may be very tedious. Thus, a mechanism to specify and use defaults for the resource demands is desirable. Most appropriate seems a hierarchical scheme that allows the inheritance of the defaults in the SDL hierarchy. This releases the user from the task of associating a resource demand with each action or transition.

Proposal for standardization We propose that resource demands are expressed by a new construct which specifies a service request similar to the approach proposed by QSDL. The new SDL construct can be integrated into the flow of control at almost any point of the behavioral description. All resource requests are blocking requests, i.e. delay the flow of control until the (indirectly) specified time delay has passed. As described above, the exact quantity of the resource demands is indirectly specified by means of service requests issued to resources. This works perfectly for all parts of the system that are explicitly specified in the SDL description, i.e. the request for computational resources to execute transitions.

Communication delays are described by introducing channel refinements. Thus, a signal sent on a channel or signal route is explicitly delayed by a service request, similar to the request for a computational resource. A disadvantage of the approach is that the SDL description may get overloaded with performance-related information. In addition, implicit communication, i.e. employed by export/import, view/reveal and remote procedure calls, cannot be modeled as there are no channels associated with it. On the other hand, modeling performance requires the explicit or implicit definition of the mapping of the communication on some resource. This is not the case with the usage of the implicit communication constructs.

In addition, we propose to use defaults to specify resource demands for all transitions of a system, block, process, service or procedure. These are also specified by means of service requests. Defaults are inherited through the inheritance tree according to the rules of SDL. The defaults apply to all transitions for which no explicit resource demand is defined.

6.2.3.5 Mapping

The different SDL entities have to be assigned to the resources in order to study the impact of resource consumption and contention. Several alternatives exist. The mapping can be specified implicitly, indirectly, or by directly naming the resource within the resource requests. However, the implicit as well as the direct mapping considerably limits flexibility.

Proposal for standardization As described above, a separate description of the mapping of the resource requests on the available resources supports modularity and flexibility of the description.

We propose that service requests are mapped on resources rather than mapping SDL entities. This is more general. It allows one to map different parts of an SDL entity, e.g. of a process on different resources. An example where this is necessary is the implementation of an SDL process in a mixed hardware/software system, where a part of the SDL process is directly implemented in hardware to speed up the execution.

We propose the use of a static description of the mapping of resource requests on available resources. Thus, a specified service request is always served by the same resource (possibly comprising several servers). This eases the mapping description. Dynamic mapping can be modeled in the application level where needed.

6.2.3.6 Performance Metrics and Sensors

Performance sensors are used to specify the performance metrics monitored and measured during the performance evaluation. QSDL provides sophisticated aggregated performance sensors while others rely on rather simple probes. Simple probes used in SPEET [StLe97] focus on providing an interface to output performance data.

With QSDL, aggregated performance sensors are special data types maintained by the underlying system during the system simulation. The sensors can be used also to dynamically influence the functional behavior of the SDL description based on load figures. This allows the dynamical adaptation of the system to meet QoS requirements or to control overload.

A different approach is known from the implementation of telecommunication systems. Load sensors are implemented by the underlying runtime system depending on the special needs of the application. Upon the detection of specific load situations, triggers in the form of SDL signals are sent to the SDL processes in charge of dealing with the problem (e.g. see [Wirt98]).

The advantage of the QSDL approach is that the sensors are part of the (extended) SDL description. Thus, sensors can be conveniently specified at the (abstract) level of SDL. However, accessing or implementing these sensors, e.g. to dynamically react to a certain state of an input queue or a processor, most probably results in polling of the respective variable. In addition, the timely reaction to specific sensor values may not always be possible.

Conversely, implementing sensors in the underlying system has the advantage that the respective sensor information is provided to the SDL description, i.e. by SDL signals, only where there is a need to react to it. Thus, the SDL description is not in charge of polling the values of these sensors. On the other hand, this approach requires the implementation of low-level constructs to monitor the sensor values and to trigger the generation of a signal where appropriate. With respect to our layered model, this represents the description of the sensor within the mapping or the resource layer.

Concerning standardization, the issue needs further study. This also holds for the question of how to output the performance metrics to be evaluated. Output may be raw data, e.g. event traces, or aggregated data, e.g. the average CPU usage, average queue size, or minimum, maximum and average response time or throughput figures.

6.2.3.7 Performance Requirements

Performance requirements are an important issue for the automatic verification of performance figures. However, their formal specification is not central for a typical performance evaluation.

Basic approaches to specifying performance requirements are

- extended MSCs [FLMS97, ScRe98],
- temporal logics used in conjunction with events [Leue95], or in conjunction with performance sensors [Dief98, QUES98], and
- the direct use of SDL to check performance requirements within the SDL description during the system simulation [QUES98].

MSCs are especially appropriate to specify response time requirements.

The direct use of SDL to specify performance requirements does not seem to be a sensible approach since this greatly interferes with the functional specification. This is especially true if performance requirements represent metrics that require the monitoring of several SDL processes. In this case, a description within SDL would require one to communicate the measured values using standard SDL communication mechanisms.

Currently, MSCs are less expressive than temporal logics. On the other hand, temporal logics are not acceptable for end users. An idea is to enhance the expressiveness of MSCs and to map these enhanced MSCs internally on a temporal logic. This allows one to support the joint verification of functional and non-functional aspects by model-checking techniques [Dief98].

6.3 Survey of Performance Engineering Concepts and Tools

After the identification of the important subtasks of performance engineering in the context of SDL (and MSC) and of the needed extra information to be integrated, we survey tools that support SDL- and MSC-based performance engineering activities. Before discussing important tools, we present a classification of the concepts on which the tools are based, and discuss the specific merits and problems of the different concepts.

6.3.1 Concepts of Tools

The main differences of the tools are the approach to specifying the needed extra information, and the underlying performance evaluation technique that is employed by the tool and on which the extended SDL specification is mapped.

6.3.1.1 Specifying Extra Information

The tools use four principal approaches to specifying the extra information needed for an SDL- or MSC-based performance evaluation:

- **Annotation:** With the annotational approach, the needed extra information is added within the SDL or MSC language by means of comments that are meaningless to the other (functional) tools but have defined semantics with respect to the performance evaluation. This approach is a workaround to real language extensions preventing interference with other tools.

- **Association:** With association, the performance-relevant information is kept outside SDL or MSC but somehow linked or associated with the respective SDL entities, e.g. by referring to SDL entities from external descriptions.

- **Integration:** With the integration approach, the needed information is added to SDL or MSC, using real extensions of the languages. Without standardization, this approach endangers the interoperability with other tools.

- **Implicit specification:** Some tools make implicit assumptions concerning performance-relevant information. For example, SDL blocks may be assumed to be the entities to be mapped on processors.

6.3.1.2 Mapping on Assessable Model

The basic approaches to performance modeling and performance evaluation have been described in sections 2.3 and 2.4, respectively. Basically all of these general approaches have been applied to the SDL- or MSC-based performance evaluation in one way or another. Evaluating the performance of extended SDL or MSC descriptions is typically done by mapping these formal descriptions onto one of the described performance models. As the performance evaluation technique is usually tied with the performance modeling approach, we deal with the two issues in a joint manner.

The following approaches, or mappings, for an SDL- or MSC-based performance evaluation have been proposed by tools:

- **Mapping on queuing network:** The SDL description is mapped on a network of queuing stations where each queuing station represents a processing or communication resource, e.g. with SPE.ED, and the LQN approach.

  Typical assumptions are exponentially distributed times and mutual independence of the stations. In order to efficiently solve queuing networks, often specific assumptions are made which do not necessarily hold in the context of SDL.

- **Mapping on Markov chain:** The SDL description is mapped on a stochastic automaton with exponentially distributed transition probabilities, i.e. on a Markov chain, and analyzed employing respective techniques. This approach is employed by Timed SDL and indirectly also by SDL-net.

  In principle, the approach is expressive and allows one to model complex features of SDL. However, with realistic problem sizes often the state explosion problem is encountered which may prevent the application of this technique.

- **Mapping on task graph:** MSCs are often mapped on task graphs where the nodes represent processing and the arcs represent communication cost. Task graphs can be analyzed using simple algorithms from graph theory to estimate the length of critical paths of the execution. Typically, these approaches are based on deterministic processing and communication times, e.g. DO-IT and SPE.ED. However, also stochastic figures may be used.

  Task graph analysis is very fast. However, mainly due to the neglect of contention on resources, the analysis is of limited use only.

- **Mapping on process graph:** The SDL as well as MSC entities may be mapped on a process graph where the nodes represent the SDL or MSC entities and the edges represent the communication relationships between the entities. Each node of the process graph represents the sum of the processing cost associated with the entity for a specific workload. The process graph may be mapped on a machine graph. This allows one to compute the sum of the load to be handled by the different resources. This approach has been taken by LASSI and the DO-IT project.

The analysis is very fast. It allows one to estimate the sum of the load caused by SDL entities or the utilization of the resources.

- **Mapping on simulation model:** The SDL or MSC description may be mapped on a simulation model and evaluated this way. The mapping of SDL on a performance model suited for simulation is employed by SPECS, HIT, the OPNET and the LQN approach. The direct simulation of SDL, or in other words, the direct mapping of the SDL description on code that forms the basis for the simulation, is supported by QUEST, ObjectGEODE and to some extent EaSy-Sim. The mapping of MSCs on a simulation model has been proposed by DO-IT, CORSAIR, and SPE.ED.

  This approach is very powerful as simulation models are typically very expressive. On the other hand simulation is very slow.

- **Emulation:** The emulation of the SDL description, or more precisely of the code derived from it, is possible as well. This is supported by the SPEET tool.

  This represents the most detailed and accurate approach to a performance evaluation. However, it requires the SDL description to be executable. In addition, due to the detailed machine model it is very slow.

6.3.2 SDL-Based Performance Evaluation

6.3.2.1 Timed SDL

Timed SDL (TSDL), due to F. Bause and P. Buchholz, is a modified version of a (relatively small) subset of SDL [BaBu91, BaBu93]. TSDL is designed for the examination of qualitative and quantitative system aspects within a single model description. TSDL essentially implements timed transition systems where rates or exponential time durations are attached to the transitions. However, it does not include the notion of resources, workload and resource requirements. A program package transforms a TSDL model into an internal representation of an equivalent finite state machine that can be treated by algorithms for qualitative and quantitative analysis. In particular, algorithms for partial state-space exploration have been shown to be very suitable. The solution algorithms employed for the performance evaluation are of Markovian type.

6.3.2.2 SDL-Net: A Petri Net Based Approach

An example for a Petri net based approach is the Ph.D. thesis of H.M. Kabutz, University of Cape Town, on SDL-net and the DNAsty tool [Kabu97]. The approach is based on QPNs (Queuing Petri Nets) that are a blend of queuing systems and place/transition nets [BKKK95].

Time is introduced by assigning stochastic durations to Petri net transitions. The SDL description itself is not changed because the external information is added at the Petri net level. Special constructs are used in the SDL-net notation to handle save, timer and channels. These constructs are represented in the QPN notation by the Save place, the Timer place and the Channel place, respectively. Other aspects of SDL, such as the structure of SDL descriptions, process creation and termination, process output, process data and process decisions are also mapped to the SDL-net.

The DNAsty toolset consists of a graphical SDL editor, an SDL to SDL-net converter, and a tool to derive the Markov chain from the SDL-net and to solve it. As an example, the well-known Initiator Responder protocol (InRes), including its failures and its performance behavior, has been investigated.

Mapping SDL descriptions to Petri nets with time-continuous stochastic time (exponential and phase-type distribution), the semantics of the original SDL description and the resulting SDL-net may differ. Another aspect concerning time is the pragmatic approximation of deterministic time durations by Coxian, i.e. phase-type distributions. The solutions proposed to resolve some of the semantic problems when transforming SDL to SDL-net are acceptable from a pragmatic viewpoint, although purists in formal specification will probably disagree.

The author points out that the usage of data types is of course very restricted for reasons that are immanent to the Petri net method. Another drawback is that the modeling approach of SDL-net does not reflect the structure of the target architecture. In particular, the concept of resources does not exist.

6.3.2.3 SPECS: SDL Performance Evaluation of Concurrent Systems

An approach using annotated SDL specifications is implemented by the SPECS tool (SDL Performance Evaluation of Concurrent Systems), which is due to P.S. Kritzinger and colleagues, University of Cape Town [BMSK96]. SPECS follows the intuitive practice of modeling SDL blocks by machines. Processes in different blocks execute concurrently, while processes in the same block execute in a multitasked way. The developer decides which processes may execute truly concurrently by placing them in different blocks. Each block is designated to run on a particular processor that has a specific processing speed. By changing the process weights and the assignment of blocks to processors, the developer can adjust relative process speeds within a block. The supplementary information needed is added separately, employing a graphical user interface. Thus, the SDL system is not changed syntactically. However, there are some semantic differences with respect to channel reliability and time.

Models built with SPECS are executed (simulated) on a virtual machine that has been derived from the SDL description and supplementary information. As stated by the authors, the approach could be improved by using similar code for the simulation (augmented by appropriate software libraries) and the implementation.

6.3.2.4 Integrating SDL with the Hierarchical Performance Modeling Tool HIT

A more general framework that uses the concept of resources has initially been proposed by E. Heck, D. Hogrefe and B. Müller-Clostermann. The research is embedded in the context of the performance tool HIT [HeHM91]. The conceptual ideas have been developed further and are now completely elaborated in the doctoral thesis of E. Heck [Heck96]. The thesis introduces a conceptual framework describing how the synthesis of formal description techniques and performance evaluation is obtained such that the properties of both worlds are preserved. The proposed framework is elaborated in detail for SDL and HIT and a prototype environment has been developed that supports the transformation of SDL descriptions into HIT models. Using HIT, full advantage can be taken of the functionality of a modern performance tool including the principle of vertical and horizontal structuring, and the principle of separating load and machine. The thesis describes how a HIT model that has been generated by transformation from an SDL description is structured. Depending on the complexity of the derived HIT model, various performance evaluation techniques supported by HIT can be applied. However, for realistic models of SDL descriptions typically only a simulation is feasible.

Templates enable the designer to transform a given SDL description into a quantitatively assessable HIT model. For different SDL structures, there are different types of templates: a model template, a partitioned-block template, a block template, and additionally, a service pool. These constructs are sufficient to transform every SDL description to a corresponding HI-SLANG representation. The original structures of the modeled system and of the SDL description (layering, frame structure of messages, role of the medium, etc.) are preserved during the modeling and transformation process. This is helpful in enabling an interpretation of the results in the original design.

The initiating paper [HeHM91] and the dissertation [Heck96] have contributed to concurrent efforts, in particular to the development of the QSDL language and the QUEST method described below.

6.3.2.5 Mapping SDL Systems to OPNET Models

A similar approach following the ideas of the SDL/HIT integration is due to J. Martins, J.-P. Hubaux, T. Saydam and S. Znatny [MHSZ96]. The authors propose to extend SDL descriptions by constructs describing delays, processing resources and workload descriptions. A translation to a performance model that is executable by the tool OPNET (Optimized Network Engineering Tool) is done manually. However, the authors state that the mapping process from augmented SDL to OPNET can be easily automated. Since OPNET is based on the EFSM paradigm, the mapping of SDL processes to OPNET can be done systematically. The performance evaluation by discrete event simulation is finally done in the OPNET environment. The approach has been applied to a small example of a TCP-like scenario.

6.3.2.6 QUEST: The Queuing SDL Tool

The QUEST approach has been developed at the University of Essen by M. Diefenbruch, J. Hintelmann and B. Müller-Clostermann [DHHM95]. QUEST is based on the adjunction of time-consuming machines that model the congestion of processes due to limited resources. By adding workload models and after defining a mapping of workload to machines, finally an assessable performance model is automatically generated. The description and construction of performance models and their evaluation is supported by the language QSDL (Queuing SDL) and the tool

QUEST. The QSDL language provides a means for the specification of load, machines and their binding. Load is modeled by QSDL processes by issuing time-consuming requests that are referred for execution to adjunct machines given by queuing stations.

QSDL processes are bound to the machines via links and pipes. Processes and machines within the same block are connected with a link.

The transformation of the QSDL description to an executable simulation program is done automatically. QUEST includes a QSDL parser, a class named SCL supporting simulation and state-space exploration, and a compiler for the translation of QSDL descriptions to a simulator in the programming language C++.

The execution of the simulation model finally yields performance values that are visualized graphically or summarized as a report. In addition to the transient behavior, the investigation of the long-term behavior of a system can be carried out by using statistical techniques. QUEST additionally supports the verification of time and performance aspects by means of model checking. The applications of QUEST range from toy examples to real-world applications [Dief97, Hint96, HiWe97].

The current version of QUEST covers SDL'88 and part of SDL'92. Some new features of SDL'92 such as priority input, value returning procedures, spontaneous transitions or different continuous signals with identical priority are supported by QUEST. Documents and software are available via ftp [QUES98].

A new version of QUEST is based on annotational extensions of SDL to integrate performance-related data. Thus, interoperability with other (functional) SDL tools is ensured.

6.3.2.7 EaSy-Sim: Coupling SDL with the SES Workbench

The EaSy-Sim environment [ScRS95], due to C. Schaffer, R. Raschhofer, and A. Simma from the University of Linz, is based on the coupling of the GEODE SDL environment with the SES simulation workbench. The basic idea is to model functional aspects of the SDL description in the SDL environment whereas all aspects related to time and non-ideal features of hardware are modeled by the SES workbench. Unlike other approaches described above that map the SDL description on a model for performance simulation, the concept of tool coupling is employed here. Thus, the SES workbench provides components to generate timeouts after a specified time, to model communication links (delay and error) and to model processing delay incurred by SDL tasks and processes.

The coupling is implemented by routing messages typically exchanged between the application-specific code and the SDL runtime support system through the SES workbench. This allows one to set a timer in the SES workbench, or to delay or scramble transmitted signals. In addition, special constructs can be inserted in the SDL description to model processing time, which are also handled by the SES workbench.

EaSy-Sim employs an extension of MSC, namely MSC/RT [Scha96], to specify real-time requirements of SDL systems. MSC/RT supports the formal specification of response time requirements, e.g. deadlines, maximum delay between two events and delay jitter. MSC/RT is employed within EaSy-Sim to automatically verify whether real-time requirements are met by the designed system.

A new version of the environment (EaSy-Sim II) is outlined in [Gerl97]. With EaSy-Sim II, the coupling with a separate performance simulation tool has been replaced by a tighter integration. This is because the coupling (co-simulation) turned out to be too inefficient to support a fast simulation.

6.3.2.8 SPEET: Performance Emulation Tool

SPEET is the SDL Performance Evaluation Tool developed at the University of Aachen by M. Steppler and M. Lott [StLe97, Step98]. The main objective of SPEET is the performance analysis of formally specified systems under real-time conditions. SPEET facilitates the functional simulation and the emulation of SDL descriptions. However, only the emulation supports a performance evaluation. The systems to be simulated or emulated can be triggered by traffic load generators and can be interconnected by transmission links which correspond to physical channels. The user can define probes within the SDL description. The data delivered by these probes during the emulation can be statistically evaluated. The strengths of the SPEET approach are the detailed workload models, e.g. the transmission models usable for the analysis of mobile communication systems, and the exact representation of existing hardware structures, e.g. microcontroller systems from Intel, Motorola and Siemens.

6.3.2.9 ObjectGEODE Performance Analyzer

The performance analyzer provided by the ObjectGEODE toolset [Veri98] represents an extension of the functional SDL simulator [Roux98]. Thus, the performance evaluation is based on a simulation and can be controlled by regular input to the simulator.

The performance simulation is based on a set of annotational extensions of SDL to specify resources, resource demands and mapping information. These extensions allow one to associate computational resources with SDL entities, e.g. blocks or processes. The resource demands (delays) are associated with actions in SDL. The delay may be a function of SDL variables. In addition, priorities may be associated with SDL processes in order to prefer some processes over others.

The statistical evaluation of the SDL description is done on the fly, controlled by special instructions provided to the simulator.

6.3.2.10 TAU Performance Library

The TAU performance library [Tele98] provides some distribution functions and other constructs that allow one to build queuing models with SDL. However, it does not really provide means for SDL-based performance evaluations.

6.3.3 MSC-Based Performance Evaluation

In this section, we describe approaches that base their performance evaluation mainly on MSC or a variant thereof. However, as MSCs are closely associated with SDL and its methodology, also SDL often plays a role with these tools.

MSCs have the advantage that they are available very early in the design process. Thus, MSC-based performance analysis can be employed in an early development phase. In addition, MSCs focus on use cases, which eases the evaluation of the performance of the system for specific cases.

6.3.3.1 LASSI/PRELASSI: MSC-Based Evaluation of Resource Utilizations

The LASSI/PRELASSI tool, due to W. Dulz and K. Hoppmann, supports a simple performance evaluation of SDL descriptions mainly based on interworkings, a synchronous variant of basic MSC.

Input to the tools is provided by an interworking specification (MSC) annotated with performance-relevant information such as relative frequency of the interworking, size of messages, and mean service times of actions. In addition, SDL is used to specify the available resources, the mapping of the SDL entities on resources, and defaults for service times of actions and of the size of messages.

The PRELASSI tool checks consistency of the interworkings with the SDL description. The performance evaluation tool LASSI analyzes the load induced by the interworkings on the SDL entities and on the available resources. LASSI derives information on the utilization of communication resources and processors. The analysis is based on deterministic times and message sizes. The analysis represents an analysis of a process graph that represents the performance aspects of the interworkings and of the SDL description.

A description of the basic ideas for the MSC-based performance analysis can be found in [Dulz96]. A detailed tool description is given in [Hopp93]. The tool is applied at Lucent Technologies to evaluate GSM systems.

6.3.3.2 DO-IT: Design and Optimization of SDL/MSC Systems

The DO-IT toolbox is a joint project of the University of Erlangen and the Technical University of Cottbus [Mits96, MiLH96]. The concept of the DO-IT toolbox is mainly due to A. Mitschele-Thiel and R. Henke. It is based on earlier work on the model-based optimization and efficient implementation of communication subsystems on parallel architectures (DSPL project) [Mits94, Mits93, Mits99] and on research on the performance evaluation of SDL/MSC systems done by W. Dulz (as described above). The goal of the DO-IT toolbox is to support performance engineering of SDL/MSC-based systems including model derivation, model-based performance evaluation and optimization. The DO-IT toolbox comprises components for the automatic derivation of the resource demands of different implementations of an SDL description, the analysis of SDL descriptions for performance bottlenecks, and the synthesis of an optimized system design deciding the issues related to software, hardware and the mapping of the load on the hardware. Central to the early integration of performance issues in the design process is the extension of MSCs to formally specify performance-relevant information.

The performance evaluation within the DO-IT toolbox is based on MSC rather than on SDL. An annotated extension of MSC (PMSC) is used to define the performance requirements including the workload, and the resource demands for different use cases [FLMS97].

The DO-IT toolbox aims at providing rather simple performance evaluation techniques based on deterministic service times. The most important techniques are a process graph analysis (to identify bottlenecks), critical path analysis and deterministic simulation. This allows the performance evaluation to be used as part of an automatic optimization process that decides on the implementation of the SDL description on a (parallel) system. The application of the DO-IT toolbox to implement a multimedia application is outlined in [MiLH96].

A major concern of the DO-IT toolbox is the efficient implementation of SDL descriptions and the process to tune the SDL description and the implementation technique to derive an optimized implementation that meets the given performance requirements [HeKM97]. The annotational extension of SDL to support this is described in [LaKö97].

6.3.3.3 CORSAIR: Rapid Prototyping of HW/SW Systems

CORSAIR (COdesign and Rapid prototyping System for ApplIcations with Real-time constraints) [DöMS00, MiSl97] is a successor of the DO-IT project conducted at the University of Erlangen. The project started in 1996 and is part of the major action of the German Research Foundation (DFG) to improve techniques and tools for the rapid prototyping of integrated control systems with stringent time requirements.

The project aims at the rapid prototyping of joint HW/SW systems to implement communication systems based on SDL and MSC. In the project, an annotated extension of SDL is used for the joint description of the application independent of its implementation in hardware or software. MSCs – also extended by means of annotations – are used to formally describe the functional and temporal requirements of the system. Based on these extended SDL [SpSD97] and MSC [FLMS97] descriptions, the partitioning of the implementation in parts to be implemented in hardware and software and other details of the implementation are decided. The decision process is mainly automatic, based on a tabu-search optimization algorithm. The decision or optimization process is based on a set of SDL- and MSC-based performance evaluation techniques. These range from MSC-based task graph analyses to quickly identify potential bottlenecks, to a more costly SDL-based performance simulation. In addition, an MSC-based schedulability analysis can be employed to verify that the required deadlines are met [SlZL98].

6.3.3.4 SPE.ED: Performance Evaluation of Software Systems

The SPE.ED tool [SmWi97], due to C.U. Smith and L.G. Williams, supports the software performance engineering approach proposed in [Smit90]. With the SPE.ED tool, MSCs accompanied by related information form the base for the performance evaluation. MSCs are mapped on execution graphs (software model), which are an extension of task graphs as introduced in section 2.3.2.2. The extensions allow one to describe loops and probabilities for conditional parts of the execution. In addition, the hierarchical structuring and refinement of execution graphs is supported. As pointed out in [SmWi97], the transformation from MSC to execution graphs is manual, but an automatic transformation should be possible.

A set of simple performance evaluation techniques is supported. The static analysis of the execution graphs allows the derivation of response time figures in the absence of other workloads. In addition, the static analysis of execution graphs does not take into account delays due to contention for resources. SPE.ED also supports the automatic creation and solution of an analytical queuing network model to derive utilization figures of the resources. The analysis technique is based on the operational analysis approach outlined in section 2.4.2.6. In addition, performance simulation is supported. For this the queuing model is mapped on a CSIM simulation model [Schw94] and evaluated.

A case study of the application of the SPE.ED tool to the design of a CAD system can be found in [WiSm98].

6.3.3.5 LQN: Mapping Traces on Layered Queuing Networks

The Model Builder as proposed by H. El-Sayed, D. Cameron and M. Woodside from Carleton University [ElCW98] transforms traces derived from the functional simulation of an SDL description on a Layered Queuing Network (LQN). The traces that form the major input of the model builder can be considered as MSCs. The transformation of the traces to the LQN is performed in a number of steps. In addition to the traces, information on the resource demands, the available resources and the mapping on the resources is needed to complete the LQN model.

LQNs [WNPM95, RoSe95] represent an adaptation of queuing networks that separate software processes, the services they implement directly, and requests to other processes. LQNs allow one to model resource contention, waiting time of messages, delays due to blocking, priorities of messages arriving at a process, and resource limitations of processes, buffers and locks. LQNs discriminate between different tasks which may represent hardware or software objects. These tasks are classified into client tasks, pure server tasks and active server tasks, depending on their specific roles in the system.

The derived LQN model can be used to calculate throughput and response time figures as well as the capacity of parts of the system at various layers. Tools support the analytic solution of the performance model, i.e. to derive the mean and variance of delays, as well as performance simulation [FHMP+95].

6.3.3.6 MSC-based Performance Debugging

Unlike most of the other approaches that focus on the early evaluation of the performance of a system, the approach pursued by F. Lemmen at the University of Erlangen [Lemm00, Lemm97, DaDL95] aims at performance measurements and debugging of implementations derived from SDL descriptions. The idea is to debug implementations based on the abstract view as provided by SDL and MSC instead of the implementation itself. The application of the approach to an SDL description of TCP/IP is described in [HoLe00].

With this approach, MSCs define specific executions of the system to be measured, e.g. a specific interaction in a communicating system such as the data processing during a file transfer or a video conference. The MSCs represent the input to automatically instrument the SDL description, i.e. add special instructions to the SDL description that, when executed, output execution information such as the current time and a label that identifies the spot in the SDL description that has prompted the output. The measured performance figures are then associated with the initial MSC on which the instrumentation has been based. Thus, the performance information is associated with the specification of the executions at a high level of abstraction. This greatly simplifies the human interpretation of the measured performance results. It is also planned to integrate the approach with the CORSAIR project described above.

6.3.3.7 MSC-based Synthesis of SDL Descriptions

The approach proposed by W. Dulz, S. Gruhl, L. Kerber and M. Söllner from the University of Erlangen and Lucent Technologies [DGKS99] supports the automatic derivation of synthetic implementations from given MSC descriptions. The synthetic implementation is executed on the target system – possibly as a small part of the system – and its performance is measured by employing some monitoring tool.

The approach works as follows. First, some given MSCs are annotated with performance information as proposed by the PMSC approach [FLMS97]. Second, SDL processes are derived from the given PMSC description. In the next step, executable code is derived from the SDL description using some code generation tool. Then the derived code is executed and measured on the target system using some monitoring tool. The approach allows to evaluate the influence of newly introduced features, i.e. described by MSCs, on the overall performance of the system. The tool has been applied to study the performance of mobile communication systems.

6.4 Summary

In the chapter, we have discussed the integration of performance engineering activities with SDL, identified the needed extra information, and provided an overview of techniques and tools that support various performance engineering activities. Support of the tools varies with respect to

- the development phase for which the tool is intended, e.g. the analysis, design, implementation or integration phase,
- the supported task, e.g. performance evaluation, performance verification, automatic design optimization, or performance monitoring, and
- the application area, e.g. a local installation or a distributed system with its interactions.

In addition, the tools differ in accuracy and the constraints concerning their applicability.

Currently, none of the tools solves all problems that may be encountered when using SDL to develop performance-critical systems. Even though there is an overlap in functionality, experience shows that different approaches (or tools) are needed for different subtasks of performance engineering.

The challenge concerning the extension of SDL and MSC with respect to time and performance is to ensure that the extensions are appropriate for a large variety of performance-related activities and tasks. In addition, as SDL and MSC are typically used in a joint manner, a harmonized approach for the extension of the two languages is important to avoid the specification of similar information in different places.

The challenge concerning tool development is the integration of different performance engineering activities in a coherent way. A tight integration of different techniques and activities is in its beginnings. For example, consider a measurement tool that provides the measured information as feedback to the SDL description which in turn is used to apply a performance evaluation with more accurate information for parts of the system.

Also missing is a complete methodology that describes how to apply which kind of performance engineering activity in which development phase and the integration of this with known methodologies, e.g. the SDL+ methodology, SOMT or similar.

Another problem concerning SDL-based performance engineering is the degree of abstraction employed by SDL, e.g. the abstraction from the runtime system. Even though this is appropriate for functional design, it complicates the application of SDL for the development of time- and performance-critical systems. Concerning this, more details concerning the aspects relevant for performance and time would help. As shown in chapter 5, the performance of the system may depend highly on the selected implementation and code generation strategy. It also seems to be useful for the SDL specification to discriminate strictly between aspects that are implemented in the final systems and aspects that are needed for other purposes, e.g. performance evaluation and verification, or functional simulation.

Chapter 7

Final Remarks

Performance problems are often quoted as being the major cause of project failure. This is especially true for complex systems such as distributed systems and telecommunication systems. Often the performance implications of decisions made during analysis and design are not obvious to the development engineers. Thus, performance problems can easily be introduced. Due to the common practice of neglecting performance issues until system testing, these problems often remain dormant until this very late phase of the development process. When detected during system testing, it is often too late to rectify the problem in an appropriate manner, i.e. with reasonable delay and cost.

In this book, we have provided a general introduction to performance engineering (chapter 2). Performance engineering is not only about building and solving performance models, e.g. Markov or other models. Instead, performance engineering is very much about understanding complex interrelationships. Thus, performance engineering has a lot in common with system development. A very good understanding of the system under development is vital to exercising performance engineering activities. For this reason, we have provided a discussion of the development of communicating systems (chapter 3), and of the implications of using SDL as a design language and as a basis for deriving implementations (chapters 4 and 5).

This laid the basis for the integration of performance-relevant aspects with the SDL methodology and the SDL language (chapter 6). The issue is currently under discussion within the ITU-T expert group responsible for the standardization and further development of SDL. To our knowledge, SDL is the only formal description technique for which such a tight integration of functional and performance aspects is pursued. However, other (non-formal) analysis and design methods also consider a tighter integration with performance issues. An example is activities in the context of UML [Doug98], which is a main competitor of SDL even in the telecommunication area.

Formal description techniques support the early specification and verification of functional aspects of the system. The integration of performance aspects with formal description techniques extends the scope of these languages by providing means to support an early evaluation of performance aspects along with functional aspects. In addition, the integration of time and performance information results in a major advantage for the functional domain, too. The introduction of bounds on time figures, especially the use of deterministic times, alleviates the state-space explosion problem well known from functional verification.

The need for the integration of performance engineering with the system engineering process is increasingly acknowledged in industry. Especially in the distributed systems and telecommunication area, the need for an early estimation of the projected performance is identified. This is also reflected by a new Workshop series on Software and Performance (WOSP), devoted to performance engineering of software systems, held for the first time in 1998 [WOSP98].

Supporting a tight integration of performance and system engineering, a number of measures are needed. The research presented in this book provides a step in this direction. The needed measures include the development of

- concepts and languages that allow the joint description of functional and non-functional aspects, i.e. include time and performance aspects, and
- methodologies and tools that support both aspects during all phases of the system development process, i.e. from analysis to system implementation, integration and testing.

The merits of such an integration are promising, i.e. the development of time- and performance-critical systems based on an abstract description technique and without fiddling with the very details of the system as is currently the case where performance issues are crucial.

However, developing performance-critical systems still requires a coordinated set of actions and interaction with humans. The idea of simply providing an abstract specification of the system and leaving the other issues to the tools will not work. Instead, engineers will have to optimize the specification and design in an appropriate way such that performance requirements are met. However, tools can support the engineer in making these decisions at various stages of the system development process.

In order to effectively develop systems based on SDL and its companions and to (automatically) derive implementations from SDL that make efficient use of the available resources and also meet given time requirements, a coordinated set of measures, applied to various phases of the development process, is needed. This comprises

- the integration of time and performance aspects with the SDL standard and its companions, i.e. to specify time requirements and the available resources, and to describe dynamic interaction of the SDL description with the underlying resources,
- techniques to support a performance analysis and the adaptation of SDL descriptions, i.e. their structure and internal behavior as well as the SDL constructs used, to provide an optimal basis for the automatic derivation of efficient code,
- adaptive code generation tools that allow one to derive efficient code from specific SDL descriptions, i.e. take into account the specific subset of SDL used to specify the application or parts thereof,
- adaptive runtime systems suited to the specific needs of the SDL description at hand, and

- high-level performance debugging that allows one to specify performance figures to be measured in SDL or MSC along with the visualization of the measured performance results at the level of SDL or MSC.

Considerable progress has been made in supporting the performance evaluation of SDL and MSC descriptions. In addition, concepts and first prototypes exist that support high-level debugging and the derivation of efficient code from SDL. However, the code generation strategies are far from being adaptive.

Extending SDL and its companions and developing tools that support the listed activities are a major challenge and will have a major impact on the role SDL plays in the future, especially the extent to which SDL can be established in the real-time and embedded systems domain.

Bibliography

[AaKo89] E.H.L. Aarts, J.H.M. Korst. Simulated Annealing and Boltzmann Machines. Wiley, 1989.

[AbPe93] M. Abbott, L. Peterson. Increasing Network Throughput by Integrating Protocol Layers. IEEE/ACM Transactions on Networking, 1(5), pp600-610, 1993.

[AdCD74] T.L. Adam, K.M. Chandy, J.R. Dickson. A Comparison of List Schedules for Parallel Processing Systems. Communications of the ACM, 17(12), Dec. 1974.

[AhBG96] B. Ahlgren, M. Björkman, P. Gunningberg. Integrated Layer Processing Can Be Hazardous to Your Performance. Protocols for High-Speed Networks V, W. Dabbous, C. Diot (eds), pp167-181, Chapman & Hall, 1996.

[AhGM96] B. Ahlgren, P. Gunningberg, K. Moldeklev. Increasing Communication Performance with a Minimal-Copy Data Path Supporting ILP and ALF. Journal of High Speed Networks, 5(2), pp203-214, 1996.

[AhHU87] A.V. Aho, J.E. Hopcroft, J.D. Ullman. Data Structures and Algorithms. Addison-Wesley, 1987.

[AjBC86] M. Ajmone Marsan, G. Balbo, G. Conte. Performance Models of Multiprocessor Systems. Series in Computer Systems, MIT Press, 1986.

[Alur90] R. Alur. Automatons for Modeling Real-Time Systems. Proc. Intl. Colloquium on Automata, Languages and Programming (ICALP), M.S. Peterson (ed.), Lecture Notes in Computer Science 443, pp332-335, Springer, 1990.

[Arth93] L.J. Arthur. Improving Software Quality – An Insider's Guide to TQM. Wiley, 1993.

[As94] H.R. van As. Media Access Techniques: The Evolution towards Terabit/s LANs and MANs. Computer Networks and ISDN Systems, 26, pp603-656, 1994.

[Axel96] J. Axelsson. Hardware/Software Partitioning Aiming at Fulfilment of Real-Time Constraints. Journal of Systems Architecture, 42(6&7), 1996.

[BaBK95] F. Bause, P. Buchholz, P. Kemper. QPN-Tool for the Specification and Analysis of Hierarchically Combined Queueing Petri Nets. Quantitative Evaluation of Computing and Communication Systems, Proc. of Performance Tools '95 and 8th GI/NTG Conference MMB '95, F. Bause, H. Beilner (eds), Lecture Notes in Computer Science, Springer, 1995.

[BaBu91] F. Bause, P. Buchholz. Protocol Analysis Using a Timed Version of SDL. 3rd Intl. Conf. on Formal Description Techniques (FORTE '90), Madrid, Spain, North-Holland, 1991.

[BaBu93] F. Bause, P. Buchholz. Qualitative and Quantitative Analysis of Timed SDL Specifications. Kommunikation in Verteilten Systemen, N. Gerner, H.G. Hegering, J. Savolvod (eds), Reihe Informatik aktuell, pp486-500, Springer, 1993.

[BaGS94] D.F. Bacon, S.L. Graham, O.J. Sharp. Compiler Transformations for High-Performance Computing. ACM Computing Surveys, 26(4), pp345-420, Dec. 1994.

[BaJL93] F. Baccelli, A. Jean-Marie, Z. Lui. A Survey on Solution Methods for Task Graph Models. Proc. Workshop on Formalisms, Principles and State-of-the-Art. N. Götz, U. Herzog, M. Rettelbach (eds), Universität Erlangen-Nürnberg, Arbeitsberichte des Instituts für Mathematische Maschinen und Datenverarbeitung, 26(14), Sept. 1994.

[BaRo87] V.R. Basili, H.D. Rombach (eds). Software Quality Assurance. Special Issue, IEEE Software, 6-9, 1987.

[BaSc95] F. Bause, M. Sczittnick. Design von Modellierungstools zur Leistungsbewertung – HIT, MACOM, QPN-Tool. it+ti: Informationstechnik und Technische Informatik, 3, 1995.

[Batt96] R. Battiti. Reactive Search: Towards Self-tuning Heuristics. Modern Heuristic Search Methods, V.J. Raywards-Smith, I.H. Osman, C.R. Reeves, G.D. Smith (eds), Wiley, 1996.

[BDPU94] R.G. Basinger, M.J. DiMario, T.C. Pingel, M.J. Urban. Total Quality Management in the Switching Systems Business Unit. AT&T Technical Journal, pp7-18, Nov./Dec. 1994.

[Beiz90] B. Beizer. Software Testing Techniques. Van Nostrand Reinhold, 1990.

[BeJe95] T. Ben Ismail, A.A. Jerraya. Synthesis Steps and Design Models for Codesign. IEEE Computer, Feb. 1995.

[BeMW88] H. Beilner, J. Mäter, N. Weißenberg. Towards a Performance Modelling Environment: News on HIT. 4th Intl. Conf. on Modelling Techniques and Tools for Computer Performance Evaluation, Palma de Mallorca, 1988.

[BHMM+97] R. Braek, O. Haugen, G. Melby, B. Moller-Pedersen, R. Sanders, T. Stalhane. TIMe – The Integrated Method, TIMe at a glance. SINTEF, Trondheim, Norway, 1997.

[BjJo82] D. Bjorner, C.B. Jones. Formal Specification and Software Development. Prentice Hall, 1982.

[BKKK95] F. Bause, H. Kabutz, P. Kemper, P. Kritzinger. SDL and Petri Net Performance Analysis of Communicating Systems. 15th Intl. Symposium on Protocol Specification, Testing and Verification (PSTV '95), P. Dembinski, M. Sredniawa (eds), North Holland, 1995.

[BMSK96] M. Bütow, M. Mestern, C. Schapiro, P.S. Kritzinger. Performance Modelling with the Formal Specification Language SDL. Joint Intl. Conf. on Formal Description Techniques for Distributed Systems and Communication Protocols (IX) and Protocol Specification, Testing and Verification (XVI) (FORTE/PSTV'96), R. Gotzhein, J. Bredereke (eds), Chapman & Hall, 1996.

[Boeh76] B.W. Boehm. Software Engineering. IEEE Transactions on Computers, 25(12), 1976.

[Boeh81] B.W. Boehm. Software Engineering Economics. Prentice Hall, 1981.

[Boeh86] B.W. Boehm. The Spiral Model of Software Development and Enhancement. ACM SIGSOFT Software Engineering Notes, 11(4), 1986.

[BoJR98] G. Booch, I. Jacobson, J. Rumbaugh. The Unified Modeling Language User Guide. Addison-Wesley, 1998.

[Bokh81] S.H. Bokhari. On the Mapping Problem. IEEE Transactions on Computers, 30(3), March 1981.

[Booc91] G. Booch. Object-Oriented Design with Applications. Benjamin/Cummings, 1991.

[Börg99] E. Börger. High Level System Design and Analysis using Abstract State Machines. Current Trends in Applied Formal Methods, D. Hutter, W. Stephan, P. Traverso, M. Ullmann (eds). Lecture Notes in Computer Science, Springer, 1999.

[Brau90] H. Braun. Massiv parallele Algorithmen für kombinatorische Optimierungsprobleme und ihre Implementierung auf einem Parallelrechner. Dissertation, Fakultät für Informatik, Universität Karlsruhe, 1990.

[Brau96] T. Braun, C. Diot. Performance Evaluation and Cache Analysis of an ILP Protocol Implementation. IEEE/ACM Transactions on Networking, 4(3), pp318-330, June 1996.

[Brau97] T. Braun. Internet Protocols for Multimedia Communications, Part II: Resource Reservation, Transport, and Application Protocols. IEEE MultiMedia, Oct.-Dec. 1997.

[BrHa93] R. Braek, O. Haugen. Engineering Real Time Systems – An Object-Oriented Methodology Using SDL. BCS Practitioner Series, Prentice Hall, 1993.

[Broo87] F.P. Brooks. No Silver Bullet, Essence and Accidents of Software Engineering. IEEE Computer, April 1987.

[Budg93] D. Budgen. Software Design. Addison-Wesley, 1993.

[Buhr84] R.J.A. Buhr. System Design with Ada. Prentice Hall, 1984.

[Butt97] G.C. Buttazzo. Hard Real-time Computing Systems. Kluwer Academic Publishers, 1997.

[BuWe96] A. Burns, A. Wellings. Real-Time Systems and Programming Languages. 2nd edition, Addison Wesley, 1996.

[Buze76] J.P. Buzen. Fundamental Laws of Computer System Performance. Proc. SIGMETRICS '76, Cambridge, MA, pp200-210, 1976.

[CCIT88] CCITT. X.208, Data Communication Networks Open System Interconnection (OSI) Abstract Syntax Notation One (ASN.1). CCITT, 1988.

[ChAb82] T.C.K. Chou, J.A. Abraham. Load Balancing in Distributed Systems. IEEE Transactions on Software Engineering, 8(4), July 1982.

[ChCo91] S. Chaumette, M.C. Counilh. A Development Environment for Distributed Systems. Lecture Notes in Computer Science 487, Springer, 1991.

[Chen76] P.P. Chen. The Entity-Relationship Model: Toward a Unified View of Data. ACM Transactions on Database Systems, 1(1), 1976.

[Cher88] D.R. Cheriton. The V Distributed System. Communications of the ACM, 31(3), pp314-333, March 1988.

[ChKi97] N. Chapman, L.G. Kirby. Performance Engineering. BT Technology Journal, pp19-25, 15(3), July 1997.

[ChYu90] G.-H. Chen, J.-S. Yur. A Branch-and-Bound-with-Underestimates Algorithm for the Task Assignment Problem with Precedence Constraint. Proc. 10th International Conference on Distributed Computing Systems, Paris, France, June 1990.

[CJRS89] D.D. Clark, V. Jacobson, J. Romkey, H. Salwen. An Analysis of TCP-Processing Overhead. IEEE Communications Magazine, pp23-29, 1989.

[Clar85] D.D. Clark. The Structuring of Systems Using Upcalls. Proc. 10th ACM Symposium on Operating Systems Principles, Oakland, CA, ACM, pp171-180, Dec. 1985.

[ClTe90] D.D. Clark, D.L. Tennenhouse. Architectural Considerations for a New Generation of Protocols. Proc. ACM SIGCOMM '90, Philadelphia, PA, ACM, pp200-208, Sept. 1990.

[DaDL95] P. Dauphin, W. Dulz, F. Lemmen. Specification-driven Performance Monitoring of SDL/MSC-specified Protocols. Proc. 8th Intl. Workshop on Protocol Test Systems, A. Cavalli, S. Budkowski (eds), Sept. 1995.

[DAPP93] P. Druschel, M.B. Abbott, M.A. Pagels, L.L. Peterson. Network Subsystem Design. IEEE Network, 7(4), pp8-17, July 1993.

[Davi91] L. Davis (ed.). Handbook of Genetic Algorithms. Van Nostrand Reinhold, 1991.

[DDKM+90] W. Doeringer, D. Dykeman, M. Kaiserswerth, B. Meister, H. Rudin, R. Williamson. A Survey of Light-Weight Transport Protocols for High-Speed Networks. IEEE Transactions on Communications, 38(1), pp2025-2039, 1990.

[DeBu78] P.J. Denning, J.P. Buzen. The Operational Analysis of Queueing Network Models, Computing Surveys, 10(3), pp225-261, 1978.

[DeMa78] T. De Marco. Structured Analysis and System Specification. Yourdon Press, 1978.

[Deo74] N. Deo. Graph Theory with Applications to Engineering and Computer Science. Prentice Hall, 1974.

[DGKS99] W. Dulz, S. Gruhl, L. Kerber, M. Söllner. Early Performance Prediction of SDL/MSC-specified Systems by Automated Synthetic Code Generation. SDL '99: The Next Millennium (Ninth SDL Forum), R. Dssouli, G.V. Bochmann, Y. Lahav (eds), pp457-471, Elsevier, 1999.

[DHHM95] M. Diefenbruch, E. Heck, J. Hintelmann, B. Müller-Clostermann. Performance Evaluation of SDL Systems Adjunct by Queuing Models. SDL '95 with MSC in CASE (Proc. Seventh SDL Forum), R. Braek, A. Sarma (eds), Elsevier, 1995.

[Dief97] M. Diefenbruch. Functional and Quantitative Verification of Time- and Resource Extended SDL-Systems with Model-Checking (in German). Proc. of Messung, Modellierung und Bewertung von Rechen- und Kommunikationssystemen, K. Irmscher (ed.), Freiberg, Germany, VDE-Verlag, 1997.

[Dief98] M. Diefenbruch. Functional and Quantitative Verification of Time- and Resource-enhanced SDL Systems with Model Checking. In [MiMR98].

[DöMS00] M. Dörfel, A. Mitschele-Thiel, F. Slomka. CORSAIR: HW/SW-Codesign von Kommunikationssystemen mit SDL. Praxis der Informationsverarbeitung und Kommunikation (PIK), 23(1), K. G. Saur Verlag, 2000.

[Doug98] B.P. Douglass. Real-Time UML – Developing Efficient Objects for Embedded Systems. Addison-Wesley, 1998.

[Dows92] K.A. Dowsland. Some Experiments with Simulated Annealing Techniques for Packing Problems. EJOR, 1992.

[Dows93] K.A. Dowsland. Simulated Annealing. Modern Heuristic Techniques for Combinatorial Problems, R. Reeves (ed.), Blackwell Scientific Publications, 1993.

[Dulz96] W. Dulz. A Framework for the Performance Evaluation of SDL/MSC-specified Systems. Proc. of the European Simulation Multiconference, ESM'96, A. Javor, A. Lehmann, I. Molnar (eds), Budapest, Society for Computer Simulation International, 1996.

[Duss98] K. Dussa-Zieger. Model-based Scheduling and Configuration of Heterogeneous Parallel Systems. Dissertation, Universität Erlangen-Nürnberg, Arbeitsberichte des Instituts für Mathematische Maschinen und Datenverarbeitung, 31(12), Dec. 1998.

[DuSW93] G. Dueck, T. Scheuer, H.-M. Wallmeier. Toleranzschwellen und Sintflut: Neue Ideen zur Optimierung. Spektrum der Wissenschaft, pp42-51, March 1993.

[EhMa85] E. Ehrig, B. Mahr. Fundamentals of Algebraic Specification. Springer, 1985.

[ElCW98] H. El-Sayed, D. Cameron, M. Woodside. Automated Performance Modelling from Scenarios and SDL Designs of Telecom Systems. Proc. of Intl. Symp. on Software Engineering for Parallel and Distributed Systems (PDSE'98), IEEE Press, 1998.

[ElHS97] J. Ellsberger, D. Hogrefe, A. Sarma. SDL – Formal Object-oriented Language for Communicating Systems. Prentice Hall, 1997.

[ElLe90] H. El-Rewini, T.G. Lewis. Scheduling Parallel Program Tasks on Arbitrary Target Machines. Journal of Parallel and Distributed Computing, 9, pp138-153, 1990.

[ETSI95] ETSI. Permanent Document MTS (95) 04, MTS – Methodologies for Standards Engineering – Specification of Protocols and Services. March 1995.

[Fent91] N. Fenton. Software Metrics: A Rigorous Approach. Chapman & Hall, 1991.

[Ferr86] D. Ferrari. Considerations on the Insularity of Performance Evaluation. IEEE Transactions on Software Engineering, 12(6), 1986.

[FHMP+95] G. Franks, A. Hubbard, S. Majumdar, D. Petriu, J. Rolia, C.M. Woodside. A Toolset for Performance Engineering and Software Design of Client-Server Systems. Journal of Performance Evaluation, 24, pp117-135, Elsevier, 1995.

[FiHa88] A.J. Field, P.G. Harrison. Functional Programming. Addison-Wesley, 1988.

[FiKe92] R.G. Fichman, C.F. Kemerer. Object-Oriented and Conventional Analysis and Design Methodologies – Comparison and Critique. IEEE Computer, Oct. 1992.

[Fisc93] J. Fischer. An Environment for SDL'92. Systems Analysis Modeling, Simulation, A. Sydow (ed.), Gordon & Breach, New York, pp107-124, 1993.

[FlCE87] A. Fleischmann, S.T. Chin, W. Effelsberg. Specification and Implementation of an ISO Session Layer. IBM System Journal, 26(3), pp255-275, 1987.

[Flei94] A. Fleischmann. Distributed Systems – Software Design and Implementation, Springer, 1994.

[FLMS97] N. Faltin, L. Lambert, A. Mitschele-Thiel, F. Slomka. An Annotational Extension of Message Sequence Charts to Support Performance Engineering. SDL '97 – Time for Testing (Eighth SDL Forum), A. Cavalli, A. Sarma (eds), pp307-322, Elsevier, Sept. 1997.

[FoFu62] L.R. Ford, D.R. Fulkerson. Flows in Networks. Princeton University Press, 1962.

[FoSc97] M. Fowler, K. Scott. UML Distilled. Addison-Wesley, 1997.

[GaHM88] D. Gantenbein, R. Hauser, E. Mumprecht. Implementation of the OSI Transport Service in a Heterogeneous Environment. HECTOR, Vol II: Basic Projects, G. Krüger, G. Müller (eds), pp215-241, Springer, 1998.

[GaJo79] M. Garey, D. Johnson. Computers and Intractability. W.H. Freeman, New York, 1979.

[GaSa79] C. Gane, T. Sarson. Structured Systems Analysis: Tools and Techniques. Prentice Hall, 1979.

[GaVa95] D.D. Gajski, F. Vahid. Specification and Design of Embedded Hardware-Software Systems. IEEE Design and Test of Computers, Spring 1995.

[Gerl97] R. Gerlich. Tuning Development of Distributed Real-Time Systems with SDL and MSC: Current Experience and Future Issues. SDL '97 – Time for Testing (Eighth SDL Forum), A. Cavalli, A. Sarma (eds), pp85-100, Elsevier, Sept. 1997.

[Gerl98] R. Gerlich. EaSySim II SDL Extensions for Performance Simulation. In [MiMR98].

[Gibb94] W.W. Gibbs. Software's Chronic Crisis. Scientific American, Sept. 1994.

[Gilb88] T. Gilb. Principles of Software Engineering Management. Addison-Wesley, 1988.

[GiLi90] D.W. Gillies, J.W.S. Liu. Scheduling Tasks with AND/OR Precedence Constraints. Proc. 2nd IEEE Symposium on Parallel and Distributed Processing, Dallas, Texas, Dec. 1990.

[GKZH95] R. German, Ch. Kelling, A. Zimmermann, G. Hommel. TimeNET – A Toolkit for Evaluating Stochastic Petri Nets with Non-Exponential Firing Times. Journal of Performance Evaluation, 24, pp69-87, Elsevier, 1995.

[GlGP99] U. Glässer, R. Gotzhein, A. Prinz. Towards a New Formal SDL Semantics based on Abstract State Machines. SDL '99: The Next Millennium (Ninth SDL Forum), R. Dssouli, G.V. Bochmann, Y. Lahav (eds), pp171-190, Elsevier, 1999.

[GlLa93] F. Glover, M. Laguna. Tabu Search. Modern Heuristic Techniques for Combinatorial Problems, R. Reeves (ed.), Blackwell Scientific Publications, 1993.

[Glov86] F. Glover. Future Paths for Integer Programming and Links to Artificial Intelligence. Computers and Operations Research, 5, pp533-549, 1986.

[GoDo95] F. Goudenove, L. Doldi. Use of SDL to specify Airbus Future Air Navigation Systems. SDL '95 with MSC in CASE (Proc. Seventh SDL Forum), R. Braek, A. Sarma (eds), Elsevier, 1995.

[Gold89] D.E. Goldberg. Genetic Algorithms in Search, Optimization, and Machine Learning. Addison-Wesley, 1989.

[Goma93] H. Gomaa. Software Design Methods for Concurrent and Real-Time Systems. SEI Series in Software Engineering. Addison-Wesley, 1993.

[Gree91] P.E. Green. The Future of Fiber-optic Computer Networks. IEEE Computer, pp78-87, Sept. 1991.

[Haas91] Z. Haas. A Protocol Structure for High-speed Communication over Broadband ISDN. IEEE Network, pp64-70, Jan. 1991

[HaFr89] F. Haist, H. Fromm. Qualität im Unternehmen: Prinzipien - Methoden - Techniken. Carl Hanser, 1989.

[HaJo90] A.J. Harget, I.D. Johnson. Load Balancing Algorithms in Loosely-Coupled Distributed Systems: a Survey. Distributed Computer Systems, H.S.M. Zedan (Ed.), Butterworths, 1990.

[Hals96] F. Halsall. Data Communications, Computer Networks and Open Systems. 4th Edition, Addison-Wesley, 1996.

[Hami86] M. Hamilton. Zero-defect Software: The Elusive Goal. IEEE Spectrum, March 1986.

[Hare87] D. Harel. Statecharts: A Visual Formalism for Complex Systems. Science of Computer Programming, 8, 1987.

[Hart93] F. Hartleb. Stochastic Graph Models for Performance Evaluation of Parallel Programs and the Evaluation Tool PEPP. Proc. Workshop on Formalisms, Principles and State-of-the-Art. N. Götz, U. Herzog, M. Rettelbach (eds), Universität Erlangen-Nürnberg, Arbeitsberichte des Instituts für Mathematische Maschinen und Datenverarbeitung, 26(14), Sept. 1994.

[Haug95] O. Haugen. Using SDL-92 Effectively. SDL '95 with MSC in CASE (Proc. Seventh SDL Forum), R. Braek, A. Sarma (eds), Elsevier, 1995.

[HCAL89] J.J. Hwang, Y.C. Chow, F.D. Anger, C.Y. Lee. Scheduling Precedence Graphs in Systems with Interprocessor Communication Times. SIAM Journal on Computing, 18(2), April 1989.

[Heat97] S. Heath. Embedded Systems Design. Butterworth-Heinemann, 1997.

[Heck96] E. Heck. Performance Evaluation of Formally Specified Systems – The Integration of SDL with HIT. Doctoral Thesis, Informatik IV, Universität Dortmund, Krehl Verlag, 1996.

[Hedd95] M. Heddes. A Hardware/Software Codesign Strategy for the Implementation of High-Speed Protocols. Ph.D. thesis, Eindhoven University of Technology, 1995.

[HeHM91] E. Heck, D. Hogrefe, B. Müller-Clostermann. Hierarchical Performance Evaluation Based on Formally Specified Communication Protocols. IEEE Transactions on Computers, 40(4), 1991.

[HeHM98] H. Hermanns, U. Herzog, V. Mertsiotakis. Stochastic Process Algebras – Between LOTOS and Markov Chains. Computer Networks and ISDN Systems, 30, pp901-924, 1998.

[Heis94] H.U. Heiss: Prozessorzuteilung in Parallelrechnern. BI-Wissenschaftsverlag, Reihe Informatik, Band 98, 1994.

[HeKM97] R. Henke, H. König, A. Mitschele-Thiel. Derivation of Efficient Implementations from SDL Specifications Employing Data Referencing, Integrated Packet Framing and Activity Threads. SDL '97 – Time for Testing (Eighth SDL Forum), A. Cavalli, A. Sarma (eds), pp397-414, Elsevier, Sept. 1997.

[HeKM98] R. Henke, H. König, A. Mitschele-Thiel. Automatische Protokollimplementierung unter Verwendung von Activity Threads. Praxis der Informationsverarbeitung und Kommunikation (PIK), 21(1), pp11-18, K. G. Saur Verlag, 1998.

[HeKö95] T. Held, H. König. Increasing the Efficiency of Computer-aided Protocol Implementations. Protocol Specification, Testing and Verification XIV, S. Vuong, S. Chanson (eds), pp387-394, Chapman & Hall, 1995.

[HeMK97] R. Henke, A. Mitschele-Thiel, H. König. On the Influence of Semantic Constraints on the Code Generation from Estelle Specifications. Formal Description Techniques and Protocol Specification, Testing and Verification (FORTE X/PSTV XVII'97), T. Mizuno, et al. (eds), pp399-414, Chapman & Hall, 1997.

[Heni80] K. Heninger. Specifying Software Requirements for Complex Systems: New Techniques and Their Applications. IEEE Transactions on Software Engineering, 6(1), Jan. 1980.

[Herz89] U. Herzog. Leistungsbewertung und Modellbildung für Parallelrechner. Informationstechnik (it), 31(1), pp31-38, 1989.

[Herz92] U. Herzog. Performance Evaluation as an Integral Part of System Design. Transputers '92: Advanced Research and Industrial Applications. M. Becker, L. Litzler, M. Trehel (eds), pp352-363, IOS Press, 1992.

[HeSo82] D.P. Heyman, M.J. Sobel. Stochastic Models in Operations Research, Vol. I. McGraw-Hill, 1982.

[Hess98] M. Hesselgrave. Avoiding the Software Performance Crisis. First Intl. Workshop on Software and Performance (WOSP 98), Tutorial Handouts, Santa Fe, New Mexico, USA, Oct. 12-16, 1998.

[Hint96] J. Hintelmann. Integration of SDL-Based QoS-Evaluation in Protocol Design. Modelling and Simulation. Proc. of the European Simulation Multiconference ESM'96, A. Javor, A. Lehmann, I. Molnar (eds). Workshop on Analytical and Numerical Modelling Techniques with Emphasis on Quality of Service Modelling, Budapest, pp. 894-898, Society for Computer Simulation International, 1996.

[HiWe97] J. Hintelmann, R. Westerfeld. Performance Analysis of TCP's Flow Control Mechanisms using Queueing SDL. SDL '97 – Time for Testing (Eighth SDL Forum), A. Cavalli, A. Sarma (eds), pp69-84, Elsevier, Sept. 1997.

[HMKL96] R. Henke, A. Mitschele-Thiel, H. König, P. Langendörfer. Automated Derivation of Efficient Implementations from SDL Specifications. Technical Report 11/96, IMMD VII, University of Erlangen-Nuremberg, Nov. 1996.

[Hoar78] C.A.R. Hoare. Communicating Sequential Processes. Communications of the ACM, 21(8), 1978.

[Hoar85] C.A.R. Hoare. Communicating Sequential Processes. Prentice Hall, 1985.

[HoAR94] E. Hou, N. Ansari, H. Ren. A Genetic Algorithm for Multiprocessor Scheduling. IEEE Transactions on Parallel and Distributed Systems, 5(2), 1994.

[HoEl98] D. Hogrefe, J. Ellsberger. The ETSI SDL Model for the Intelligent Network Application Protocol. Proc. First Workshop of the SDL Forum Society on SDL and MSC, Y. Lahav, A. Wolisz, J. Fischer, E. Holz (eds), Informatik-Bericht Nr. 104, Humboldt-Univ. zu Berlin, 1998.

[HoLe00] R. Hofmann, F. Lemmen. Specification-driven Monitoring of TCP/IP. Proc. of the 8th Euromicro Workshop on Parallel and Distributed Processing (EURO-PDP 2000), IEEE Computer Society, Rhodos, Greece, Jan. 2000.

[Holl75] J.H. Holland. Adaptation in Natural and Artificial Systems. University of Michigan Press, 1975.

[Holz91] G. Holzman. Design and Validation of Computer Protocols. Prentice-Hall, 1991.

[Hopp93] K. Hoppmann. Lastanalysator für hierarchische Signalisiernetze. Diplomarbeit, Universität Erlangen-Nürnberg, IMMD VII, April 1993.

[HoSa78] E. Horowitz, S. Sahni. Fundamentals of Computer Algorithms. Computer Science Press, 1978.

[HoTa85] J.J Hopfield, D.W. Tank. 'Neural' Computation of Decisions in Optimization Problems. Biological Cybernetics, 52, pp141-152, 1985.

[Hump91] W.S. Humphrey. Managing the Software Process. Addison-Wesley, 1991.

[HuPe91] N.C. Hutchinson, L.L. Peterson. The x-Kernel: An Architecture for Implementing Network Protocols. IEEE Transactions on Software Engineering, 17(1), pp64-76, Jan. 1991

[HWWL+96] E. Holz, M. Wasowski, D. Witaszek, S. Lau, J. Fischer, P. Roques, K. Verschaeve, E. Mariatos, J.-P. Delpiroux. The INSYDE Methodology. Deliverable INSYDE/WP1/HUB/400/v2, ESPRIT Ref: P8641, Jan. 1996.

[HwXu90] K. Hwang, J. Xu. Mapping Partitioned Program Modules onto Multicomputer Nodes Using Simulated Annealing. Proc. 1990 Intl. Conference on Parallel Processing, vol. II, Pennsylvania State University Press, University Park, PA, 1990.

[INMO88] INMOS Limited. The Transputer Databook. INMOS Data Series, 1988.

[IRSK96] E. Inocencio, M. Ricardo, H. Sato, T. Kashima. Combined Application of SDL-92, OMT. MSC and TTCN. Joint Intl. Conf. on Formal Description Techniques for Distributed Systems and Communication Protocols (IX) and Protocol Specification, Testing and Verification (XVI) (FORTE/PSTV'96), R. Gotzhein, J. Bredereke (eds), Chapman & Hall, 1996.

[ISO84] ISO. International Standard 7498, Information Processing Systems – Open Systems Interconnection – Basic Reference Model, 1984.

[ISO89] ISO. International Standard 9074, Estelle – A Formal Description Technique Based on an Extended State Transition Model, 1989.

[ISO96] ISO/IEC. Conformance Testing Methodology and Framework, Part 3: The Tree and Tabular Combined Notation (TTCN). ISO/IEC IS 9646-3, 1996.

[ITU93] ITU-T. Z.100, CCITT Specification and Description Language (SDL). ITU, 1993.

[ITU93a] ITU-T. Z.100, Appendix I. ITU, SDL Methodology Guidelines. ITU, 1993.

[ITU95] ITU/ISO. Open Distributed Processing – Reference Model, Part 1 to 4. ISO Intl. Standard 10746 1-4 and ITU-T Recommendations X.901 - X.904, 1995.

[ITU95a] ITU-T. Z.105, SDL Combined with ASN.1 (SDL/ASN.1). 1995.

[ITU97] ITU-T. Z.120, Message Sequence Chart (MSC). ITU, 1997.

[ITU97a] ITU-T. SDL+ Methodology: Manual for the use of MSC and SDL (with ASN.1). Supplement 1 to Z.100, 1997.

[ITU97b] ITU-T. Z.100 Addendum 1, Corrections to Recommendation Z.100, 1997.

[ITU99] ITU-T. Draft New Recommendation Z.109: Languages for Telecommunications Applications – SDL Combined with UML. June 1999.

[ITU99a] ITU-T. Draft Revised Recommendation Z.100: Languages for Telecommunications Applications – Specification and Description Language. Nov. 1999.

[JaBR99] I. Jacobson, G. Booch, J. Rumbaugh. The Unified Software Development Process. Addison-Wesley, 1999.

[Jack75] M.A. Jackson. Principles of Program Design. Academic Press, 1975.

[Jain91] R. Jain. The Art of Computer Systems Performance Analysis - Techniques for Experimental Design, Measurement, Simulation, and Modeling. Wiley, 1991.

[JAMS89] D.S. Johnson, C.R. Aragon, L.A. McGeoch, C. Schevon. Optimization by Simulated Annealing: An Experimental Evaluation; Part I, Graph Partitioning. Operations Research, 37, pp865-892, 1989.

[JCJÖ92] I. Jacobson, M. Christerson, P. Jonsson, G. Övergaard. Object-Oriented Software Engineering – A Use Case Driven Approach. Addison-Wesley, 1992.

[Kabu97] H.M. Kabutz. Analytical performance evaluation of concurrent communicating systems using SDL and stochastic Petri nets. Doctoral Thesis, Department of Computer Science, University of Cape Town, Republic of South Africa, 1997.

[KaGa93] B. Kao, H. Garcia-Molina. Subtask Deadline Assignment for Complex Distributed Soft Real-time Tasks. Stanford University, Report No. STAN-CS-93-1491, 1993.

[KaLe93] A. Kalavade, E.A. Lee. A Hardware-Software Codesign Methodology for DSP Applications. IEEE Design and Test of Computers, Sept. 1993.

[Kant92] K. Kant. Introduction to Computer System Performance Evaluation. McGraw-Hill, 1992.

[KDHM+95] R. Klar, P. Dauphin, F. Hartleb, R. Hofmann, B. Mohr, A. Quick, M. Siegle. Messung und Modellierung paralleler und verteilter Rechensysteme. Teubner, 1995.

[KeLi70] B.W. Kernighan, S. Lin. An Efficient Heuristic Procedure for Partitioning Graphs. Bell Systems Technical Journal, 49(2), pp291-308, 1970.

[KiGV83] S. Kirkpatrick, C.D. Gellat and M.P. Vecchi. Optimization by Simulated Annealing. Science, 220, pp671-680, 1983.

[Klei75] L. Kleinrock. Queueing Systems, Vol. 1, Theory. Wiley, 1975.

[Klei82] W. Kleinöder. Stochastische Bewertung von Aufgabenstrukturen für hierarchische Mehrrechnersysteme. Dissertation, Universität Erlangen-Nürnberg, Arbeitsberichte des Instituts für Mathematische Maschinen und Datenverarbeitung, 15(10), Aug. 1982.

[KLKZ96] H. Kameda, J. Li, C. Kim, Y. Zhang. Optimal Load Balancing in Distributed Computer Systems. Springer, 1996.

[Koho89] T. Kohonen. Self-organization and Associative Memory. 3rd edition, Springer, 1989.

[KöKr96] H. König, H. Krumm. Implementierung von Kommunikationsprotokollen. Informatik-Spektrum, 19, pp. 316-325, Springer, 1996.

[KrMS90] U. Krieger, B. Müller-Clostermann, M. Sczittnick. Modeling and Analysis of Communication Systems Based on Computational Methods for Markov Chains, IEEE Journal on Selected Areas in Communications, 8(9), pp. 1630-1648 (Special Issue on Modeling and Analysis of Telecommunication Systems), 1990.

[LaKö97] P. Langendörfer, H. König. Improving the Efficiency of Automatically Generated Code by Using Implementation-specific Annotations. Proc. 3rd Intl. Workshop on High Performance Protocol Architectures (HIPPARCH'97), Uppsala, 1997.

[LaKö99] P. Langendörfer, H. König. COCOS – A Configurable SDL Compiler for Generating Efficient Protocol Implementations. SDL '99: The Next Millennium (Ninth SDL Forum), R. Dssouli, G.V. Bochmann, Y. Lahav (eds), pp259-274, Elsevier, 1999.

[LaSW97] K.G. Larson, B. Steffen, C. Weise. Continuous Modeling of Real-Time and Hybrid Systems: From Concepts to Tools. Intl. Journal on Software Tools for Technology Transfer. Springer, 1(1+2), Dec. 1997.

[Lave83] S. Lavenberg. Computer Performance Modelling Handbook. Academic Press, 1983.

[LCPM85] B.M. Leiner, R. Cole, J. Postel, D. Mills. The DARPA Internet Protocol Suite. IEEE Communications Magazine, 23, pp29-34, March 1985.

[Lemm97] F. Lemmen. Objektorientierte Instrumentierung einer SDL'92-Spezifikation im Kontext des SDL/MSC-gesteuerten Monitoring. Proc. 9. ITG/GI-Fachtagung Messung, Modellierung und Bewertung von Rechen-und Kommunikationssystement (MMB'97), K. Irmscher, Ch. Mittasch, K. Richter (eds), pp217-229, Sept. 1997.

[Lemm00] F. Lemmen. Spezifikationsgesteuertes Monitoring zur Integration der Leistungsbewertung in den formalen Entwurf von Kommunikationssystemen. Dissertation, Universität Erlangen-Nürnberg, Arbeitsberichte des Instituts für Mathematische Maschinen und Datenverarbeitung, 33(2), Febr. 2000.

[Leng90] T. Lengauer. Combinatorial Algorithms for Integrated Circuit Layout. Wiley, 1990.

[LeOe96] S. Leue, P. Oechslin. On Parallelizing and Optimizing the Implementation of Communication Protocols. IEEE/ACM Transactions on Networking, 4(1), 1996.

[Leue95] S. Leue. Specifying Real-Time Requirements for SDL Specifications – A Temporal Logic-Approach. 15th Intl. Symposium on Protocol Specification, Testing and Verification (PSTV '95), P. Dembinski, M. Sredniawa (eds), North Holland, 1995.

[Lind94] C. Lindemann. DSPNexpress: a Software Package for the Efficient Solution of Discrete and Stochastic Petri Nets. Performance Evaluation, 20, July 1994.

[Lind98] C. Lindemann. Performance Modelling with Deterministic and Stochastic Petri Nets. John Wiley & Sons, 1998.

[Ling94] R.C. Linger. Clean Room Process Model. IEEE Software, March 1994.

[Lipo82] N. Lipow. Number of Faults per Line of Code. IEEE Transactions on Software Engineering, July 1982.

[LiSw80] B. Lientz, E. Swanson. Software Maintenance Management. Addison-Wesley, 1980.

[LSIV97] C. Loftus. E. Sherratt, E. Inocencio, P. Viana. The Unification of OMT, SDL and IDL for Service Creation. SDL '97 – Time for Testing (Eighth SDL Forum), A. Cavalli, A. Sarma (eds), pp443-458, Elsevier, Sept. 1997.

[LWFH98] Y. Lahav, A. Wolisz, J. Fischer, E. Holz (eds). Preface. Proc. First Workshop of the SDL Forum Society on SDL and MSC, Informatik-Bericht Nr. 104, Humboldt-Univ. zu Berlin, 1998.

[LZGS84] E.D. Lazowska, J. Zahorjan, G.S. Graham, K.C. Sevcik. Quantitative System Performance – Computer System Analysis Using Queueing Network Models. Prentice Hall, 1984.

[MaRa99] N. Mansurov, A. Ragozin. Using Declarative Mappings for Automatic Code Generation from SDL and ASN.1. SDL '99: The Next Millennium (Ninth SDL Forum), R. Dssouli, G.V. Bochmann, Y. Lahav (eds), pp275-290, Elsevier, 1999.

[McCl89] C. McClure. Software Automation in the 1990s and Beyond. Proc. of the Fall 1989 CASE Symposium. Andover, MA, Digital Consulting, Sept. 1989.

[MHSZ96] J. Martins, J.-P. Hubaux, T. Saydam, S. Znatny. Integrating Performance Evaluation and Formal Specification. Proc. of Intl. Conf. on Communications (ICC96), IEEE Press, 1996.

[MiLH96] A. Mitschele-Thiel, P. Langendörfer, R. Henke. Design and Optimization of High-Performance Protocols with the DO-IT Toolbox. Formal Description Techniques IX – Theory, Application and Tools (Proc. Joint Intl. Conf. FORTE/PSTV'96), R. Gotzhein, J. Bredereke (eds), pp45-60, Chapman & Hall, 1996

[Miln80] R. Milner. A Calculus of Communicating Systems. Lecture Notes in Computer Science 92, Springer, 1980.

[MiMR98] A. Mitschele-Thiel, B. Müller-Clostermann, R. Reed (eds). Proc. of Workshop on Performance and Time in SDL and MSC. Report IMMD VII-1/98, University of Erlangen, 1998.

[MiMü99] A. Mitschele-Thiel, B. Müller-Clostermann. Performance Engineering of SDL/MSC Systems. Computer Networks, A. Cavalli (ed.), 31(17), pp1801-1815, Elsevier, June 1999.

[MiSl97] A. Mitschele-Thiel, F. Slomka. Codesign with SDL/MSC. CONSYSE'97: Intl. Workshop on Conjoint Systems Engineering, K. Buchenrieder, A. Sedelmeier (eds), pp1-18, Applied Computer Science and Technology Series, IT-Press, June 1999.

[Mits93] A. Mitschele-Thiel. Automatic Configuration and Optimization of Parallel Transputer Applications. Transputer Applications and Systems '93, R. Grebe et al. (eds), vol. 2, IOS Press, Sept. 1993.

[Mits94] A. Mitschele-Thiel. Die DSPL-Entwicklungsumgebung – Ein automatischer Ansatz zur Abbildung und effizienten Realisierung dedizierter Anwendungen auf parallele Systeme. Dissertation, Fortschritt-Berichte VDI, Reihe 10: Informatik/Kommunikationstechnik, Nr. 315, 152pp, VDI Verlag, 1994.

[Mits96] A. Mitschele-Thiel. Methodology and Tools for the Development of High Performance Parallel Systems with SDL/MSCs. Software Engineering for Parallel and Distributed Systems, I. Jelly, I. Gorton, P. Croll (eds), Chapman & Hall, 1996.

[Mits98] A. Mitschele-Thiel. Performance Evaluation of SDL Systems (Invited Talk). Proc. First Workshop of the SDL Forum Society on SDL and MSC, Y. Lahav, A. Wolisz, J. Fischer, E. Holz (eds), Informatik-Bericht Nr. 104, Humboldt-Univ. zu Berlin, pp35-44, 1998.

[Mits99] A. Mitschele-Thiel. Integrating Model-Based Optimization and Program Transformation to Generate Efficient Parallel Programs. Journal of Systems Architecture, 45, pp465-482, Elsevier, 1999.

[MMOP+94] A.B. Montz, D. Mosberger, S.W. O'Malley, L.L. Peterson, T.A. Proebsting, J.H. Hartman. Scout: A Communications-Oriented Operating System. Technical Report TR 94-20, Department of Computer Science, The University of Arizona, Tucson, AZ, 1994.

[Moll82] M.K. Molloy. Performance Analysis Using Stochastic Petri Nets. IEEE Transactions on Computers, 31(9), pp913-917, Sept. 1982.

[Morl95] T.I. Morley. The Application of SDL to Control Systems for Industrial Machines. SDL '95 with MSC in CASE (Proc. Seventh SDL Forum), R. Braek, A. Sarma (eds), Elsevier, 1995.

[MRCM96] N.N. Mansurov, A.S. Ragozin, A.V. Chernov, I.V. Mansurov. Tool Support for Algebraic Specifications of Data in SDL-92. Joint Intl. Conf. on Formal Description Techniques for Distributed Systems and Communication Protocols (IX) and Protocol Specification, Testing and Verification (XVI) (FORTE/PSTV'96), R. Gotzhein, J. Bredereke (eds), Chapman & Hall, 1996.

[Myer79] G.J. Myers. The Art of Software Testing. Wiley, 1979.

[Nils77] N.J. Nilsson. Problem Solving Methods in Artificial Intelligence. McGraw-Hill, New York, 1977.

[NiSi91] X. Nicollin, J. Sifakis. An Overview and Synthesis on Timed Process Algebras. Real-Time: Theory in Practice, Lecture Notes in Computer Science 600, Springer, 1991.

[NoTh93] M.G. Norman, P. Thanisch. Models of Machines and Computation for Mapping on Multicomputers. ACM Computing Surveys, 25(3), pp263-302, Sept. 1993.

[OFMR+94] A. Olsen, O. Faergemand, B. Moller-Pedersen, R. Reed, J.R.W. Smith. Systems Engineering Using SDL-92. North Holland, 1994.

[Olse93] N.C. Olsen. The Software Rush Hour. IEEE Software, Sept. 1993.

[OMG99] OMG Unified Modeling Language Specification, Version 1.3, June 1999.

[OmLe90] P.W. Oman, T.G. Lewis (eds). Milestones in Software Engineering. IEEE Press, 1990.

[OsGl95] O. Oskarsson, R.L. Glass. An ISO 9000 Approach to Building Quality Software. Prentice Hall, 1995.

[PaFi89] M. Paterok, O. Fischer. Feedback Queues with Preemptive-Distance Priorities. Proc. ACM SIGMETRICS/Performance '89, Performance Evaluation Review, 17(1), 1989.

[Page88] M. Page-Jones. The Practical Guide to Structured Systems Design. 1988.

[Paul95] M.C. Paulk. The Evolution of the SEI's Capability Maturity Model for Software. Software Process – Improvement and Practice, 1, pp3-15, Wiley, 1995.

[PCCW93] M.C. Paulk, B. Curtis, M.B. Chrissis, C.V. Weber. Capability Maturity Model for Software, Version 1.1. Technical Report, CMU/SEI-93-TR-24, Software Engineering Institute, Carnegie Mellon University, 1993.

[PeDa96] L. Peterson, B. Davie. Computer Networks – A Systems Approach. Morgan Kaufmann Publishers, San Francisco, 1996.

[PeDe97] V. Perrier, N. Dervaux. Combining Object-Oriented and Real-Time Programming from an OMT and SDL Design. SDL '97 – Time for Testing (Eighth SDL Forum), A. Cavalli, A. Sarma (eds), pp459-472, Elsevier, Sept. 1997.

[PeSö93] C. Peterson, B. Söderberg. Artificial Neural Networks. Modern Heuristic Techniques for Combinatorial Problems, R. Reeves (ed.), Blackwell Scientific Publications, 1993.

[Pete77] J.L. Peterson. Petri Nets. ACM Computing Surveys, 9(3), pp223-252, Sept. 1977.

[PlAt91] B. Plateau, K. Atif. Stochastic Automata Network for Modeling Parallel Systems. IEEE Transactions on Software Engineering, 17(10), Oct. 1991.

[Pool98] R. Pooley. The Unified Modelling Language and Performance Engineering. First Intl. Workshop on Software and Performance (WOSP 98), Tutorial Handouts, Santa Fe, New Mexico, USA, Oct. 12-16, 1998.

[Pres97] R.S. Pressman. Software Engineering – A Practitioner's Approach, 4th edition, McGraw-Hill, 1997.

[PrSa90] C.-C. Price, M.A. Salama. Scheduling of Precedence-Constrained Tasks on Multiprocessors. The Computer Journal, 33(3), 1990.

[Prot92] Protocol Engines Inc.: Xpress Transfer Protocol Definition. Revision 3.6, 1992.

[QUES98] QUEST and QSDL Homepage. http://www.cs.uni-essen.de/Fachgebiete/SysMod/Forschung/QUEST/, 1998.

[RaES88] J. Ramanujam, F. Ercal, P. Sadayappan. Task Allocation by Simulated Annealing. Proc. Intl. Conf. on Supercomputing, vol. III, Boston, MA, May 1988.

[Ramc74] C. Ramchandani. Analysis of Asynchronous Concurrent Systems by Timed Petri Nets. Ph.D. Thesis, MIT, Report MAC-TR-120, 1974.

[RaPa97] S. Raman, L.M. Patnaik. Optimization via Evolutionary Processes. Advances in Computers, vol. 45, M.V. Zelkowitz (ed.), Academic Press, 1997.

[RBPE+91] J. Rumbaugh, M. Blaha, W. Premerlani, F. Eddy, W. Lorensen. Object-Oriented Modeling and Design. Prentice Hall, 1991.

[Reed93] R. Reed (ed.). SPECS. North-Holland, 1993.

[Reed96] R. Reed. Methodology for Real Time Systems. Computer Networks and ISDN Systems, 28, pp1685-1701, 1996.

[Reed97] R. Reed. SDL and MSC in International Organizations: ITU-T. SDL '97 – Time for Testing (Eighth SDL Forum), A. Cavalli, A. Sarma (eds), pp231-241, Elsevier, Sept. 1997.

[Reev93] C.R. Reeves. Genetic Algorithms. Modern Heuristic Techniques for Combinatorial Problems, R. Reeves (ed.), Blackwell Scientific Publications, 1993.

[Reev93a] C.R. Reeves (ed.). Modern Heuristic Techniques for Combinatorial Problems, Blackwell Scientific Publications, 1993.

[Robi92] P.J. Robinson. Hierarchical Object-Oriented Design. Prentice Hall, 1992.

[RoBu95] J. Rozenblit, K. Buchenrieder (eds). Codesign – Computer-Aided Software/Hardware Engineering. IEEE Press, 1995.

[RoSc77] D.T. Ross, K.E. Schoman Jr. Structured Analysis for Requirements Definition. IEEE Transactions on Software Engineering, 3(1), 1977.

[RoSe95] J.A. Rolia, K.C. Sevcik. The Method of Layers. IEEE Transactions on Software Engineering, 21(8), Aug. 1995.

[Roux98] J.L. Roux. SDL Performance Analysis with ObjectGEODE. In [MiMR98].

[Royc70] W.W. Royce. Managing the Development of Large Software Systems: Concepts and Techniques. Proc. WESCON, Aug. 1970.

[RuGG96] E. Rudolph, P. Graubmann, J. Grabowski. Tutorial on Message Sequence Charts. Computer Networks and ISDN Systems, 28, 1629-1641, 1996.

[RuJB98] J. Rumbaugh, I. Jacobson, G. Booch. The Unified Modeling Language Reference Manual. Addison-Wesley, 1998.

[SaTr87] R. Sahner, K. Trivedi. Performance Analysis and Reliability Analysis Using Directed Acyclic Graphs. IEEE Transactions on Software Engineering, 13(10), Oct. 1987.

[Scha96] C. Schaffer. MSC/RT: A Real-Time Extension to Message Sequence Charts (MSCs). Interner Bericht TR140-96, Institut für Systemwissenschaften, Johannes Kepler Universität Linz, 1996.

[Schr00] R. Schröder. Weiterentwicklung von SDL-Datenkonzepten. Dissertation, Humboldt Universität zu Berlin, in preparation.

[Schw94] H. Schwetman. CSIM17: A Simulation Model-Building Toolkit. Proc. Winter Simulation Conference, Orlando, 1994.

[Schw97] M. Schwehm. Globale Optimierung mit massiv parallelen Genetischen Algorithmen. Dissertation, Universität Erlangen-Nürnberg, Arbeitsberichte des Instituts für Mathematische Maschinen und Datenverarbeitung, 30(1), Jan. 1997.

[ScRe98] I. Schieferdecker, A. Rennoch. Usage of Timed MSCs for Test Purpose Definition. In [MiMR98].

[ScRS95] C. Schaffer, R. Raschhofer, A. Simma. EaSy-Sim: A Tool Environment for the Design of Complex, Real-Time Systems. Proc. Intl. Conf. on Computer Aided Systems Technologies, Innsbruck, Springer, 1995.

[ScSu93] D.C. Schmidt, T. Suda. Transport System Architecture Services for High-Performance Communication Systems. IEEE Journal on Selected Areas in Communications, 11(4), pp489-506, May 1993.

[SeBr88] C. Sechen, D. Braun, A. Sangiovanni-Vincetelli. Thunderbird: A Complete Standard Cell Layout Package. IEEE Journal of Solid-State Circuits, 23, pp410-420, 1988.

[Shen92] H. Shen. Self-adjusting Mapping: A Heuristic Mapping Algorithm for Mapping Parallel Programs on to Transputer Networks. The Computer Journal, 35(1), 1992.

[SiLe90] G.C. Sih, E.A. Lee. Scheduling to Account for Interprocessor Communication within Interconnected-Constrained Processor Networks. Proc. 1990 International Conference on Parallel Processing, vol. I, Pennsylvania State University Press, University Park, PA, 1990.

[Simu90] Simulog. New Users' Introduction into QNAP2, Version 7.0. 1990 (see also www.simulog.fr for recent information).

[SlKr87] M. Sloman, J. Kramer. Distributed Systems and Computer Networks. Prentice Hall, 1987.

[SlZL98] F. Slomka, J. Zant, L. Lambert. Schedulability Analysis of Heterogeneous Systems for Performance Message Sequence Charts. Proc. 6th Intl. Workshop on Hardware/Software Codesign. IEEE Computer Society Press, 1998.

[Smit90] C.U. Smith. Performance Engineering of Software Systems. SEI Series in Software Engineering, Addison-Wesley, 1990.

[Smit90a] M.F. Smith. Software Prototyping: Adoption, Practice and Management. McGraw-Hill, 1990.

[SmWi97] C.U. Smith, L.G. Williams. Performance Engineering Evaluation of Object-Oriented Systems with SPE.ED. Intl. Conf. on Computer Performance Evaluation: Modelling Techniques and Tools, Lecture Notes in Computer Science 1245, Springer, 1997.

[SpSD97] S. Spitz, F. Slomka, M. Dörfel. SDL* – An Annotated Specification Language for Engineering Multimedia Communication Systems. Proc. 6th Open Workshop on High Speed Networks. Institut für Nachrichtenvermittlung und Datenverarbeitung (IND), Universität Stuttgart, pp77-84, Oct. 1997.

[Step98] M. Steppler. Performance Analysis of Communication Systems Formally Specified in SDL. Proc. First Intl. Workshop on Software and Performance (WOSP 98), Santa Fe, New Mexico, USA, ACM, Oct. 12-16, 1998.

[StJG93] C. Steigner, R. Joostema, C. Groove. PAR-SDL: Software Design and Implementation for Transputer Systems. Transputer Applications and Systems '93 (Proc. of the 1993 World Transputer Congress, Aachen, Germany), vol. 2, IOS Press, 1993.

[StLe97] M. Steppler, M. Lott. SPEET - SDL Performance Evaluation Tool. SDL '97 – Time for Testing (Eighth SDL Forum), A. Cavalli, A. Sarma (eds), pp53-67, Elsevier, Sept. 1997.

[StMC74] W.P. Stevens, G.J. Myers, L.L. Constantine. Structured Design. IBM Systems Journal, 13, 1974.

[Svob89] L. Svobodova. Implementing OSI Systems. IEEE Journal on Selected Areas in Communications, 7(7), pp1115-1130, 1989.

[Tane96] A. Tanenbaum. Computer Networks. Third Edition, Prentice Hall, 1996.

[Teic97] J. Teich. Digitale Hardware/Software-Systeme. Springer, 1997.

[Tele98] Telelogic Malmö AB: Tau User's Manual. 1998.

[ThAS93] D.E. Thomas, J.K. Adams, H. Schmit. A Model and Methodology for Hardware-Software Codesign. IEEE Design and Test of Computers, Sept. 1993.

[Thom93] B. Thome. Definition and Scope of Systems Engineering. Systems Engineering – Principles and Practice of Computer-based Systems Engineering, B. Thome (ed.), Wiley, 1993.

[Trau93] H. Trauboth. Software-Qualitätssicherung. Handbuch der Informatik 5.2, R. Oldenbourg Verlag, 1993.

[TrSo89] J.P. Tremblay, P.G. Sorenson. The Theory and Practice of Compiler Writing. McGraw-Hill, 1989.

[VaAa88] P.J.M. Van Laarhoven, E.H.L. Aarts. Simulated Annealing: Theory and Applications. Kluwer, 1988.

[Veri98] Verilog. ObjectGEODE – Technical Documentation, 1998.

[WaGr97] T. Walter, J. Grabowski. A Proposal for a Real-Time Extension of TTCN. Kommunikation in Verteilten Systemen, M. Zitterbart (ed.), Reihe Informatik aktuell, pp182-196, Springer, 1997.

[Walc94] M. Walch. Evaluating Parallel Processing of Communication Protocols. GMD-Bericht Nr. 238, R. Oldenbourg Verlag, 1994.

[WaMa87] R.W. Watson, S.A. Mamrak. Gaining Efficiency in Transport Services by Appropriate Design and Implementation Choices. ACM Transactions on Computer Systems. 5(2), pp97-120, 1987.

[WaMe85] P.T. Ward, S.J. Mellor. Structured Development for Real-Time Systems. Vol 1-3, Yourdon Press, 1985.

[Warn74] J.P. Warnier. Logical Construction of Programs. Van Nostrand Reinhold, 1974.

[WaTs88] L. Wang, W. Tsai. Optimal Assignment of Task Modules with Precedence for Distributed Processing by Graph Matching and State-Space Search. BIT, 28, 54-68, 1988.

[WeEL97] T. Weigert, A. Ek, Y. Lejeune. References in SDL. ITU Telecommunications Standardisation Sector, Temporary Document TDL 613, Study Group 10, Question 6 (SDL), Oct. 1997.

[Wier98] R. Wieringa. A Survey of Structured and Object-Oriented Software Specification Methods and Techniques. ACM Computing Surveys, 30(4), pp459-527, Dec. 1998.

[Wirt98] K. Wirth. Overload Control in GSM – Handling the Problem in the Context of SDL. In [MiMR98].

[WiSm98] L.G. Williams, C.U. Smith. Performance Evaluation of Software Architectures. Proc. First Intl. Workshop on Software and Performance (WOSP 98), Santa Fe, New Mexico, USA, ACM, Oct. 12-16, 1998.

[WNPM95] C.M. Woodside, J.E. Neilson, D. Petriu, S. Majumdar. The Stochastic Rendezvous Network Model for Performance of Synchronous Client-server-like Distributed Software. IEEE Transactions on Computers, 44(1), pp20-34, Jan. 1995.

[WoMo89] C.M. Woodside, J.R. Montealegre. The Effect of Buffering Strategies on Protocol Execution Performance. IEEE Transactions on Communications, 37(6), pp545-554, June 1989.

[Wood95] C.M. Woodside. A Three-View Model for Performance Engineering of Concurrent Software. IEEE Transactions on Software Engineering, 21(9), pp754-767, Sept. 1995.

[WOSP98] Proc. First Intl. Workshop on Software and Performance (WOSP 98), Santa Fe, New Mexico, USA, ACM, Oct. 12-16, 1998.

[Your92] E. Yourdan. Decline and Fall of the American Programmer. Prentice Hall, 1992.

[Zave84] P. Zave. The Operational Versus the Conventional Approach to Software Development. Communications of the ACM, Feb. 1984.

Index

RECEIVED
MAY 1 7 2001